CAMBRIDGE EARTH SCIENCE SERIES
Editors:
A.H. Cook, W.B. Harland, N.F. Hughes
A. Putnis, J.G. Sclater and M.R.A. Thomson

A geologic time scale

This reference book is a concentrated review of the time scales used in geology in order to date stratigraphic sequences and to define the geologic epochs. The text presents, discusses and evaluates the state of chronostratigraphic, chronometric and other scales; this includes a revised calibration in years of the standard stratigraphic scale.

There are chapters with (1) introductory and historical background, (2) stratigraphic time scale and correlation charts for each period or other time interval with commentaries, (3) a new way of presenting the data for age estimates of each boundary (chronograms) which attempts to depict the strengths and weaknesses of the calibration, (4) a new magnetostratigraphic time scale and (5) a summary of world events in the form of charts. This book with its profusion of tables and data will serve as a fundamental reference source in the Earth Sciences.

The cover

The panels on the cover summarise the provisional time scale adopted in this book.

The panel on the front cover is plotted to a linear scale of approximately 20 million years (20 Ma) to a millimetre. To the left is a scale of chronometric divisions defined numerically; to the right are shown some chronostratic divisions of higher rank abstracted from the tabulation on the back cover.

The two panels on the back cover summarise the chronostratic time scale as developed in the book. The numerical values are provisional estimates. The divisions are spaced for typographic convenience.

In this series

PALAEOMAGNETISM AND PLATE TECTONICS
M. W. McElhinny

JURASSIC ENVIRONMENTS
A. Hallam

PALAEOBIOLOGY OF ANGIOSPERM ORIGINS
N. F. Hughes

MESOZOIC AND CENOZOIC PALEOCONTINENTAL MAPS
A. G. Smith and J. C. Briden

MINERAL CHEMISTRY OF METAL SULFIDES
D. J. Vaughan and J. R. Craig

EARTHQUAKE MECHANICS
K. Kasahara

PHANEROZOIC PALEOCONTINENTAL WORLD MAPS
A. G. Smith, A. M. Hurley and J. C. Briden

EARTH'S PRE-PLEISTOCENE GLACIAL RECORD
M. J. Hambrey and W. B. Harland

A HISTORY OF PERSIAN EARTHQUAKES
N. N. Ambraseys and C. P. Melville

A GEOLOGICAL TIME SCALE
W. B. Harland, A. Cox, P. G. Llewellyn, C. A. G. Pickton, A. G. Smith and R. Walters

THE GREAT TOLBACHIK FISSURE ERUPTION
S. A. Fedotov and Ye. K. Markhinin

FOSSIL INVERTEBRATES
U. Lehmann and G. Hillmer

GENERAL HYDROGEOLOGY
E. V. Pinneker

THE DAWN OF ANIMAL LIFE
M. E. Glaessner

ATLAS OF CONTINENTAL DISPLACEMENT
200 MILLION YEARS TO THE PRESENT
H. G. Owen

A geologic time scale

W. B. HARLAND
A. V. COX
P. G. LLEWELLYN
C. A. G. PICKTON
A. G. SMITH
R. WALTERS

assisted by K.E. Fancett

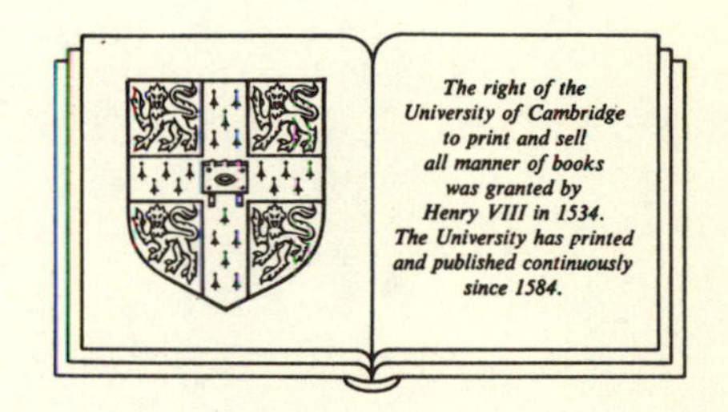

CAMBRIDGE UNIVERSITY PRESS
Cambridge
New York New Rochelle Melbourne Sydney

Published by the Press Syndicate of the University of Cambridge
The Pitt Building, Trumpington Street, Cambridge CB2 1RP
32 East 57th Street, New York, NY 10022, USA
10 Stamford Road, Oakleigh, Melbourne 3166, Australia

First published 1982
Reprinted 1984, 1985, 1987

Printed in Great Britain at the University Press, Cambridge

Library of Congress catalogue card number: 82-4333

British Library cataloguing in publication data

A geologic time scale.–(Cambridge Earth Science Series)

1. Geological time
I. Harland, W. B.
551.7′01 QE508

ISBN 0 521 24728 4 hard covers
ISBN 0 521 28919 X paperback

Contents

Figures

Tables

Charts

Preface

Geologic time scales have been published at frequent intervals for more than 100 years. Many think existing scales are good enough and are discouraged by the diminishing returns that further effort would appear to yield. On the other hand new levels of precision are possible and scales in current use are so uneven that improved constraints at a few critical points could be of great use. Each advance in precision opens up new applications: so, for the foreseeable future, the activity is worthwhile. In this perspective we make no apology for a 1982 version. We believe it will be useful; we know it will be superseded. Its particular value may be to display some aspects of how the geologic time scale is constructed, and so to make evident its short-comings, and to marshal material for a further improved scale. The 1964 time-scale volume of the Geological Society of London was commonly accepted uncritically; we hope this will not prove to be the case with this publication which our initial optimism led us to suppose might be published in 1981, 100 years after the International Geological Congress (IGC) in Bologna which set out as one of its targets the standardisation of the chronostratigraphic scale.

We were all only too well aware of the limitations of the *Phanerozoic time-scale* volumes of 1964 and 1971. WBH in his Cambridge Arctic Shelf Programme (CASP) work realised that there was urgent need for an updated scale that could be used and so CAGP wrote an internal report circulated to all CASP subscribers. This CASP scale was used in the IGCP Pre-Pleistocene Tillite Project and appears on pages 12 and 13 of the resulting volume (Hambrey & Harland 1981). Quite independently AGS needed an updated scale to improve his computer map sequences. PGL had also constructed a time scale for British Petroleum (BP) in 1975, revised in due course by RW in October 1980 and circulated as an internal report by BP. When BP received the CASP report the overlap of our intentions was immediately apparent.

Shortly before this WBH had been invited to join a Geological Society of London Working Group to plan a Geological Time Scale Symposium and at first it seemed that the outlet for his interest in revising the scale was to participate actively in that. However, it was decided at an early stage that the Society would organise independent speakers for a wide-ranging symposium rather than undertake a cooperative investigation. As with the Sydney Symposium (Cohee, Glaessner & Hedberg 1978) valuable papers would accrue but no agreed time scale for general use was then planned.

We had intended somehow to revise the magnetic reversal scale in our work. By chance, AVC happened to be in Cambridge and also needed an improved time scale for magnetic reversals, hence Chapter 4. Chapters 1 and 2 were the prime responsibility of WBH and RW. AGS (and AVC) worked on Chapter 3 and although CAGP was not able to attend the meetings he agreed to produce the index. KEF helped WBH with the initial planning and assembly of material and later helped with editing and assembling the reference list, a task taken on by T.A. Brewer when KEF left Cambridge.

It was decided from the outset that a wallchart would be published so as to display the essentials of the time scale argued more fully in the book. RW and PGL undertook to coordinate its design and execution and Chapter 5 describes some aspects of the chart. Similarly WBH coordinated the assembly of material for the book.

The authors are indebted to the Directors of the British Petroleum Company Limited for providing facilities for the complete drawing of all the illustrations for this work and the wallchart, and for allowing PGL and RW to participate. Also they thank A.A. Miles and his colleagues at BP for undertaking the drawing.

The reference list shows the literature on which we have depended, but one reference deserves particular mention as we have used it *in toto*, i.e. Armstrong's invaluable list (in Cohee *et al.* 1978).

We are indebted for stratigraphic consultations to a number of colleagues: N.F. Hughes generally and for Mesozoic classification; R.W. Hey and N.J. Shackleton for considerable help with the Quaternary time scale but most especially M.J. Hambrey, who compiled the material, drafting much of the text and all of Chart 2.17; R.T. Wu for assistance with problems of Triassic stratigraphy and with other Chinese columns; W.H.C. Ramsbottom, President of the IUGS Subcommission on Carboniferous Stratigraphy, for unpublished material and advice and also D.G. Smith, who contributed to our understanding; P.F. Friend, who checked on Devonian work; R.B. Rickards for bringing us up to date with Silurian stratigraphic developments and C.P. Hughes and J.W. Cowie, who respectively advised on Ordovician and Cambrian classification. For the Precambrian time scale we acknowledge help from G. Vidal and E. Nisbet. We are indebted to A.J. Fleet for a state-of-the-art summary on global sea-level changes and for compiling a composite Phanerozoic sea-level curve (Fig. 5.1).

Each author was constrained by other duties which prevented the whole book being completed with all authors participating. We foresaw this and accepted the consequence that a less than perfect work would result if we hurried on to complete it as best we could. We planned that our work, even if not published, would be available to the contributors to the Geological Society Symposium in 1982, hoping it would be useful and accepting that it would thereby shortly be bettered.

In selecting spelling and typographic conventions we have had in mind the need to promote international stratigraphic standards that will be convenient for scientists of many language communities. This has led us, when adopting a name that is rooted in a language other than English, for the chronostratic scale, to opt for the simpler form as in Visean (without accents) or Paleozoic (without diphthong). The latter practice was extended, we hope consistently, to some other terms (e.g. paleomagnetic).

In order to keep the book compact for convenient use some material has been printed in smaller type.

A. V. Cox
School of Earth Sciences
Stanford University
Stanford
California 94305
USA

W. B. Harland
A. G. Smith
Department of Earth Sciences
Cambridge University
Downing Street
Cambridge CB2 3EQ

P. G. Llewellyn
R. Walters
Exploration and Production Division
British Petroleum Research Centre
Chertsey Road
Sunbury-on-Thames
Middlesex TW15 7LN

C. A. G. Pickton
Deminex UK Oil and Gas Ltd
Bowater House
68 Knightsbridge
London SW1X 7LD

Abbreviations and acronyms used in this book

Abbreviations for chronostratic names are given in Appendix 3

AGI	American Geological Institute
BP	Before Present (1950)
BP	British Petroleum Company
CASP	Cambridge Arctic Shelf Programme
CGMW	Commission for the Geological Map of the World
CUP	Cambridge University Press
DSDP	Deep Sea Drilling Project
GSC	Geological Survey of Canada
GSL	Geological Society of London
IAGA	International Association of Geomagnetism and Aeronomy
IAU	International Astronomical Union
IGC	International Geological Congress
IGCP	International Geological Correlation Programme
INQUA	International Quaternary Association
ISSC	International Subcommission on Stratigraphic Classification (of IUGS)
IUGS	International Union of Geological Sciences
NZGS	New Zealand Geological Survey
RSS	Regional Stratigraphic Scales
SGCS	Standard Global Chronostratigraphic Scale
SI	Système International d'Unités
SSS	Standard Stratigraphic Scale
TSS	Traditional Stratigraphic Scale
Unesco	United Nations Educational Scientific and Cultural Organisation
USGS	United States Geological Survey

1
Introduction

1.1 Objective

A geologic time scale is composed of standard stratigraphic divisions based on rock sequences and calibrated in years. It is thus the joining of two different kinds of scale. A chronometric scale is based on units of duration – the standard second – hence a year. A chronostratic scale is now conceived as a scale of rock sequence with standardised reference points selected in sections, each particularly complete at the boundary and known as a boundary stratotype. The chronostratic scale is a convention to be agreed rather than discovered, while its calibration in years is a matter for discovery rather than agreement. Whereas the chronostratic scale once agreed should generally stand unchanged, its evaluation will be subject to repeated revision. For this reason no geologic time scale can be final and our particular attempt must be qualified by 1982, its year of publication.

1.2 The traditional stratigraphic time scale

The prodigious stratigraphic labours of the nineteenth century resulted in innumerable competing stratigraphic schemes. To impose some order the first International Geological Congress (IGC) in Paris in 1878 set as its objective the production of a standard stratigraphic scale. Suggestions were made for standard colours (Anon. 1880, pp.70–82), uniformity of geologic nomenclature (pp.82–4) and the adoption of uniform subdivisions (pp.85–7). There was also a review of several regional stratigraphic problems. In the succeeding congress at Bologna in 1881 many of the above suggestions were taken substantially further, i.e. international maps were planned with standard colours for stratigraphic periods and rock types (e.g. Anon, 1882, pp.297–411) and annexes contained national contributions towards standardisation of stratigraphic classification etc. (pp.429–658).

In spite of this promising start the IGCs did not have the continuing organisation to carry these proposals through, except for the commissions set up to produce international maps. It was not until the establishment of the International Union of Geological Sciences (IUGS) around 1960 that the promise had a means of fulfilment, through the IUGS's Commission on Stratigraphy and its many subcommissions.

By 1878 the early belief that the stratigraphic systems and other divisions being described in any one place were natural chapters of Earth history was fading and the need to agree some convention was widely recognised. Even so, the practice continued of describing stratal divisions largely as biostratigraphic units, and even today it is an article of faith for many that divisions of the developing international stratigraphic scale are defined by the fossil content of the rocks. To follow this through, however, leads to difficulties: boundaries may change with new fossil discoveries; boundaries defined by particular fossils will tend to be diachronous; there will be disagreement as to which taxa shall be definitive. So the traditional stratigraphic scale is of necessity evolving into a new kind of standard stratigraphic scale.

1.3 Development of the standard stratigraphic scale

At the 1948 IGC one of the first attempts to standardise artificially a stratigraphic boundary (Pliocene–Pleistocene boundary at the Calabrian base in Italy) was made on the basis that such a decision had to be an agreed convention and that it was necessary to standardise divisions at their boundaries only, and in only one locality. The international procedure to standardise such a boundary at a single point in a reference section was worked out by the Silurian–Devonian Boundary Working Group. Their procedure was first to agree the approximate position in the biostratigraphic sequence that would do least violence to existing usage and then to find a succession anywhere in the world where the Silurian–Devonian boundary was represented in fossiliferous rock with the best characters for correlation.

If we take this procedure as a guide, the requirements for the standard global stratigraphic scale may be summarised as follows.

(1) A sequence of agreed reference points in continuous sections of uniform (marine) sedimentary facies selected with suitable characters for international correlation, preservation and access. The point in the boundary stratotype section is then conceived as representing the point in time when that part of the rock was formed. Pairs of such points then define the intervening time span.

(2) The procedure has a significant consequence in the conception of chronostratic divisions. Before the standardisation just described, the intervals were conceived as being the time equivalent of rocks already defined. Thus systems (series, stages or chronozones) were first described and the geologic periods (epochs, ages, chrons) were derived as the corresponding time intervals. The new procedure of defining boundary points effectively reverses the derivation. The time division (period etc.) is defined by its initial and terminal points while the corresponding rock (system etc.) can only be estimated by correlation. This generally yields a well-dated main body of the rock division, but with uncertain and often unidentifiable boundaries. Because of the primacy of time in a time scale we use Early, Middle and Late rather than Lower, Middle and Upper. To avoid usage such as 'early Early' for subdivisions it is well also to seek names for all

epochs (i.e. rank below period and above age).

(3) Various names have been proposed for this newly standardised scale. The Geological Society of London (GSL) used Standard Stratigraphic Scale (SSS) in relation to the traditional stratigraphic scale (TSS) and regional stratigraphic scales (RSS) out of which it was evolving (George *et al.* 1967). The International Subcommission of Stratigraphic Classification (ISSC) referred to it as the Standard Global Chronostratigraphic Scale (SGCS) in the *International stratigraphic guide* (Hedberg 1976). Unfortunately both the American stratigraphic code and the ISSC *Guide* into which it grew confuse the matter somewhat. They divided the standard scale of periods and systems as described here into two parts. *Geochronologic units* refer to periods etc. and *Chronostratigraphic units* to systems etc. It is obvious that time and rock are different (e.g. as indicated by the words period and system), but when defined they both derive from the same standard reference points. It is really one time scale of such points and everything that happened in the time intervals so defined has to be correlated and described as well as may be. The two apparently distinct disciplines (geochronology and chronostratigraphy in Hedberg's terminology), are likewise different aspects of the single discipline of time-correlation.

(4) It is both traditional and convenient to use a hierarchy of names for these intervals (era, period, epoch, age, chron) in such a way that the boundaries of the successively larger divisions are coterminous, i.e. they coincide. There is no difference in procedure for defining boundaries of larger or smaller divisions. Era boundaries are defined the same way as chrons and indeed the same boundary definition may serve several coterminous divisions. The use of the hierarchy is largely a matter of habit but it has its uses in both economy of description and in describing events of different duration or uncertainty of correlation.

(5) The names for the spans are generally those favoured from classic sections. Once selected for the SGCS, however, they cease to have local reference and must be used internationally for the time span defined by the limiting points. It is convenient to retain familiar names but when redefined at some distance from the eponymous locality the local geologists must accept that their name has acquired a new meaning and be careful only to use other lithostratigraphic unit names for the rocks they describe.

(6) The above principles developed for the global scale can be applied to standard regional scales as a step in the process of correlation, but the multiplication of scales is not generally helpful. The work of standardisation is considerable and need not be multiplied. Until such a global time span is standardised all regional scales may be regarded as competing for acceptance as global boundary points.

1.4 The geochronometric scale

The proposal of a chronometric scale is quite different. The scale is periodic and is compounded of units of equal duration. Therefore all that is necessary is to define a standard unit (a second based on cesium, and so derive one year, or on the alternative International Astronomers' year). In the same way that a linear scale is constructed from unit lengths and is so defined, the chronometric scale exists by virtue of the definition of a unit of duration.

There is a further matter of convention, namely, to compound the units into longer named intervals. Such a scheme of millennia (10^6 a), gigennia (10^9 a), etc. is by no means essential but, as with the higher ranks of the chronostratic hierarchy, they may be convenient in general expressions of age. Unlike the chronostratic divisions they will be defined not by reference points in rock but by initial and terminal points, each defined by a finite number of units of duration BP (Before Present – conventionally before 1950). These matters are taken further in Appendix 4.

There are those who think that there is some advantage in treating Precambrian history as sufficiently different from Phanerozoic history as to require the use of named chronometric divisions only for Precambrian time. The Subcommission on Precambrian Stratigraphy of the IUGS agreed in 1976 that the boundary between Archean and Proterozoic should be defined at 2500 Ma (exactly); moreover other subdivisions of Precambrian time are also being proposed along the same lines, as will be seen in the discussion in Chapter 2. An alternative is to extend named chronostratic divisions backwards into Precambrian time. Moreover, a parallel forward development of named chronometric divisions through Phanerozoic time cannot be discounted. The chronostratic scale is extended here (Chapter 2) as seemingly an inevitable development already in progress. Chronometric divisions for all geologic time in 500 Ma intervals have also been proposed (e.g. Harland 1975, 1978) and named from Latin rather than Greek roots thus: Priscotime (to 4000 Ma); Antiquotime (4000 to 2500 Ma), Mediotime (2500 to 1500 Ma); Novotime (1500 Ma onwards). This has not found favour, but, in the absence of other recommendations on the initial Archean (=Antiquo) boundary, Priscoan is applied in Chapter 2 for pre-4000 Ma time, i.e. defined chronometrically.

1.5 Statement of age

The two artificially devised scales outlined above (chronostratic and chronometric) do not in themselves enable us to date or to time-correlate rocks one with another. Their function is to provide common bases for comparison of ages. They reduce the number of ways in which geologic ages are generally stated (i.e. to two conventions; one verbal and one numerical as it happens). The two do not define each other and so they are both needed. According to circumstances some rocks can be dated chronometrically more precisely than chronostratically and for others more precise ages can be given chronostratically. Only if the conversion of one scale to the other were altogether more precise than it now is could all ages be usefully given in years.

So on the one hand it is not sensible to abandon either of the time scales because of the uncertainty and loss of information that would result; on the other hand it is unnecessary to use more than two scales because the ages of rocks can be usefully stated in terms of one or the other.

1.6 Natural chronologies

There are unlimited scales or chronologies that could be, and indeed have been, derived from natural phenomena.

There is the simplest binary scale of magnetic reversals in which only two alternate states are recorded; this is the subject of Chapter 4. There are scales with degrees between two extreme states such as climatic curves – e.g. between glacial and inter-glacial or cold and warm, or between high and low sea level, or between greater and lesser tectonic activity (as outlined in Chapter 5). Then there are decay scales such as in radioactive series, or cooling curves, and finally because of the multiplicity of biological evolutions there are as many biostratigraphic scales as there are useful groups of taxa within any time span; some are listed in Chapter 2. Each of these kinds of natural sequence has its own distinctive properties and value for correlation. All depend on interpretation of rock whose age is best expressed in one or other of the two time scales. While it is necessary to define the SGCS it is impossible so to agree any natural chronology except its terminology.

It is, however, the interest in natural phenomena that motivates science. The geologic time scale is only a tool or language in the interpretation of Earth history; moreover the time scale has no application without time-correlation which is entirely dependent on the interpretation of natural phenomena.

1.7 Local rock units

Rock is the ultimate objective reference for both the study of the natural phenomena of geologic history and for the evidence of age. There is a well-established geologic convention for describing and classifying rock in named units, i.e. formations combined into groups, supergroups and complexes or divided into members and beds (e.g. Hedberg 1976).

All stratigraphic units as originally described were in effect local rock units even if they were intended to have regional or global significance. There is therefore most confusion in the eponymous areas of the SGCS. The original systems, series and stages were initially described as bodies of rock and in many cases this usage persists explicitly, as for example in the hierarchy of South African stratigraphy (e.g. Kent & Hugo 1978).

1.8 Geologic time scales

To return to the point at which we began. A geologic time scale is really a dual scale: two scales – chronostratic and chronometric – side by side, fitted to each other more or less successfully. Table 1.1 shows one of the earliest attempts at a geologic time scale constructed in 1893 before radiometric methods were conceived. H. S. Williams was one of many who attempted this. He used geochrones as his unit; his geochrone being the duration of a well-known period: the Eocene geochrone. Charts 1.1 and 1.2 show the chronostratic divisions and classification adopted here with a sequence of ages (in Ma BP) from a number of earlier scales for comparison with our own. These are from Holmes 1937; Holmes (B scale) 1947; Holmes 1959; Kulp 1961; the GSL *Phanerozoic time-scale* (Harland, Smith & Wilcock 1964); Lambert in *The Phanerozoic time-scale – a supplement* (Harland *et al.* 1971); Van Eysinga 1975; Armstrong in *The geologic time-scale* 1978 from the IGC Sydney Symposium; the BP internal report by R. Walters 1980 (after a BP Alaska internal report by P. G. Llewellyn 1975); the CASP internal report by C. A. G. Pickton 1980, the table from which was published in Hambrey & Harland (1981, pp.12 and 13); and this work.

These time scales are made by interpolation between, and extrapolation from, tie-points that relate to particular rock samples in which a fortunate combination of characters allows, for example, radiometric determinations on rocks closely related to those with fossils that can be used to correlate with the stratotype. Methods other than radiometric of obtaining comparative durations are relative thicknesses, rhythmites, numbers of similar biozones and rates of ocean spreading. Other methods of making correlation feasible (besides biostratigraphic) are lithostratigraphic, paleomagnetic and paleoclimatic. Indeed the best points are those in rocks with the most characters and determinations. But it is a matter of chance where in the geologic column such useful rocks are found, and this leads to some parts of the combined scale being far more effective than others.

Some statement of qualification of uncertainty of each calibration is useful. There are several elements to be considered. Experimental error, usually expressed as standard error, informs only that the same rock unit gave such a scatter of determinations. The environmental history of the mineral in the rock and of the rock itself modify the closed system on which the determinations are based. Therefore it is well not to refer to radiometric ages as absolute ages but rather as *apparent*, i.e. distinct from *true* age. There are uncertainties introduced by interpolation between determined points. There are uncertainties due to correlation precision by paleontological methods, or whatever other methods are used; one

Table 1.1 *Standard time scale of geochronology, on the basis of the Eocene Period for a time unit or geochrone (H. S. Williams 1893, p.295)*

Recent Quaternary	1	
Pliocene Miocene	1	3
Eocene	1	
Cretaceous	4	
Jurassic	3	9
Triassic	2	
Carboniferous	6	
Devonian	5	
Upper Silurian	4	45
Lower Silurian or Ordovician	15	
Cambrian	15	

Charts 1.1 and 1.2. Chronostratic scale with successive chronometric calibrations including our own summarised from Chapters 2 and 3. Values based on old constants are printed in italics.

Chart 1.1. Holocene to Permian.

Eon	Era	Sub-era Period	Epoch	Age	Abbrev.	Holmes1937	Holmes1947B	Holmes1959	Kulp 1961	Geol. Soc. London 1964 constants *old*	Geol. Soc. London 1964 constants new	GSL. 1971	Van Eysinga 1975	Armstrong 1978	BP1980 (internal)	Pickton in Hambrey & Harland 1981	(Pickton source)	This work 1982 Ma	Age	Duration	
Phanerozoic	Cenozoic	Quaternary(Q) or Pleistogene	Holocene		Hol								*0·01*			0·01		0·01	Hol	·01	2·0
			Pleistocene		Ple		*1*	*1*	*1*	*1·5/2*	1·5/2		*1·8*			2·0		2·0	Ple	1·99	
		Tertiary / Neogene	Pliocene (Pli) 2	Piacenzian	Pia														Pia	3·1	22·6
			Pliocene (Pli) 1	Zanclian	Zan	*16*	*12*	*11*	*13*	*7*	7		*5*		5	5	Labrecque *et al* 1978	5·1	Zan		
			Miocene 3	Messinian	Mes														Mes	6·2	
			Miocene 3	Tortonian	Tor													11·3	Tor		
			Miocene 2	Serravallian	Srv														Srv	3·1	
			Miocene 2	Langhian - Late	Lan2													14·4	Lan2		
			Miocene 1	Langhian - Early	Lan1														Lan1		
			Miocene 1	Burdigalian	Bur														Bur	10·2	
		(Ng)	Miocene 1	Aquitanian	Aqt	*32*	*26*	*25*	*25*	*26*	27		*22·5*		22·5	22·5		24·6	Aqt		
		Paleogene	Oligocene 2	Chattian	Cht										32		Hardenbol + Berggren 1978	32·8	Cht	8·2	40·4
			Oligocene 1	Rupelian	Rup	*48*	*38*	*40*	*36*	*37/38*	38/39		*38*			37		38·0	Rup	5·2	
			Eocene 3	Priabonian	Prb				*45*						40			42·0	Prb	4	
			Eocene 2	Bartonian	Brt										44				Brt	8·5	
			Eocene 2	Lutetian	Lut				*52*						49			50·5	Lut		
			Eocene 1	Ypresian	Ypr	*68*	*58*	*60*	*58*	*53/4*	54/5		*55*		53·5	53·5		54·9	Ypr	4·4	
			Paleocene 2	Thanetian	Tha										60			60·2	Tha	5·3	
	(Cz)	(TT) (Pg)	Paleocene (Pal) 1	Danian	Dan			*70*	*63*	*65*	67		*65*	65	65	65		65	Dan	4·8	
	Mesozoic	Cretaceous	K2 Senonian	Maastrichtian	Maa				*72*	*70*	72			72	70	73	Odin 1978	73	Maa	8	79
			K2 Senonian	Campanian	Cmp					*76*	78			84	78	84		83	Cmp	10	
			K2 Senonian	Santonian	San				*84*	*82*	84			88	82	86		87·5	San	4·5	
			K2 Senonian	Coniacian	Con					*88*	90		*80*	90	86	88		88·5	Con	1	
			K2	Turonian	Tur				*90*	*94*	96			92	92	90		91	Tur	2·5	
			K2	Cenomanian	Cen				*110*	*100*	102	*95*	*100*	106	100	95		97·5	Cen	6·5	
			K1	Albian	Alb				*120*	*106*	109			116	108	107	Van Hinte 1978 a	113	Alb	15·5	
			K1	Aptian	Apt					*112*	115			123	115	115		119	Apt	6	
			K1	Barremian	Brm					*118*	121		*118*	127	121	121		125	Brm	6	
			K1 Neocomian	Hauterivian	Hau					*124*	127			130	126	126		131	Hau	6	
			K1 Neocomian	Valanginian	Vlg					*130*	133			136	131	131		138	Vlg	7	
		(K)	K1 Neocomian	Berriasian	Ber	*108*	*127*	*135*	*135*	*(136)*	(139)	*(125)*	*141*	143	135	135		144	Ber	6	
		Jurassic	J3 Malm	Tithonian	Tth					*(146)*	(150)	*(145)*		149	138	141	Van Hinte 1978 b	150	Tth	6	69
			J3 Malm	Kimmeridgian	Kim					*151*	155			157	143	143		156	Kim	6	
			J3 Malm	Oxfordian	Oxf					*157*	161		*160*	162	149	149		163	Oxf	7	
			J2 Dogger	Callovian	Clv					*162*	166			166	156	156		169	Clv	6	
			J2 Dogger	Bathonian	Bth				*166*	*167*	171				165	165		175	Bth	6	
			J2 Dogger	Bajocian	Baj								*176*	177	171	171		181	Baj	6	
			J2 Dogger	Aalenian	Aal					*172*	176				1·74	174		188	Aal	7	
			J1 Lias	Toarcian	Toa					*178*	182			188	178	178		194	Toa	6	
			J1 Lias	Pliensbachian	Plb					*183*	187			198	183	183		200	Plb	6	
			J1 Lias	Sinemurian	Sin					*188*	192				189	189		206	Sin	6	
		(J)	J1 Lias	Hettangian	Het	*145*	*152*	*180*	*181*	*192*	196		*195*	211	192	192		213	Het	7	
		Triassic	Tr3	Rhaetian	Rht									220		(197)	Banks 1973	219	Rht	6	35
			Tr3	Norian	Nor									228		(202)		225	Nor	6	
			Tr3	Carnian	Crn				*200*	*205*	210			234		(207)		231	Crn	6	
			Tr2	Ladinian	Lad									238		(214)		238	Lad	7	
			Tr2	Anisian	Ans					*215*	220			242		(221)		243	Ans	5	
			Scythian Tr1	Olenekian – Spathian	Spa											(224)			Spa	1¼	
			Scythian Tr1	Olenekian – Smithian	Smi											(228)			Smi	1¼	
			Scythian Tr1	Induan – Dienerian	Die											(231)			Die	1¼	
	(Mz)	(Tr)	Tr1 (Scy)	Induan – Griesbachian	Gri	*193*	*182*	*225*	*230*	*225*	230	*235*	*230*	247	225	235		248	Gri	1¼	
	Paleozoic	Permian	P2	Tatarian	Tat					*230*	235			252		(239)		253	Tat	5	38
			P2	Kazanian	Kaz					*240*	245			259		(242)			Kaz	2½	
			P2	Ufimian	Ufi								*251*		240	(255)		258	Ufi	2½	
			P1	Kungurian	Kun					*256½*	262½			269		(266)		263	Kun	5	
			P1	Artinskian	Art					*266½*	272½			278		277		268	Art	5	
			P1	Sakmarian	Sak											(248)			Sak	9	
	(Ph)	(P)	P1	Asselian	Ass	*227*	*203*	*270*	*280*	*280*	286	*>276*	*280*	288	290	290		286	Ass	9	

Chart 1.2. Pre-Permian.

Eon	Era	Period / Sub-period	Epoch	Age		Abbrev.	Holmes 1937	Holmes 1947B	Holmes 1959	Kulp 1961	Geol. Soc. London 1964 constants old	new	G.S.L. 1971	Van Eysinga 1975	Armstrong 1978	BP 1980 (internal)	Pickton in Harland & Hambrey 1981	This work 1982 Ma	Age	Duration
Phanerozoic	Palaeozoic	Permian P_1		Asselian		Ass	227	203	270	280	280-	286	>276	280	288	290	290	286	Ass	9
		Carboniferous, Pennsylvanian	Gzelian	Noginskian	C	Gze													Gze	?
			Kasimovian	Krevyakinsk.	B, A (Stephanian)	Kas													Kas	?
			Moscovian	Myachkov.	Ctb, D, C, B, A (Westphal.)	Mos					292½	296·5			307	300	300	296	Mos	?
																	308			34
		(Pen)	Bashkirian	Yeadonian	C, B (Namurian)	Bsh					312½	319·5			330	315	(320)	315	Bsh	?
																	(328)	320?		
		Mississippian	Serpukhovian		A	Spk													Spk	13 ?
											325	332		325	341	325	(335)	333		
			Visean	Holkerian		Vis													Vis	19 (40)
										320	337½	344·5			356	335	(350)	352		
		(C)(Mis)	Tournaisian	Ivorian, Hastarian		Tou													Tou	8
							275	255	350	345	345	352	360	345	368	345	365	360		
		Devonian	D_3	Famennian		Fam					353	360			379		373	367	Fam	7
				Frasnian		Frs				365	359	366			385	358	381	374	Frs	7
			D_2	Givetian		Giv									391		388	380	Giv	6
				Eifelian		Eif				390	370	377		370	396	370	395	387	Eif	7 (48)
			D_1	Emsian		Ems					374	381			401		401	394	Ems	7
				Siegenian		Sig					390	398			406		406	401	Sig	7
		(D)		Gedinnian		Ged	313	313	400	405	395	403	405	395	(417/425)	405	411	408	Ged	7
		Silurian	Pridoli			Prd											415	414	Prd	6
			Ludlow			Lud									432		422	421	Lud	7
			Wenlock	Sheinwoodian		Wen								423	440		428	428	Wen	7 (30)
		(S)	Llandovery			Lly													Lly	10
							341	350	440	425	435	443		435	446	435	440	438		
		Ordovician	Ashgill			Ash													Ash	10
															445		447	448		
			Caradoc	Costonian		Crd													Crd	10 (67)
										445	445	453		450	465		469	458		
			Llandeilo			Llo													Llo	10
															477		477	468		
			Llanvirn			Lln													Lln	10
															492		485	478		
			Arenig			Arg													Arg	10
							392	430	500	500	500	509		500	500	500	495	488		
		(O)	Tremadoc			Tre													Tre	17
															510		505	505		
		Cambrian	Merioneth (Mer)	Dolgellian		Dol													Dol	9
				Maentwrogian		Mnt													Mnt	9
										530	515	524		515	524	515	520	523		
			St. David's (St D)	Menevian		Men													Men	9
				Solvan		Sol													Sol	8 (85)
											540	549		540	545	540	540	540		
			Caerfai	Lenian		Len													Len	15
				Atdabanian		Atb													Atb	15
(Ph)	(Pz)	(Є)	(Crf)	Tommotian		Tom	470	510	600		570	579	574	570	574	570	570	570	Tom	20
																	590	590		
Proterozoic	Sinian	Vendian	Ediacaran			Edi											640		Edi	40 (80)
			Varangian (Var)			Mortensnes											700	650	Mor	20
		(V)				Smalfjord													Sma	20
	(Z)	Sturtian (U)				U												670	U	130
	Riphean	?																800		
																		~1050		
		Yurmatin (Y)				Y												~1350	Y	~300
	(Rip)	Burzyan (B)				B												~1650	B	~300
																		~2100		
(Pt) 2500		Huronian (H)				H												~2400	H	~300
																		~2630		
		Randian (Ran)				Ran												~2800	Ran	170
Swazian																			Sw	950
																		~3750		
		Isuan (I)				I													I	~150
Hadean		lunar sequence (Hde)				Hde													Hde	

Sources noted beside the Pickton column: Bouroz 1978, Gordon+Manet 1978; McKerrow 1978; Ross *et al* 1978; Cowie+Cribb 1978.

N.B. These divisions are not established internationally. They are selected from works as discussed in chapter 2 so as to illustrate how the Precambrian chronostratic scale may develop around good successions anywhere.

Chronometric definitions (Ga)

Proterozoic (Pt)	Pt_3	1·0	Hadrynian	Z 0·8
	Pt_2	1·8	Helikian	Y
	Pt_1		Aphebian	X
— 2·5 —				
Archean (Ar)	Ar_3	2·9?		W
	Ar_2	3·5?		
	Ar_1			
— 4·0 —				
Priscoan (Pr)				

might by analogy refer here also to apparent rather than (indeterminable) true age by time-correlation. There are structural uncertainties as to whether the relationships between rocks determined are assumed. No way has yet been devised of expressing these uncertainties concisely. The expression $\pm x$ is misleading if given without qualification. One major cause of discrepancy between numerical values for the same rock has been resolved by international co-operation, namely, the agreement to use the same decay constants. Appendix 1 provides the means to convert older values to those based on constants agreed in 1976.

It is the main purpose of this work to show not so much our conclusions for a preferred time scale for 1982 – which it will be seen does not differ greatly from previous efforts – but rather to show as clearly as we can how such a scale has been constructed. Believing, as we do, that a scale is in need of frequent revision to take account of new results from many disciplines, we hope this work will encourage improvement of it by displaying its limitations. One unintended example of this is worth noting. Ideally the chapters would have been written in the order in which they appear, so that after defining the chronostratic scheme our best efforts at calibrating it in years would be made and this calibration would then be applied to the magnetostratigraphic scale. As it happened the authors were unable to work in that sequence and the values used in Chapter 3 were arrived at and the drawings were completed before a final draft of Chapter 2 was available.

The final draft of Chapter 2 differed from the earlier one in many minor but in some major respects. Significant changes in Carboniferous classification had been made. It will be seen from Chart 1.2 that radiometric ages had been related to European divisions of largely continental facies but it is a simple matter to convert to an alternative scale. A different problem emerged for the Triassic Period. The Scythian Epoch had been treated as a single age unit but was then divided into four ages. On this basis its duration has therefore probably been underestimated. Conversely the Rhaetian Age appears to be characterised by only one biozone of the kind of which six are listed for the Norian Age. So the duration of Rhaetian has probably been overestimated. But, because a tie-point occurs between Scythian and Rhaetian ages we cannot simply increase one at the expense of the other.

The number of subdivisions or their duration is unimportant when all their boundaries are well defined if they can be correlated or calibrated in time. Even if Rhaetian spanned only one-sixth of Norian time we see no reason, as some do, to suppress it or reduce its rank on these grounds provided it fulfils a distinctive correlation function. Moreover, there is no principle that can be adduced to argue that effective biozones have or should have equal duration. Nevertheless, in constructing a geologic time scale the fact is that very few boundaries are well calibrated at tie-points. Such tie-points are discussed in detail in Chapter 3 and are indicated by thicker lines in the right-hand column of Charts 1.1 and 1.2. For longer intervals between tie-points we have used a rough and ready method treating ages as being of approximately equal duration and so interpolating them between tie-points. An obvious next refinement is to consider a number of different aspects of the history encompassed by each age to assess more likely comparative durations. This we had no time to do.

So already before publication it is clear that the dual scale offered here is in need of revision. Indeed this underlines our initial statement that this kind of dual time scale is expected to improve but can never be final.

2
The chronostratigraphic scale

2.1 Introduction

In this chapter we attempt to set out a chronostratigraphic (chronostratic) scale. Some parts of it are already well established by international decision and/or conventional usage; other parts are not. Where no standard has been agreed we have taken the liberty to suggest one. This will irritate some readers; but it may be useful to have an interim scheme for the whole scale and it may hasten the day when such a scale is established. Although the first steps towards such an international convention were taken at the International Geological Congress (IGC) in Bologna in 1881 (Anon. 1882) it was only by slow degrees that the nature of the task was understood to require the precise definition of boundaries. An attempt to agree a Pliocene–Pleistocene boundary in 1948 at the IGC in London was an important first step. A thorough application of the principle awaited the decision of the International Union of Geological Sciences (IUGS) to establish the Silurian–Devonian boundary at Klonk in Bohemia – finally decided in 1972 at the IGC in Montreal (McLaren 1977). Under the same authority there are now active groups working on nearly every remaining boundary, so that within five or ten years we may expect the main points of the scale to be established.

The essential requirement of such a scale is a sequence of reference points defined in (boundary) stratotype sections which have a good correlation potential. A sufficient spread of these, named (e.g. Klonk for Silurian–Devonian boundary) or otherwise labelled, would alone define a chronostratic scale (Hughes *et al.* 1967). Of course, when such a scale is established it does no more than provide a single agreed standard for calibration or against which to correlate.

In order to convert the traditional stratigraphic names and classifications to a single standard stratigraphic scale for general use, three kinds of decision or agreement need to be made, and made by a single authority (i.e. IUGS): (1) a scheme of divisions with appropriate classification together with (2) agreed names for each division that shall correspond to the time spans between the boundaries and (3) agreed standardisation of the boundaries. We consider these three elements of a chronostratic scale.

(1) The classification has developed traditionally on a hierarchical basis with eons (e.g. Phanerozoic), eras (e.g. Mesozoic), periods (e.g. Jurassic), epochs (e.g. Late Jurassic), ages (e.g. Oxfordian) and chrons (e.g. Mariae). The number of ranks is not a matter of principle but an accident of history. There have been attempts to standardise, but sub-eras, sub-periods or sub-epochs have intervened. The rock formed anywhere in such intervals constitutes respectively the eonothem, erathem, system, series, stage and chronozone. A convenient feature of the hierarchical classification is that boundaries of divisions of higher rank shall coincide with those of lower rank; this is the convention of coterminosity. A hierarchy is not essential, and if one is used it matters little whether the distinct ranks in the hierarchy are strictly maintained. Provided each span is properly defined, then usage will consolidate a suitable number of names. Higher ranks have descriptive advantages for slower or uncertain timing and they avoid the need to remember names of lower ranks.

(2) The names for the ages mostly began as formational names of which some were used more widely as regional stages. They mostly have original body-stratotype localities and sections. In constructing the standard scale for a particular span of time of a given rank one name (and no more) is needed. This may be one most familiar for that interval or it may be newly coined. If it be a familiar name – as most are – then boundaries will be defined, possibly elsewhere, and then it may cease to name exactly the original interval. So in pressing the claims of a favourite name for international use, its supporters must accept that its boundaries, which really define the named time span, could well be fixed in better localities elsewhere.

(3) Definition and standardisation is a two-fold process. Agreement is first necessary as to the approximate time in Earth history that would make a boundary convenient and acceptable. This may well be decided on a biostratigraphic basis before the locality for the reference point in the boundary stratotype is decided.

In this chapter we divide geologic time, as is traditional, into periods (after Precambrian time); each is introduced historically and there is a commentary on a corresponding chart. In each case the charts distinguish the chronostratic time scale on the left by lower-case lettering, while the rock units or regional stages are printed to the right in capital letters.

An attempt is made to present a single standard scale with its hierarchy of time divisions and, if one is not yet agreed, a tentative scheme is suggested. On the right-hand side of the chart a selection of local successions or regional stages is selected from amongst the multitude available. These are included to give an indication of distant equivalents and exact correlation is not intended.

For epochs we prefer stratigraphic names to the formal qualifications Early, Middle and Late but generally these are not available. The advantage of such a name is that it can be defined unambiguously, whereas the terms Early, Middle and Late have often been used in different senses and are also liable to confusion with the informal usage (early, middle and late), which is perhaps more generally useful. We have not attempted to standardise the ending of epoch names. For ages on the other hand the suffix '-ian' is generally applied.

For time scale divisions we have used a system of abbreviations which has been tested for uniqueness and may have wider use (see Appendix 3). Abbreviations for eon, era and period names have largely been determined internationally and we have taken into ac-

Chart 2.1. Precambrian time scales.

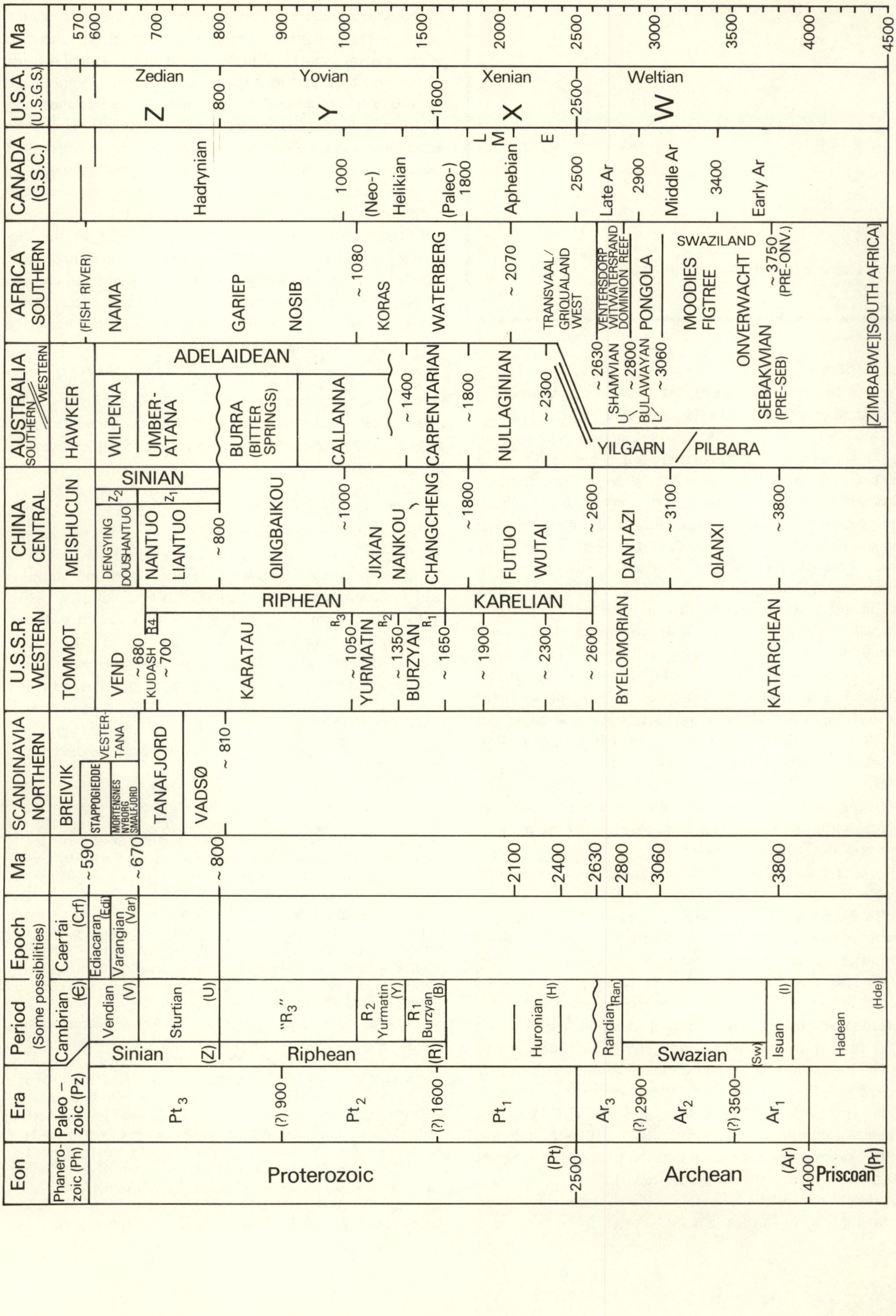

Column	Entries (top to bottom)
Eon	Phanerozoic (Ph); Proterozoic (Pt); 2500; Archean (Ar); 4000; Priscoan (Pr)
Era	Paleo-zoic (Pz); Pt_3; (?) 900; Pt_2; (?) 1600; Pt_1; Ar_3; (?) 2900; Ar_2; (?) 3500; Ar_1
Period (Some possibilities)	Cambrian (Є); Sinian (Z): Vendian (V), Sturtian (U); Riphean (R): "R_3", R_2 Yurmatin (Y), R_1 Burzyan (B); Huronian (H); Randian (Ran); Swazian (Sw); Isuan (I); Hadean (Hde)
Epoch	Caerfai (Crf); Ediacaran (Edi); Varangian (Var)
Ma	~ 590; ~ 670; ~ 800; 2100; 2400; 2630; 2800; 3060; 3800
SCANDINAVIA NORTHERN	BREIVIK; STAPPOGIEDDE; MORTENSNES, NYBORG, SMALFJORD; VESTERTANA; TANAFJORD; VADSØ; ~ 810
U.S.S.R. WESTERN	TOMMOT; VEND; ~ 680; KUDASH; R4; ~ 700; KARATAU; R_3; ~ 1050; YURMATIN; R_2; ~ 1350; BURZYAN; R_1; ~ 1650; ~ 1900; ~ 2300; ~ 2600; BYELOMORIAN; KATARCHEAN; RIPHEAN; KARELIAN
CHINA CENTRAL	MEISHUCUN; SINIAN: DENGYING, DOUSHANTUO (Z_2), NANTUO, LIANTUO (Z_1); ~ 800; QINGBAIKOU; ~ 1000; JIXIAN; NANKOU; CHANGCHENG; ~ 1800; FUTUO; WUTAI; ~ 2600; DANTAZI; ~ 3100; QIANXI; ~ 3800
AUSTRALIA SOUTHERN / WESTERN	HAWKER; WILPENA; UMBER-ATANA; BURRA (BITTER SPRINGS); CALLANNA; ADELAIDEAN; ~ 1400; CARPENTARIAN; ~ 1800; NULLAGINIAN; ~ 2300; ~ 2630; U SHAMVIAN; ~ 2800; L BULAWAYAN; ~ 3060; SEBAKWIAN (PRE-SEB); YILGARN / PILBARA; [ZIMBABWE]
AFRICA SOUTHERN	(FISH RIVER); NAMA; GARIEP; NOSIB; ~ 1080; KORAS; WATERBERG; ~ 2070; TRANSVAAL/ GRIQUALAND WEST; VENTERSDORP; WITWATERSRAND; DOMINION REEF; PONGOLA; SWAZILAND: MOODIES, FIGTREE, ONVERWACHT; ~ 3750 (PRE-ONV.); [SOUTH AFRICA]
CANADA (G.S.C.)	Hadrynian; 1000; (Neo-) Helikian; (Paleo-) 1800; Aphebian (L, M, E); 2500; Late Ar; 2900; Middle Ar; 3400; Early Ar
U.S.A. (U.S.G.S.)	Zedian Z; 800; Yovian Y; 1600; Xenian X; 2500; Weltian W
Ma	570; 600; 700; 800; 900; 1000; 1500; 2000; 2500; 3000; 3500; 4000; 4500

count the symbols used on the Unesco *Geological world atlas* (Choubert & Faure-Muret 1976). For the rest we have introduced a three-letter system after finding that two letters gave insufficient scope for recognition and uniqueness. We prefer the numerical subscript to indicate early, middle and late. Two, three or more divisions can so be noted, and it avoids the unfortunate ambiguity of the abbreviation 'L' for Late or Lower. We recommend generally to write such words in full, also to show by the initial if it is formal (with capital) or informal (with lower-case initial).

The chronometric column headed Ma (SI for millions of years) is added for convenience. The figures are transferred from Chapter 3 and are not part of the argument of this chapter. As it happens the work on Chapters 3 and 4 preceded that on Chapter 2 and the numerical calibration (and the drawing of figures in Chapters 3 and 4) were completed on an earlier chronostratic scale. If there had been time to consider the whole work together we should have applied the knowledge potentially available from Chapter 2 to give better relative durations for the different ages.

Where necessary transliteration of names has been revised: Chinese in the contemporary Pinyin, and Russian according to the standard PCGN/BGN system (Permanent Committee on Geographic Names, US Board on Geographic Names as used, e.g. in *The Times Atlas*). As already observed we have accepted American spelling to avoid diphthongs in names such as Paleozoic. This will not find favour with some British colleagues but we make this choice deliberately, believing that we are setting out a time scale for international use, the names of which should not need transliterating within the English language. It is the privilege of those whose ancestry and mother tongue is English to contribute to the evolving international language of science.

2.2 Precambrian time scales

The confusing historical sequence of names and classifications for Precambrian rocks and divisions has been recounted many times (e.g. Wilmarth 1925, Harland 1974). In the end, of the many early names proposed, only *Proterozoic* and *Archean* achieved the approval of the IUGS. The consolidated name Precambrian was also agreed in 1972 by the IUGS against the alternative pre-Cambrian (cf. pre-Vendian or pre-Pleistocene). It was thought by some to be an advance that recognised a supposed unity of Precambrian rocks and studies in contrast to Phanerozoic stratigraphy. But identical principles and procedures apply to all rocks even though the evolution of the Earth resulted in widely different rocks forming at different times and at any one time.

2.2.1 Chronometric divisions

Many students of Precambrian rocks think of dividing Precambrian time at boundaries defined only by numbered multiples of duration of a standard year. It was first attempted to discover natural divisions of Earth or regional history and to define the age of the event in years; for example the Geological Survey of Canada (GSC) introduced the names Hadrynian, Helikian and Aphebian for tectonic divisions of the Canadian shield with boundaries at 880, 1640 and 2390 Ma (Stockwell 1964) and successively rounded off to 1000, 1800 and 2500 Ma (Douglas 1980). At the same time the United States Geological Survey (USGS) accepted the principle of artificial boundaries in round numbers for Precambrian W, X, Y and Z (James 1972), even though the figures chosen also approximated to convenient American map divisions, so making it more likely that the scheme would remain a regional one.

The Subcommission on Precambrian Stratigraphy of the IUGS agreed by a majority at its Fifth Meeting in Duluth, Minnesota, 15–19 September 1979 to recommend provisionally to divide the Proterozoic Eon into three eras divided at 900 and 1600 Ma. Competing opinions favoured a boundary at 1500 Ma and/or use of four divisions. The Proterozoic–Archean boundary at 2500 Ma had already been decided. It was also suggested to divide Archean time at 2900 Ma and 3500 Ma (a three-fold division) 'to fit the Archean geologic record as now known in most parts of the world'. These recommendations and suggestions were to be reviewed at the next meeting in 1982 (Sims 1980).

Precambrian time may be conceived as extending indefinitely backwards before 4600 Ma and for that matter in so far as it has meaning before 10 000 Ma. The oldest commonly used division of Precambrian time is Archean. It originally referred to the oldest known rocks. According to the 1980 *AGI Glossary of Geology* (Bates & Jackson 1980) Archeozoic is 'the earlier part of Precambrian time corresponding to Archean rocks'. Archeozoic was indeed introduced earlier than Archean (Harland 1974). In 1976 the Subcommission on Precambrian Stratigraphy treated Archean as a standard time division terminating at 2500 Ma, no initial boundary was established. However, its relation to known terrestrial rocks persists in the minds of most geoscientists because in nearly every case attempts to divide Archean time into Early, Middle and Late; 1, 2 and 3 etc. result in divisions such that Middle and Late Archean generally refer to rocks approximately in the intervals 3500 to 3000 Ma and 3000 to 2500 Ma. Moreover the oldest rocks so far recorded from southern West Greenland and from southern Africa are commonly referred to as 'earliest Archean'.

With this in mind and with an increasing need to refer to the still earlier history of the Earth and Solar System, a solution that some of us have suggested is to define the initial Archean boundary at (say) 4000 Ma. The Eon would then span exactly 4000 to 2500 Ma. This would do no violence to majority usage of the name. On this basis we name pre-Archean time Priscoan from Prisco time (Latin priscus = former, previous, olden of times) which was suggested for this purpose in 1975 (Harland). Others of us, however, would prefer to extend Archean time back to 5000 Ma and redefine its divisions. The consequence of this alternative procedure is that all rocks in the Solar System, including meteorites are likely to be Archean in age and no new names are required. Archean time could be subdivided into five 500 Ma units, provisionally named Archean 1–5. In any case no international convention can be adopted until settled by the IUGS Committee on Stratigraphy. Until that time we make these suggestions to test opinion; for it seems that, because of prevailing preoccupation with terrestrial Archean rocks, little thought has been given to earlier history as an extension of the same fundamental stratigraphic discipline. The widespread colloquial use of adjectives as nouns (e.g. 'the Precambrian', 'the Archean') may suggest that the record in rock is in mind and some extension is needed.

2.2.2 Chronostratic divisions

The alternative and equally valid approach to a Precambrian time scale is to follow the principles being applied successfully to define Phanerozoic divisions by reference to supracrustal rocks. On the coterminosity principle the terminal Proterozoic boundary coincides with the initial Phanerozoic, Paleozoic and Cambrian boundary. This is expected (in 1983 or 1984) to be defined in a stratotype. It will have the effect of making the Proterozoic division a hybrid – beginning with a chronometric boundary and ending with a chronostratic one.

Similar boundaries will be defined to limit divisions where there are good stratal characteristics for correlation. The five columns listed on Chart 2.1 represent successions in Scandinavia, USSR, China, Australia and Africa; each indicates a possible chronostratic standard for some part of Precambrian time. These columns were selected and simplified as follows. The Scandinavian column is based on that in Finnmark by Edwards & Føyn (1981). The USSR column is from Keller (1979, p.421). The China column is composite and based on Wang & Liu (1980). In it the Sinian units are from the Yangtze Gorge and south China region and the older units are from north China. The southern Australia column is based on Coats (1981) and Rutland *et al.* (1981) and the units from western Australia are plotted approximately according to Hallberg & Glikson (1981). The southern Africa column is composite, some names being selected from those proposed by Kent & Hugo (1978) for international chronostratic use: Swazian, Randian, Vaalian, Mogolian (from River Mogol in Griquatown West) and Namibian. The rock sequences are also adapted from Anhaeusser & Wilson (1981) for Swazian rocks, and Button *et al.* (1981) for Randian through Mogolian. Button *et al.* included the Dominion Reef as well as the West Rand and Central Rand Groups in the Witwatersrand Supergroup but we identify it as one of the familiar 'Witwatersrand Triad'. The Zimbabwe sequence is from Nisbet, Wilson & Bickle (1981).

No international decisions have yet been made, so it might be premature to select from the above successions those that may provide international standards. Nevertheless in the period and epoch columns of our chart the suggestions made are indicative of the approach which will develop by degrees as the chronostratic scale extends backwards in time.

The Sinian Era. With the impending definition of the initial Cambrian boundary the question arises as to what scheme of classification and nomenclature shall be applied to the preceding chronostratic time divisions. The ambiguity of the name Eo-Cambrian, both in definition and classification, has effectively ruled it out for international use. It may, however, still have a value in informal use to indicate an uncertain age at about the Phanerozoic threshold. Sinian and Vendian are two contenders for the formal division to precede Cambrian. Sinian has priority and was extensively used in the USSR as well as China for many years. Then when Precambrian studies in the USSR advanced beyond those in China the name Vendian was introduced and has been widely used. It seems almost to have been adopted internationally. There are however some difficulties, not least that good type sequences with reference points have not been established, nor has the classification scheme achieved agreement even in the USSR. In the meantime research in China has surged and the claims of Sinian are being pressed again. Sinian had two widely differing meanings – including or excluding the thick northern succession of Jixian altogether older than about 800 Ma. Now that sequence has been excluded, Sinian is again restricted to rocks younger than about 800 Ma as originally so named from the widespread platform succession typified in the Yangtze Gorge (as on Chart 2.1). A more complete geosynclinal succession in Xinjiang may well provide a better standard with four distinct tillite horizons that facilitate international correlation elsewhere. These are the Bayishi, Altungol, Tereeken and Hangeerqiaoke of Wang *et al.* (1981). The suggestion tabulated in Chart 2.1 is to establish a Sinian Era or Sub-era (Bayishi to Hangeerqiaoke) divided into two periods, Sturtian and Vendian. So Vendian is retained in the restricted sense commonly used so as to include two epochs (e.g. Harland & Herod 1975), i.e. Varangian (Kulling 1951) and Ediacaran (Cloud 1972). This would be appropriate if a Xinjiang section for Sinian correlation were used and might avoid some difficult problems of correlation between Vendian and latest Riphean (e.g. Vidal 1979). Within this span there are relatively good palynological and climatological bases for correlation. Vidal suggested a three-fold division of Vendian namely Valdaian, Varangerian and an early division (Vidal 1981). Certainly northern Europe has good successions for correlation in this interval but Valdaian is not used here (it is already established in Quaternary nomenclature). Varangian has priority over Varangerian. Vetternian was suggested by Vidal for the early division (personal communication 1981); it is related to a small and isolated outcrop area in Sweden.

The classification favoured here is designed to satisfy the aspirations of the Australian, Chinese, Russian and Scandinavian communities who have each contributed significantly to this latest Precambrian scale in describing key successions. The Ediacaran division has been proposed for some time e.g. Cloud (1972), Glaessner (1977) and most recently Jenkins (in press) who renewed the case for an 'Ediacaran Period'. On this basis at least two ages are proposed within it, namely (later) Poundian and (earlier) Wonokan. Varangian is also divided into at least two ages, (later) Mortensnes and (earlier) Smalfjord from the two tillite horizons in the Varanger Fjord area. Sturtian is similarly capable of divisions that have a correlation potential combining biostratigraphic and climatic characters.

Pre-Sinian eras. If Riphean is to be used for the next earlier era (as shown on Chart 2.1) then a radical restriction of R_3 would be needed as Karatavian (based on Karatau) overlaps Sturtian. Confusion has long existed in the later Riphean classification and on the basis suggested here, if (and only if) a Russian-based international chronostratic division were wished by Soviet colleagues would a new definition be necessary.

Still earlier chronostratic divisions are merely suggested here to indicate the way in which the classification will develop. The Huronian sequence with its four potential epochs (three of them being glacial cycles) has a distinct correlation potential and would provide a good standard for this part of Earth history.

Earlier names in the column are culled from Kent & Hugo (1978) who with good reason have argued that African Precambrian successions provide good standards for international use. Of these we adopt the Archean sequences: Randian (including the Witwatersrand) and Swazian to include the rich sequences of Swaziland and Zimbabwe. Isuan (from West Greenland) is suggested for the earlier terrestrial record. Hadean is adopted from Cloud (1976) for the pre-Isuan sequence whose record may not be preserved on Earth but is perhaps best known from the Moon.

2.3 The Phanerozoic Eon

The *Palaeozoic Series* was first proposed by Sedgwick in 1838 for the rocks overlying the *Primary Stratified Groups*; J. Phillips in 1840 applied Paleozoic to those 'transitional' rocks up to and including the Old Red Sandstone, and in 1841 he extended it to include all rocks from Cambrian to Permian, when he also introduced the names *Mesozoic* and *Cainozoic* in the modern sense.

'Post-Pre-Cambrian' time was the alternative to expressing Cambrian through Holocene time until Chadwick in 1930 proposed *Phanerozoic* for this span (Paleozoic + Mesozoic + Cenozoic) and *Cryptozoic* was his Proterozoic + Archeozoic (e.g. Harland 1974).

All the above names carry a descriptive meaning related to the evolution of animal life; but they have also become conventional names in a stratigraphic hierarchy and that is how we use them. In other words these boundaries are coterminous with successively lower ranks at the defined point. Thus the initial Phanerozoic boundary will be defined at the point that will define the initial Cambrian boundary and it is not to begin according to successive discoveries or opinions about animal evolution.

2.4 The Cambrian Period

The history of the definition and classification of Cambrian rocks in the British Isles, especially in North Wales, from the initial publication of the name in 1835 by Adam Sedgwick to the ultimate resolution of the conflict about the Cambrian–Silurian boundary, through establishment of the Ordovician System by Lapworth in 1879, has been outlined by Stubblefield (1956) and Cowie, Rushton & Stubblefield (1972). Cambria is a variant of Cumbria (ancient British kingdom, in present day north-west England) latinised from the Welsh *Cymry* (= fellow country-man, compatriot against invading Anglo–Saxons). The Celtic *Cymru* survives only in Wales and Cambrian already pertained to Wales when Sedgwick adopted it. Detailed stratigraphic information on the various divisions and units proposed appear in the *Lexique stratigraphique international,* not only for Cambrian rocks of the British Isles, but for rocks of all ages and many regions. These volumes will not generally be referred to here and are now mainly useful for the history of stratigraphic names. However, Vol. 1, Europe Fasc. 3aIII by Stubblefield in 1959 gives much definitive information on the Cambrian strata of the British Isles.

Sedgwick had first used the group names Lower and Upper. His Lower Cambrian was divided into the Bangor Group (with Llanberis Slates and Harlech Grits) below and the Ffestiniog Group (with Lingula Flags, Tremadoc Slates and Arenig Slates) above; his Upper Cambrian included Bala strata and extended up to the base of the Woolhope Limestone. Since 1879 the Cambrian System has comprised only Sedgwick's Lower Cambrian and not all of that. Arenig was the lowest series of the new Ordovician System and Tremadoc has long properly been a part of the Cambrian System, as argued by Whittington & Williams (1964) for example. However, we accept the prevailing international practice, whether or not based on a misunderstanding, to make Tremadoc the first Ordovician epoch.

The difficulties of Cambrian classification and correlation arise partly from the paucity of appropriate fossils. A consequence of this led in North America to the early placing of *Olenellus*-bearing rocks above, rather than below, those with the more abundant *Paradoxides.* These relationships were hardly cleared up until about 1890 (Cowie *et al.* 1972).

In central Europe Barrande (in 1859) listed the old primordial fauna with *Paradoxides* and Brøgger showed the Olenellid *Holmia* to be older than *Paradoxides* in central Norway (e.g. Harland 1974).

In North America in the meantime the name Acadian, introduced by J. W. Dawson in 1868, was proposed for strata characterised by *Paradoxides* etc. as 'the oldest member of the Paleozoic of America' by Walcott in 1891 when he also introduced the name Potsdam for Upper Cambrian. The order was sorted out but not the nomenclature. In 1903 Walcott replaced Potsdam by Saratogan and then in 1912 by St Croixian. He had used Georgian originally from 1882 (Hitchcock in 1861) for Lower Cambrian and in 1912 Walcott recommended Waucoban as a provincial rock series and applied the name Taconian for general use for the Early Cambrian Epoch. The Taconic System had been proposed as early as 1842 by Emmons for the earliest system beneath the New York System (Wilmarth 1925).

In due course in North America these and other names have had regional significance often contrasted between Appalachian and Cordilleran usage, i.e. Georgian, Acadian and Potsdam (or Saratogan) in the east (e.g. Blackwelder 1912) and somewhat later, in the west, Waucoban, Albertan and Croixian. Usage now tends to favour the western nomenclature in North America.

This leaves us today with a general acceptance of three Cambrian epochs, but no internationally accepted names for them. Because of the widely different usages of the names Early, Middle and Late Cambrian, and also because of the need to refer informally to parts of epochs (e.g. early Early Cambrian or worse) a nomenclature defined by a stable classification and standardised in type areas, not necessarily near the eponymous localities, is to be preferred. The choice made here is for names long used unambiguously for British Cambrian rocks and/or recently proposed by the Geological Society of London (GSL) Working Group (Cowie *et al.* 1972).

2.4.1 The Caerfai Epoch

Caerfai is used here in preference to Comley for the Early Cambrian Epoch because Comley, however well defined (Cowie *et al.* 1972), could mislead so long as the Upper Comley Group of Middle Cambrian age be so named. There is not a definitive sequence in South Wales to justify

Chart 2.2. Cambrian chronostratic scale and correlation.

Pd.	Epoch	Age	Biostratigraphic correlation	Ma	WALES North	WALES South	NORWAY C. NORWAY	NORWAY Finnmark	SIBERIA	CHINA	AUSTRALIA	N. AMERICA
O	Tremadoc		*Dictyonema flabelliforme*	505	DICTYOMEMA BAND & TYNLLAN				OLENTIAN (KAZAKHSTAN)		DATSONIAN	TREMPEALEAUAN
Cambrian (€)	Merioneth (Late Cambrian) (Mer) (€3)	Dolgellian (Dol)	*Acerocare*		'LINGULA FLAGS' GROUP				SHIDERTINIAN	FENGSHAN	PAYNTONIAN	CROIXIAN
			Peltura scarabaeoides, *Peltura minor*, *Protopeltura praecursor*		DOLGELLY					CHANGSHAN	'PRE-PAYNTONIAN'	FRANCONIAN
			Leptoplastus									
			Parabolina spinulosa		FFESTINIOG						'POST-IDAMEAN'	
		Maentwrogian (Mnt)	*Olenus & Agnostus obesus*		MAENTWROG				TUORIAN	GUSHAN	IDAMEAN	DRESBACHIAN
			Agnostus pisiformis	525							MINDYALLAN	
	St David's (Middle Cambrian) (St D) (€2)	Menevian (Men): Late (Men3)	↳ *Paradoxides forchhammeri*: *Lejopyge laevigata*		CLOGAU	MENEVIAN			MAYAN	ZHANGXIA		ALBERTIAN
			Solenopleura brachymetopa									
			Ptychagnostus lundgreni & P.(G) nathorsti									
		Middle (Men2)	↳ *Paradoxides paradoxissimus*: *Ptychagnostus punctuosus*							XUZHUANG		
			Hypagnostus parvifrons									
		Early (Men1)	*Tomagnostus fissus & Ptychagnostus atavus*		CEFN COCH				AMGAN			
		Solvan (Sol): Late & Middle (Sol 2 & 3)	*Ptychagnostus gibbus*		GAMLAN	SOLVA				MAOZHUANG	TEMPLETONIAN	
		Early (Sol1)	↳ *E. oelandicus*: *Eccaparadoxides oelandicus pinus*		BARMOUTH							
			Eccaparadoxides insularis	540	HAFOTTY						ORDIAN	
	Caerfai (Early Cambrian) (Crf) (€1)	Lenian (Len)	*Anabaraspis*, *Lermontovia dzevanovskii & Paramicmacca*, *Bergeroniellus expansus*, *Bergeroniellus micmacciformis*		HARLECH: RHINOG	CAERFAI	EVJEVIK LST	DVOLBBADASGAISSA	LENAN	LONGWANGMIAO		WAUCOBAN
		Atdabanian (Atb)	*Judomia & Dipharus attleborensis*		LLANBEDR		HOLMIA SHALE		PETROTSVET	CANGLANGPU	'LOWER CAMBRIAN'	
			Judomia (& Fallotaspis)									
		Tommotian (Tom)	*Dokidocyathus lenicus*		DOLWEN		BRASTAD SH & LST	BREIVIK ?		QIONGZHUSI		
			↳ *Dokidocyathus regularis*: *L. bella*									
			L. tortuosa									
			Ajacicyathus sunnaginicus	590 ?								
P€	Ediacaran	Poundian (Pou)	*Anabarites trisulcatus*				RINGSAKER	STOPPOGIEDE	YUDOMA			

using the name Caerfai; that could well be a composite one from extra British regions. But Caerfai has, for example, been selected by the USGS (Cohee 1970) as the European standard name and appears also on Van Eysinga's (1975) chart. So it is already familiar internationally.

2.4.2 The St David's Epoch

St David's (?Davidian by analogy with Croixian) is the name suggested for the Middle Cambrian Epoch (Cowie *et al.* 1972). This seems to be entirely appropriate. Rocks with *Paradoxides* were first identified in the St David's area; and the divisions Menevian and Solvan from the same area in South Wales have a long history in Middle Cambrian stratigraphy.

2.4.3 The Merioneth Epoch

Merioneth (a former county in North Wales, now included in Gwynedd) is the name for the restricted Late Cambrian Epoch and such a name is necessary to make it clear that Tremadoc is excluded. The form of these names is left as proposed without adding the familiar '-ian' suffix and is also in conformity with some British practice for Ordovician and Silurian epochs. This practice also serves to distinguish epochs from age divisions which are generally preferred to end with an '-ian'.

The sequence of ages listed here are taken from Cowie (in press) and we have not the information to attempt to justify them. Their identity is clear to some extent from the column of correlative fossils listed also in that work.

The initial Cambrian boundary has been discussed many times and one such paper (Harland 1974) contributed to the formation of the Precambrian–Cambrian Boundary Working Group Project No.29 of the International Geological Correlation Programme (IGCP) (Leader: J. W. Cowie), which is expected to establish an internationally agreed initial Cambrian boundary in 1983/4. It is premature to discuss where the reference boundary stratotype will be; however, the Siberian successions are amongst the most serious contenders and it is not surprising that the three ages proposed by Cowie (in press) for the Early Cambrian (Caerfai) Epoch are Siberian (Tommotian, Atdabanian and Lenian).

For Middle Cambrian (St David's) ages, Cowie (in press) uses three *Paradoxides* zones as ages. In Chart 2.2 we have included these and retained Solvan and Menevian which have a long tradition, and all are South Welsh names like St David's.

In the Late Cambrian (Merioneth) Epoch Maentwrogian and Dolgellian units are retained but the Ffestiniog stage that once separated them (in a three-fold division – e.g. Cohee *et al*. 1967) has been dropped. All are North Welsh names.

The columns in Chart 2.2 are modified as explained above from Cowie (in press) for the left-hand side of the chart. On the right the Norway column is from Martinsson (1974) and gives only Early Cambrian divisions because thereafter all the Scandinavian sequences are described in terms of a complex of biostratigraphic names. The China column is from Wang & Liu (1980). The other columns are from Cowie *et al.* (1972).

2.5 The Ordovician Period

The Ordovician System was founded by Lapworth in 1879 and solved the Murchison–Sedgwick conflict with overlapping claims for their Silurian and Cambrian systems respectively. It was therefore instrumental in defining the three periods of the Early Paleozoic Era. The history was outlined by Whittington & Williams (1964) and detailed British sequences were charted by Williams *et al.* (1972).

Lapworth cited 3600 m of volcanic ash and sedimentary rocks in the Arenig–Bala district and the boundaries of the original system were clearly defined there – above the Tremadoc (Amnodd Shales), i.e. beginning with the Basal Grit of Sedgwick's Arenig Group and ending below the Llandovery (Cwm yr Aethnen Mudstones) at the top of the Foel y Ddinas Mudstone which was the youngest formation of Sedgwick's Bala Group.

Whittington & Williams (1964) recounted the complex history of error and confusion that led to various departures from the above simple definitions and so gave some support for including Tremadoc rocks in the Ordovician System. In short, while priority and the logic of the initial definition would exclude Tremadoc rocks, the fact is that a majority of geologists throughout the world have allowed these mistakes to pass unchecked into general use and so established the convention that we now follow by beginning the Ordovician Period at the point that would make the Tremadoc Group also Ordovician. There is, after all, no truth to be established – only a convention to be agreed. In fact, *Dictyonema flabelliforme* has commonly been taken as the index fossil in correlation of earliest Ordovician rocks, though a point in a reference section somewhere will be needed to standardise the initial point of the Ordovician Period.

The divisions of the Ordovician Period are therefore based on names and successions of rocks in Britain. These successions are, however, scattered and do not comprise a single or even two or three standard sequences so that no column of British rock units has been attempted in Chart 2.3. They are clearly depicted in the correlation charts in Williams *et al.* (1972).

Lapworth used two rather than three Ordovician divisions which corresponded to Sedgwick's original Arenig and Bala. This two-fold division was reaffirmed by Whittington & Williams (1964). We adopt the same Late Ordovician Subperiod bringing back the name Bala for it. Bala happens to coincide with Upper Cambrian as defined by Sedgwick in 1852 and 1873. For the earlier part of Ordovician time there is some difficulty. Sedgwick's Arenig has since been divided into Arenig, Llanvirn and Llandeilo so cannot now be used in a different sense. A three-fold division of the Period is also commonly employed, especially in America, and although Whittington & Williams argued against it, now that we have added the Tremadoc Epoch this makes the three-fold division seem more reasonable to us. The name Canadian if defined as coterminous with Tremadoc plus Arenig, would be an unambiguous name for our Early Ordovician Subperiod in spite of Canadian facies being totally foreign to the type successions. We are after all seeking names for time

Chart 2.3. Ordovician chronostratic scale and correlation.

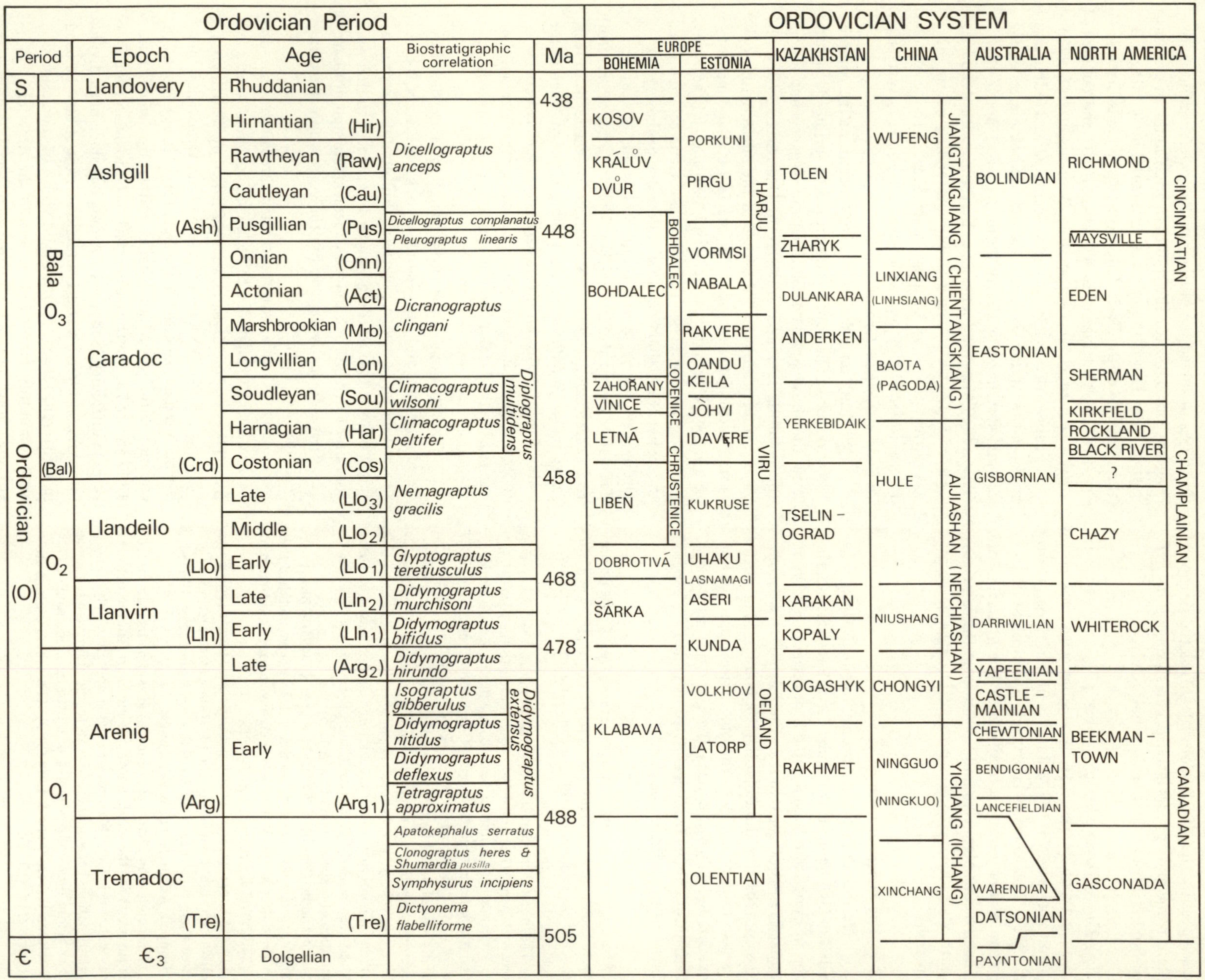

Ordovician Period						ORDOVICIAN SYSTEM					
Period		Epoch	Age	Biostratigraphic correlation	Ma	EUROPE: BOHEMIA	EUROPE: ESTONIA	KAZAKHSTAN	CHINA	AUSTRALIA	NORTH AMERICA
S		Llandovery	Rhuddanian		438						
Ordovician (O)	Bala O_3	Ashgill	Hirnantian (Hir)	*Dicellograptus anceps*		KOSOV	HARJU: PORKUNI	TOLEN	JIANGTANGJIANG (CHIENTANGKIANG): WUFENG	BOLINDIAN	CINCINNATIAN: RICHMOND
			Rawtheyan (Raw)			KRÁLŮV DVŮR	PIRGU				
			Cautleyan (Cau)								
		(Ash)	Pusgillian (Pus)	*Dicellograptus complanatus*	448			ZHARYK			MAYSVILLE
		Caradoc	Onnian (Onn)	*Pleurograptus linearis*		BOHDALEC: BOHDALEC	VORMSI		LINXIANG (LINHSIANG)		EDEN
			Actonian (Act)	*Dicranograptus clingani*			NABALA	DULANKARA			
			Marshbrookian (Mrb)				RAKVERE	ANDERKEN			
			Longvillian (Lon)				VIRU: OANDU		BAOTA (PAGODA)	EASTONIAN	
			Soudleyan (Sou)	*Climacograptus wilsoni* (*Diplograptus multidens*)		LODENICE: ZAHOŘANY	KEILA				SHERMAN
			Harnagian (Har)	*Climacograptus peltifer* (*Diplograptus multidens*)		VINICE	JÕHVI	YERKEBIDAIK			KIRKFIELD ROCKLAND
	(Bal)	(Crd)	Costonian (Cos)			LETNÁ	IDAVERE				BLACK RIVER
	O_2	Llandeilo	Late (Llo_3)	*Nemagraptus gracilis*	458	CHRUSTENICE: LIBEŇ	KUKRUSE	TSELIN – OGRAD	AIJIASHAN (NEICHIASHAN): HULE	GISBORNIAN	CHAMPLAINIAN: ?
			Middle (Llo_2)								CHAZY
		(Llo)	Early (Llo_1)	*Glyptograptus teretiusculus*		DOBROTIVÁ	UHAKU				
		Llanvirn	Late (Lln_2)	*Didymograptus murchisoni*	468	ŠÁRKA	LASNAMAGI ASERI	KARAKAN	NIUSHANG	DARRIWILIAN	WHITEROCK
		(Lln)	Early (Lln_1)	*Didymograptus bifidus*				KOPALY			
	O_1	Arenig	Late (Arg_2)	*Didymograptus hirundo*	478		OELAND: KUNDA			YAPEENIAN	
			Early	*Isograptus gibberulus* (*Didymograptus extensus*)		KLABAVA	VOLKHOV	KOGASHYK	CHONGYI	CASTLE – MAINIAN	
				Didymograptus nitidus (*Didymograptus extensus*)						CHEWTONIAN	CANADIAN: BEEKMAN – TOWN
				Didymograptus deflexus (*Didymograptus extensus*)			LATORP	RAKHMET	YICHANG (ICHANG): NINGGUO (NINGKUO)	BENDIGONIAN	
		(Arg)	(Arg_1)	*Tetragraptus approximatus* (*Didymograptus extensus*)						LANCEFIELDIAN	
		Tremadoc		*Apatokephalus serratus*	488		OLENTIAN		XINCHANG	WARENDIAN	GASCONADA
				Clonograptus heres & Shumardia pusilla							
				Symphysurus incipiens							
		(Tre)	(Tre)	*Dictyonema flabelliforme*						DATSONIAN	
€		$€_3$	Dolgellian		505					PAYNTONIAN	

divisions only. But we are at a loss for a name (short of inventing one) for Llanvirn and Llandeilo. This is not so serious because if Middle Ordovician be used then that is what we intend.

Referring now to the finer division of Ordovician time into epochs and ages we first note that pre-Bala epochs are not yet formally divided into named ages whereas eleven Bala ages have been established for some time. The ***Tremadoc*** Epoch is unambiguously related to the Tremadoc Group in North Wales with *Dictyonema;* on historical grounds it is Cambrian as explained above but by popular preference we take it as Ordovician. ***Arenig*** was reduced to its present span by separation of Murchison's ***Llandeilo*** (e.g. by Ramsay and Salter in 1866) and from this in due course Hicks in 1881 introduced the ***Llanvirn*** together with Llandeilo between Arenig and Bala. ***Caradoc*** was Murchison's name first used for Bala by Ramsay and Salter in 1866 from which in due course Marr (in 1905) separated ***Ashgill.*** All these epochs are now defined approximately in terms of graptolite zones as shown in Chart 2.3. The Ashgill–Llandovery boundary referred to below in Scotland may be related to the Welsh section where Hirnantian age is related to the Hirnantian fauna of the Foel y Ddinas Mudstone but the Dobb's Linn (south Scotland) sequence may be better as a standard for correlation.

The chart shows the major British graptolite zones used for correlation as in Williams *et al.* (1972). The consolidation of two biozones to *Diplograptus multidens* is based on common usage. The biozones are generally of longer duration than the ages, whose finer division is possible using fossil assemblages. The Tremadoc biostratigraphic correlation is based on Scandinavian biozones taken from Cowie *et al.* (1972). The post-Tremadoc divisions are from Williams *et al.* (1972) as are the columns for Bohemia, Estonia and Kazakhstan. The China section is from Sheng Shen-Fu (1980). The Australian stages are from Webby *et al.* (1981). The post-Canadian North American sequence is from Sweet & Bergstrom (1976, p.134), and the earlier part of the sequence is based on Sweet & Bergström's 1975 column, as given in Sheng Shen-Fu (1980).

2.6 The Silurian Period

The Silurian System was named by Murchison in 1839 for the Welsh Borderland tribe the Silures, and first included rocks that were claimed as Cambrian by Sedgwick. Silurian was subsequently used in two senses – to include or exclude what are now recognised as Ordovician strata, so the less ambiguous name Gotlandian was proposed in 1893 for the post-Ordovician period and competed with the name Silurian until in 1960 at the IGC in Copenhagen, Silurian was officially adopted in its restricted sense, i.e. Murchison's Upper Silurian.

Murchison included Llandovery, Wenlock and Ludlow in his Upper Silurian in 1859. These names are unambiguous and are preferred to Early, Middle and Late Silurian epochs for that reason. Lapworth in 1879 and 1880 used Valentian (approximately = Llandovery) and Salopian (approximately = Wenlock + Ludlow) but this usage was discontinued (e.g. by O. T. Jones in 1929) when a fourth division (the Downton) came into use (including part of Murchison's Tilestones that he had included with what later became the Old Red Sandstone). Whittard (1961) and Cocks *et al.* (1971) summarised the confused nomenclatural history of the Period.

The four epochs (S_1 – S_4) have more recently been recombined informally into two sub-periods (Early Silurian = $S_1 + S_2$ and Late Silurian = $S_3 + S_4$) when the divisions of the Wenlock (S_2) and Ludlow (S_3) epochs were recommended (as here) by the IUGS Subcommission on Silurian Stratigraphy (Holland 1980).

2.6.1 The Llandovery Epoch

The initial Llandovery (Ordovician–Silurian) boundary is not well standardised in the Llandovery district itself and so is referred to the Dobb's Linn section near Moffat, Scotland where Lapworth conceived the Ordovician System. It is taken at the base of beds bearing a *Glyptograptus persculptus* zone assemblage.

The ages plotted in our chart were named by Cocks, Toghill & Ziegler (1970) in the type area in Wales. These four ages may informally constitute in pairs early and late Llandovery. The equivalence of graptolitic and shelly facies has long been a question in these rocks, with problems of correlation between Scotland and Wales. The lettered symbols in the biostratigraphic correlation column refer to a sequence A, B and C for a superseded usage of Early, Middle and Late Llandovery and are based upon a sequence of Welsh shelly (largely brachiopod) faunas (Cocks *et al.* 1970).

2.6.2 The Wenlock Epoch

The name Wenlock was first used by Murchison in 1839 from the type area of Wenlock Edge in the Welsh Borderlands. A scheme of formations and members was developed over the years. However, a GSL initiative led to a reinvestigation of these rocks and proposals for an international scale that could be correlated by the graptolite sequence (Bassett *et al.* 1975). This is adopted in the accompanying classification (without detailing the sequence of evolving classifications) and the definitions are abstracted below.

The initial ***Sheinwoodian*** (Llandovery–Wenlock) boundary is seen in the standard section (National Grid SO 5688 9839) in Hughley Brook which lies 200 m SE of Leasowes Farm and 500 m NE of Hughley Church. The marker point for the boundary is taken in the left (N) bank of the stream at the base of unit G which is the base of the Buildwas Formation and immediately above the Purple Shales in a measured section (Bassett *et al.* 1975, p.13).

The ***Homerian*** Age was divided into the two chrons – Whitwell and Gleedon – because of the possibility of international correlation by the graptolite zones and the likely ease of recognition of their bases elsewhere by correlation with the standard section. The spacing of these graptolite zones in the chart follows that of Bassett *et al.* (1975, p.2) presumably being the best time scale interpolated by thicknesses in the 'Wenlock Shales' – mainly Coalbrookdale Formation.

Chart 2.4. Silurian chronostratic scale and correlation.

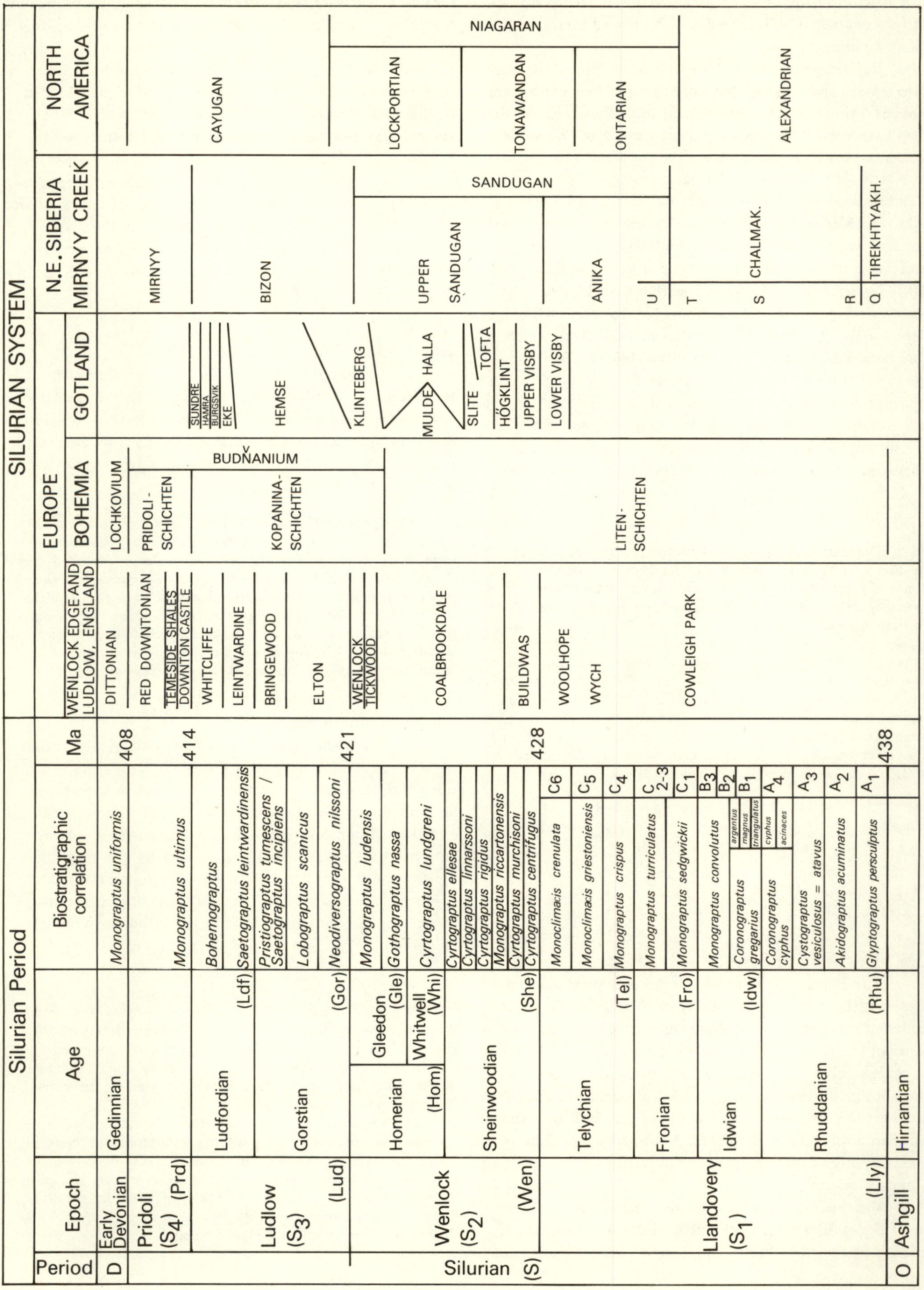

Period	Epoch	Age	Biostratigraphic correlation		Ma	WENLOCK EDGE AND LUDLOW, ENGLAND (EUROPE)	BOHEMIA (EUROPE)	GOTLAND (EUROPE)	N.E. SIBERIA MIRNYY CREEK	NORTH AMERICA
		Silurian Period				SILURIAN SYSTEM				
D	Early Devonian	Gedinnian	*Monograptus uniformis*		408	DITTONIAN	LOCHKOVIUM			
Silurian (S)	Pridoli (S_4) (Prd)		*Monograptus ultimus*		414	RED DOWNTONIAN TEMESIDE SHALES DOWNTON CASTLE	PRIDOLI-SCHICHTEN (BUDŇANIUM)	SUNDRE HAMRA BURGSVIK	MIRNYY	CAYUGAN
	Ludlow (S_3)	Ludfordian	*Bohemograptus*			WHITCLIFFE	KOPANINA-SCHICHTEN (BUDŇANIUM)	EKE	BIZON	
		(Ldf)	*Saetograptus leintwardinensis*			LEINTWARDINE				
		Gorstian	*Pristiograptus tumescens* / *Saetograptus incipiens*			BRINGEWOOD		HEMSE		
			Lobograptus scanicus			ELTON				
	(Lud)	(Gor)	*Neodiversograptus nilssoni*		421					
	Wenlock (S_2)	Homerian — Gleedon	*Monograptus ludensis*			WENLOCK TICKWOOD		KLINTEBERG	UPPER SANDUGAN (SANDUGAN)	LOCKPORTIAN (NIAGARAN)
		Gleedon (Gle)	*Gothograptus nassa*			COALBROOKDALE	LITEN-SCHICHTEN	MULDE / HALLA		
		Homerian (Hom) — Whitwell (Whi)	*Cyrtograptus lundgreni*							
		Sheinwoodian	*Cyrtograptus ellesae*					SLITE / TOFTA		TONAWANDAN (NIAGARAN)
			Cyrtograptus linnarssoni							
			Cyrtograptus rigidus							
			Monograptus riccartonensis					HÖGKLINT		
			Cyrtograptus murchisoni			BUILDWAS		UPPER VISBY		
	(Wen)	(She)	*Cyrtograptus centrifugus*		428					
	Llandovery (S_1)	Telychian	*Monoclimacis crenulata*	C6		WOOLHOPE		LOWER VISBY	ANIKA (SANDUGAN)	
			Monoclimacis griestoniensis	C_5		WYCH				ONTARIAN (NIAGARAN)
		(Tel)	*Monograptus crispus*	C_4						
		Fronian	*Monograptus turriculatus*	C_{2-3}					U	
		(Fro)	*Monograptus sedgwickii*	C_1		COWLEIGH PARK			T — CHALMAK.	ALEXANDRIAN
		Idwian	*Monograptus convolutus*	B_3						
			Coronograptus gregarius — *argentus*	B_2						
		(Idw)	*Coronograptus gregarius* — *magnus*, *triangulatus*	B_1					S — CHALMAK.	
		Rhuddanian	*Coronograptus cyphus* — *cyphus*, *acinaces*	A_4						
			Cystograptus vesiculosus = atavus	A_3						
			Akidograptus acuminatus	A_2					R	
	(Lly)	(Rhu)	*Glyptograptus persculptus*	A_1	438				Q — TIREKHTYAKH.	
O	Ashgill	Hirnantian								

The initial ***Whitwell*** boundary is located in the 'small side-stream (National Grid SO 6194 0204) to the tributary of Sheinton Brook which flows through Whitwell Coppice, 500 m north of Homer. The marker point . . . is within a more or less continuous section of olive to grey-green mudstones, blocky-fractured and thin-bedded, its exact position coinciding with the point at which the *ellesae/lundgreni* biozone boundary cuts the right (north) bank of the stream.' More biostratigraphic details were given for international correlation by Bassett *et al.* (1975).

The initial ***Gleedon*** boundary is defined (at National Grid SO 5016 8999) on the SE side of the track 182 m E of Eaton Church and coincides with the point where the *lundgreni/nassa* biozone boundary cuts the track.

2.6.3 The Ludlow Epoch

The Ludlow Epoch has been redivided (Holland 1980) by combining four ages used earlier (e.g. Cocks *et al.* 1971) into two that are now better established as follows:

Holland (1980)	Cocks *et al.* (1971)
Ludfordian	Whitcliffian
	Leintwardinian
Gorstian	Bringewoodian
	Eltonian

The initial ***Gorstian*** (Wenlock–Ludlow) boundary was defined by Holland, Lawson & Walmsley (1963) at Pitch Coppice in the Ludlow anticline in the standard section in the Old Quarry (National Grid SO 4726 7301) on the S side of the Ludlow/Wigmore road, about 2 km NE of Aston Church. This definition is in younger strata than was sometimes earlier supposed.

The initial ***Ludfordian*** boundary was taken at the base of the beds with *Saetograptus leintwardinensis* (Holland 1980) but was defined precisely by Holland *et al.* (1963) in the section on the NW face of Sunnyhill Quarry in Mary Knoll Valley 2.8 km SSW of Ludlow (National Grid SO 4953 7255).

2.6.4 The Pridoli Epoch

This epoch approximates to post-Ludlow, pre-Gedinnian time but its name is not yet internationally established. Downton is the name used in the Welsh Borderland sequence and would rationally follow on the sequence Llandovery, Wenlock and Ludlow. Graptolites have not been found in this facies. The alternative preferred in this work is Pridoli from Pridoli-Schichten (labelled E332 20–80 m in Barrande's section in Bohemia). It is preferred partly because the classic Silurian–Devonian boundary is established there, as is also the Pridoli–Lochkovian boundary. The Ludlow–Pridoli boundary has not been defined in any locality, but it approximates to the base of the *Monograptus ultimus* biozone.

The epochs, ages and biostratigraphic correlation columns on Chart 2.4 are based on Cocks *et al.* (1971) with modifications from Bassett *et al.* (1975) and Holland (1980) as detailed above. R. B. Rickards kindly provided details of the zone fossils. The Wenlock Edge and Ludlow, Bohemia and Gotland columns are also from Cocks *et al.* (1971); the north-east Siberia column is based on Oradovskaya & Sobolevskaya (1979) and the North American column is from Norford *et al.* (1970, p.604).

2.7 The Devonian Period

The Devonian System was established in Devon, England by Sedgwick and Murchison in 1839, after a controversy from 1834 to 1840 as to whether the rocks were Silurian or Carboniferous (Rudwick 1979) They were shown to be coeval with the Old Red Sandstone. The division of the System on the basis of marine faunas, however, came to be established through the work of Dumont, Beyrich, Roemer and many others in the Ardenne–Rhenish area where more recent attempts to standardise the time divisions of the period have been based (Ziegler 1979). For a time the name Rhenian was preferred by some to Devonian.

2.7.1 The Early Devonian Epoch

Gedinnian was first defined by Dumont in 1848 from Gedinne in Belgium as the initial Lower Devonian stage. Its initial boundary is also the Silurian–Devonian boundary which was the first to be fully agreed by the Commission on Stratigraphy of the IUGS (at Montreal 1972). They defined the boundary stratotype section at Klonk near Prague in the Barrandian area of Bohemia, Czechoslovakia. The reference point selected is just below a bed with the first and abundant occurrence of *Monograptus uniformis* and *M. uniformis angustidens,* i.e. in bed No. 20 (7–10 cm thick), described by Chlupáč, Jaeger & Zikmundova (1972). This horizon was approximately the base of the Gedinnian stage so that the initial point of the Gedinnian Age is now thereby standardised (McLaren 1977). The boundary being established in Bohemia (Pridolian–Lochkovian boundary) gives some claim to Lochkovian as a stage name approximating to Gedinnian but Gedinnian is internationally established.

Siegenian (named by Kayser in 1881, from Siegen in Germany) was originally the Coblentzian (Coblenzian) of Dumont 1848 and the lower Coblenzian of Gosselet 1880–88. The Gedinnian–Siegenian boundary would correspona to the base of the type Siegenian. However, that is poorly fossiliferous and a better standard for the age boundary might be established in Bohemia.

Emsian was introduced by Dorlodot in 1900 to clarify confusion on nomenclature (discussed in detail by Ziegler (1979)). The initial boundary of the Emsian Age might be taken, for example, at the base of Grès de Vireaux (Ziegler 1979).

2.7.2 The Middle Devonian Epoch

Eifelian, named from Eifel in Germany, was first applied by Dumont in 1848 to older and younger rocks. The standard stage as now understood was restricted to the present meaning in 1937 (see Richter 1942, Ziegler 1979). The name Couvinian (from Couvin in Belgium named by Dupont in 1885) has been widely used by French-speaking geologists; it corresponds to all present Eifelian and part of the upper Emsian strata and so it is not recommended here for the standard scale. The initial Eifelian boundary might be taken at the base of the Eifel Stufe in the Welleldorf standard Eifel section at the boundary between Heisdorf and Lauch beds.

Chart 2.5. Devonian chronostratic scale and correlation.

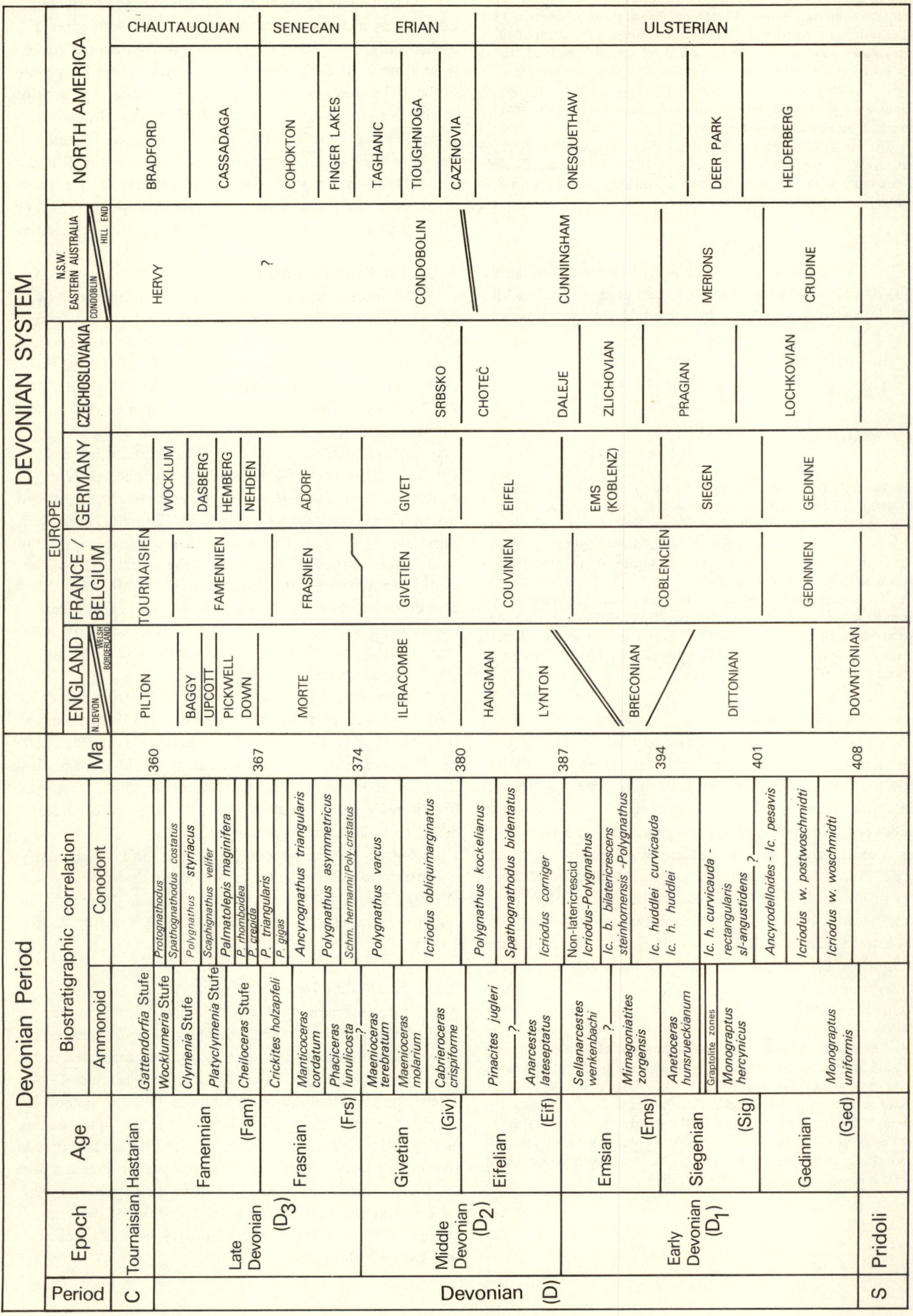

Period	Epoch	Age	Ammonoid	Conodont	Ma	England (N. Devon / Welsh Borderland)	France / Belgium	Germany	Czechoslovakia	N.S.W. Eastern Australia (Condoblin / Hill End)	North America	North America
Devonian Period			Biostratigraphic correlation			DEVONIAN SYSTEM: Europe						
C	Tournaisian	Hastarian	*Gattendorfia* Stufe			PILTON	TOURNAISIEN			HERVY	BRADFORD	CHAUTAUQUAN
Devonian (D)	Late Devonian (D_3)	Famennian (Fam)	*Wocklumeria* Stufe	*Protognathodus*	360							
				Spathognathodus costatus				WOCKLUM				
			Clymenia Stufe	*Polygnathus styriacus*		BAGGY						
			Platyclymenia Stufe	*Scaphignathus velifer*		UPCOTT	FAMENNIEN	DASBERG				
				Palmatolepis maginifera		PICKWELL DOWN		HEMBERG			CASSADAGA	
			Cheiloceras Stufe	*P. rhomboidea*				NEHDEN				
				P. crepida								
		Frasnian (Frs)	*Crickites holzapfeli*	*P. triangularis*	367					?		
				P. gigas								
			Manticoceras cordatum	*Ancyrognathus triangularis*		MORTE	FRASNIEN	ADORF			COHOKTON	SENECAN
			Phaciceras lunulicosta	*Polygnathus asymmetricus*							FINGER LAKES	
			?	*Schm. hermanni/Poly cristatus*	374							
	Middle Devonian (D_2)	Givetian (Giv)	*Maenioceras terebratum*	*Polygnathus varcus*		ILFRACOMBE	GIVETIEN	GIVET			TAGHANIC	ERIAN
			Maenioceras molarium	*Icriodus obliquimarginatus*						CONDOBOLIN	TIOUGHNIOGA	
			Cabrieroceras crispiforme						SRBSKO		CAZENOVIA	
		Eifelian (Eif)	*Pinacites jugleri*	*Polygnathus kockelianus*	380	HANGMAN	COUVINIEN	EIFEL	CHOTEČ			ULSTERIAN
			?	*Spathognathodus bidentatus*								
			Anarcestes lateseptatus	*Icriodus corniger*		LYNTON						
	Early Devonian (D_1)	Emsian (Ems)	*Sellanarcestes wenkenbachi*	Non-latericrescid *Icriodus-Polygnathus*	387			EMS (KOBLENZ)	DALEJE	CUNNINGHAM	ONESQUETHAW	
			?	*Ic. b. bilatericrescens steinhornensis -Polygnathus*					ZLICHOVIAN			
			Mimagoniatites zorgensis	*Ic. huddlei curvicauda*; *Ic. h. huddlei*		BRECONIAN	COBLENCIEN					
		Siegenian (Sig)	*Anetoceras hunsrueckianum*		394			SIEGEN	PRAGIAN	MERIONS	DEER PARK	
			Graptolite zones									
			Monograptus hercynicus	*Ic. h. curvicauda - rectangularis sl-angustidens* ?		DITTONIAN						
		Gedinnian (Ged)		*Ancyrodelloides - Ic. pesavis*	401				LOCHKOVIAN		HELDERBERG	
				Icriodus w. postwoschmidti			GEDINNIEN	GEDINNE		CRUDINE		
			Monograptus uniformis	*Icriodus w. woschmidti*		DOWNTONIAN						
S	Pridoli				408							

Givetian (named from Calcaire de Givet, France, by Gosselet in 1879) has been used in approximately the present sense from its inception. The initial boundary of the age is taken by modern German authors at or near the lower range of *Stringocephalus burtini* (Ziegler 1979).

2.7.3 The Late Devonian Epoch

Frasnian from Frasnes, Belgium (Gosselet in 1880) is used approximately in its original sense: Adorfian, Manticoceras Stufe, as well as the North American Senecan have also been proposed. The initial boundary would be placed at approximately the base of the Assise de Fromelennes or alternatively at the base of the Assise de Frasnes. In House *et al.* (1977, p.8) the boundary is taken 'at the base of the goniatite *Phaciceras lunulicosta* zone of the Manticoceras Stufe'. This is the boundary of the German orthochronologic scale which ammonoid workers have used throughout this century and which conodont workers have also sought to use.

Famennian from Famenne in Belgium was first used by Dumont in 1885 for his earlier (1848) Système de Condroz. Gosselet in 1879 used it in the present sense. The initial boundary of the age would be near the base of the Schistes de Senzeilles but, as with all the other boundaries within the Devonian Period, has yet to be defined. The terminal boundary is related to the top of the Wocklumeria Stufe (with *Wocklumeria sphaeroides)* and below the Gattendorfia Stufe (see discussion under Tournaisian (Section 2.8.1)).

The time divisions and the two biostratigraphic correlation columns of Chart 2.5 are from House *et al.* (1977) upon which the four European columns have also been based. Details of eastern Australian rocks are taken from Hill (1967) and those of North America (Appalachian Basin) from Oliver *et al.* (1967).

2.8 The Carboniferous Period

Because of their economic value and good outcrops in Britain, Carboniferous rocks were amongst the first to be classified. The name Coal Measures was proposed by Farey in 1807 and 1811, Millstone Grit by Whitehurst in 1778 and with the Mountain or Carboniferous Limestone these three major divisions (listed by William Phillips in 1818) constituted the Medial or Carboniferous Order (the latter including the Old Red Sandstone) as set out by Conybeare & Phillips (1822). In a detailed historical discussion Ramsbottom showed that Carboniferous was Conybeare's creation in that work (140 pages of detailed description in England and Wales). It was the first system to be established though it was referred to variously as an order, formation, group or series before system was applied by John Phillips in 1835 (Ramsbottom 1981). The three British units correspond to the three north European divisions. Green and others in 1878 grouped the upper two units as Upper Carboniferous and a two-fold division was again used in Belgium by d'Omalius d'Halloy in 1808, the upper one being the Terrain Houiller (Zittel 1901, Ramsbottom *et al.* 1978).

The three divisions, Namurian (proposed by Purves in 1883), Westphalian and Stephanian (both by de Lapparent in 1893), were systematised by Jongmans in 1928 when Westphalian was divided into A, B and C based on goniatite biozones and Westphalian D based on floras. Westphalian E was later referred to Stephanian.

The two-fold division of the Period was recognised by the Heerlen Conference in 1935 when Dinantian was introduced for Lower Carboniferous and the IUGS Subcommission on Carboniferous Stratigraphy proposed Silesian for the upper part in 1960. The same Subcommission in 1972 (George & Wagner) regarded Dinantian and Silesian as sub-systems, Namurian, Westphalian and Stephanian being ranked as series.

The definition of the initial Silesian Sub-period marker-point corresponds to the earliest occurrence of *Cravenoceras leion* (Heerlen Conference 1958) and the section at Little Mearley Clough, Pendle Hill, Lancashire, England was suggested as a boundary stratotype for deciding a reference point to mark also the initial Pendleian Age.

The division of the 'Carboniferous or Pennine' System into Coal Measures or Pennsylvanian (above) and Lower Carboniferous or Mississippian is attributed to H. S. Williams in 1891 (Wilmarth 1925). Williams also included in the Pennsylvanian the 'Coal Measure Conglomerate' or Millstone Grit (Pottsville Formation). Mississippian had already been in use since Winchell proposed it in 1869 as the designation of the Carboniferous or Mountain Limestone of the United States. By 1891 it already encompassed three groups of several formations. Ulrich in 1911 divided the Mississippian into Waverleyan and Tennesseean 'systems'. Anthracolithic was also used for Carboniferous by the USGS (Wilmarth 1925).

The Russian sequences have more to offer in the way of marine facies which are richly fossiliferous through to Permian rocks. Unlike those of western Europe and North America, Soviet Carboniferous rocks were classified into Lower, Middle and Upper which led to much ambiguity because Lower Carboniferous and Upper Carboniferous each span quite different times in each region.

Uralian (Ouralien) was proposed by de Lapparent in 1893 as the latest Carboniferous division intended as a marine equivalent of Stephanian but discontinued when many supposedly Uralian rocks proved to be Permian. Orenburgian was also proposed as the latest Carboniferous division in the southern Urals, and even recently this was shown as a Carboniferous stage above the Gzelian, but it was later abandoned as being part Permian (Sherlock 1948, p.14).

The conclusion of a long story in which three areas (western Europe, Russia and North America) seemed to be competing equally for the role of providing names and possibly stratotypes for a global Carboniferous scale seems now to be approaching and the proposal made here anticipates it. Chart 2.6 combines the recent thinking of some authorities and we are reinforced in this step by an impression

Chart 2.6. Carboniferous chronostratic scale and correlation.

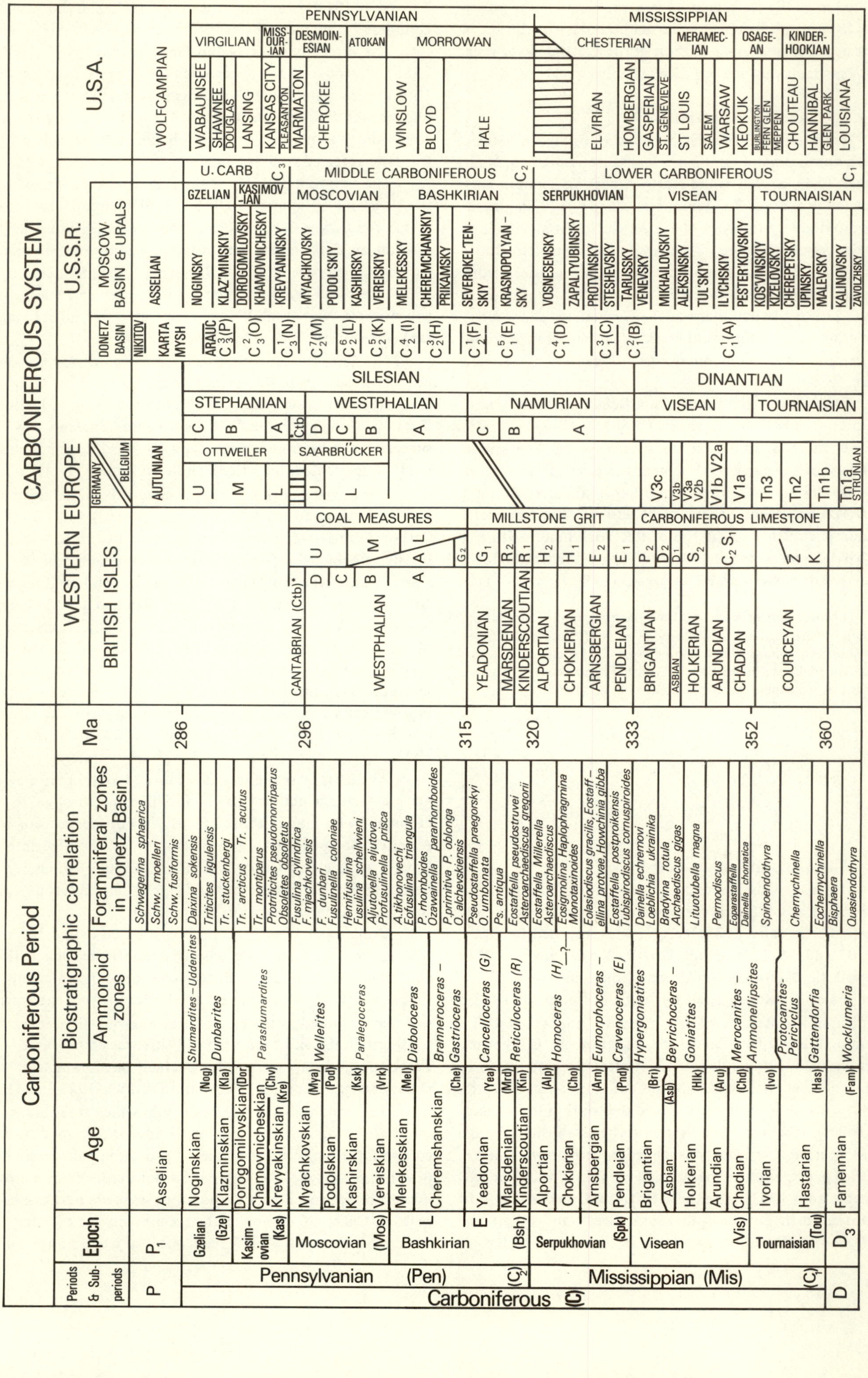

Carboniferous Period: Periods & Sub-periods	Epoch	Age	Biostratigraphic correlation: Ammonoid zones	Biostratigraphic correlation: Foraminiferal zones in Donetz Basin	Ma	CARBONIFEROUS SYSTEM, Western Europe: British Isles	British Isles zones	British Isles lithostratigraphy	Germany / Belgium	Western Europe stages	Western Europe series	U.S.S.R.: Donetz Basin	U.S.S.R.: Moscow Basin & Urals	U.S.S.R. stages	U.S.A.	U.S.A. series
P	P_1	Asselian		Schwagerina sphaerica; Schw. moelleri; Schw. fusiformis	286				AUTUNIAN			NIKITOV; KARTA MYSH	ASSELIAN		WOLFCAMPIAN	
Carboniferous (C); Pennsylvanian (Pen) (C₂)	Gzelian (Gze)	Noginskian (Nog)	Shumardites – Uddenites	Daixina sokensis					OTTWEILER U	C	STEPHANIAN; SILESIAN	ARAUC C_3^3(P)	NOGINSKY	GZELIAN; U. CARB C_3	WABAUNSEE	VIRGILIAN; PENNSYLVANIAN
		Klazminskian (Kla)	Dunbarites	Triticites jigulensis; Tr. stuckenbergi					M	B			KLAZ'MINSKIY		SHAWNEE; DOUGLAS	
	Kasimovian (Kas)	Dorogomilovskian (Dor)	Parashumardites	Tr. arcticus, Tr. acutus								C_3^2(O)	DOROGOMILOVSKY	KASIMOVIAN	LANSING	
		Chamovnicheskian (Chv)		Tr. montiparus					L	A			KHAMOVNICHESKY		KANSAS CITY	MISSOURIAN
		Krevyakinskian (Kre)		Protriticites pseudomontiparus; Obsoletes obsoletus	296							C_3^1(N)	KREVYANINSKY		PLEASANTON	
	Moscovian (Mos)	Myachkovskian (Mya)	Wellerites	Fusulina cylindrica; F. mjachkovensis		CANTABRIAN (Ctb)*; WESTPHALIAN	D	U; COAL MEASURES	SAARBRÜCKER U	Ctb; D	WESTPHALIAN	C_2^7(M)	MYACHKOVSKY	MOSCOVIAN; MIDDLE CARBONIFEROUS	MARMATON	DESMOINESIAN
		Podolskian (Pod)		F. dunbari; Fusulinella coloniae			C		L	C		C_2^6(L)	PODOL'SKIY		CHEROKEE	
		Kashirskian (Ksk)	Paralegoceras	Hemifusulina; Fusulina schellwieni			B	M		B			KASHIRSKY			
		Vereiskian (Vrk)		Aljutovella aljutova; Profusulinella prisca								C_2^5(K)	VEREISKIY			ATOKAN
	Bashkirian (Bsh) L	Melekesskian (Mel)	Diaboloceras	A. tikhonovechi; Eofusulina triangula			A	A; L		A		C_2^4(I)	MELEKESSKY	BASHKIRIAN	WINSLOW	MORROWAN
		Cheremshanskian (Che)	Branneroceras – Gastrioceras	P. rhomboides; Ozawainella pararhomboides; P. primitiva P. oblonga; O. alchevskiensis	315			G_2				C_2^3(H)	CHEREMCHANSKIY; PRIKAMSKY		BLOYD	
	E	Yeadonian (Yea)	Cancelloceras (G)	Pseudostaffella praegorskyi; O. umbonata		YEADONIAN	G_1	MILLSTONE GRIT		C	NAMURIAN	C_2^1(F)	SEVEROKEL'TEN-SKIY		HALE	
		Marsdenian (Mrd)	Reticuloceras (R)	Ps. antiqua		MARSDENIAN	R_2			B		C_1^5(E)	KRASNOPOLYAN-SKY			
		Kinderscoutian (Kin)		Eostaffella pseudostruvei; Asteroarchaediscus gregorii	320	KINDERSCOUTIAN	R_1							C_2		
Mississippian (Mis) (C₁)	Serpukhovian (Spk)	Alportian (Alp)	Homoceras (H)	Eostaffella Millerella; Asteroarchaediscus		ALPORTIAN	H_2			A		C_1^4(D)	VOSNESENSKY	SERPUKHOVIAN; LOWER CARBONIFEROUS		CHESTERIAN; MISSISSIPPIAN
		Chokierian (Cho)		Eosigmolina Haplophragmina; Monotaxinoides		CHOKIERIAN	H_1						ZAPALTYUBINSKY			
		Arnsbergian (Arn)	Eumorphoceras –	Eolasiodiscus gracilis, Eostaffellina protvae, Howchinia gibba		ARNSBERGIAN	E_2					C_1^3(C)	PROTVINSKY; STESHEVSKY		ELVIRIAN	
		Pendleian (Pnd)	Cravenoceras (E)	Eostaffella postproikensis; Tubispirodiscus cornuspiroides	333	PENDLEIAN	E_1					C_1^2(B)	TARUSSKY			
	Visean (Vis)	Brigantian (Bri)	Hypergoniatites	Dainella echremovi; Loeblichia ukrainika		BRIGANTIAN	P_2; D_2	CARBONIFEROUS LIMESTONE	V3c		VISEAN; DINANTIAN		VENEVSKY	VISEAN	HOMBERGIAN; GASPERIAN	
		Asbian (Asb)	Beyrichoceras – Goniatites	Bradyina rotula; Archaediscus gigas		ASBIAN	D_1		V3b				MIKHAILOVSKIY		ST. GENEVIEVE	
		Holkerian (Hlk)		Lituotubella magna		HOLKERIAN	S_2		V3a; V2b				ALEKSINSKY		ST LOUIS	MERAMECIAN
		Arundian (Aru)		Permodiscus		ARUNDIAN	C_2 S_1		V1b V2a			C_1^1(A)	TUL'SKIY		SALEM; WARSAW	
		Chadian (Chd)	Merocanites – Ammonellipsites	Eoparastaffella; Dainella chomatica	352	CHADIAN			V1a				ILYCHSKIY; PESTER'KOVSKIY		KEOKUK	OSAGEAN
	Tournaisian (Tou)	Ivorian (Ivo)		Spinoendothyra		COURCEYAN	Z		Tn3		TOURNAISIAN		KOS'VINSKIY; KIZELOVSKY	TOURNAISIAN	BURLINGTON; FERN GLEN; MEPPEN	
		Hastarian (Has)	Protocanites-Pericyclus; Gattendorfia	Chernychinella; Eochernychinella	360		K		Tn2; Tn1b				CHEREPETSKY; UPINSKY; MALEVSKY		CHOUTEAU; HANNIBAL; GLEN PARK	KINDERHOOKIAN
D	D_3	Famennian (Fam)	Wocklumeria	Bisphaera; Quasiendothyra					Tn1a STRUNIAN				KALINOVSKY; ZAVOLZHSKY	C_1	LOUISIANA	

that the Carboniferous Subcommission of IUGS may be formalising some such scheme (e.g. Ramsbottom 1981). We follow Bouroz and others in 1977 (Rotai 1979) in adopting Mississippian and Pennsylvanian as sub-systems for the international scale partly because of priority over Silesian but mainly because their mutual boundary coincides approximately with that of the Lower–Middle Carboniferous boundary in Russia. We propose then to use a largely European scheme for Mississippian divisions and a largely Russian scheme for the major part of Pennsylvanian. Apart from acknowledging by name the three regional communities that have contributed, this scheme also makes use of the detailed preparatory work in defining ages as well as epochs in the two Special Reports of the Geological Society of London (George *et al.* 1976, and Ramsbottom *et al.* 1978) and in the symposium: *The Carboniferous of the USSR* (Wagner, Higgins & Meyen 1979). We outline its epochs and ages in order.

2.8.1 The Mississippian Sub-period

The initial ***Tournaisian*** boundary is likely to be defined at a horizon approximating to the top of the Etroeungt Limestones and the base of the Hangenberg Kalk in Germany with the appearance of *Gattendorfia subinvoluta.*

'The level chosen represents an attempt at closest possible conformity with the current definition of the boundary, namely at the base of the *Gattendorfia* zone as recommended by the 1935 Heerlen Congress. It is at the first appearance of the conodont *Siphonodella sulcata* within the evolutionary lineage from *S. praesulcata* to *S. sulcata.* This is immediately below the lowermost record of *Gattendorfia* in Hönnetal. It is now essential to begin the search for the section best suitable as the boundary stratotype . . .' (from the recommendation and invitation of the Working Group on the Devonian–Carboniferous Boundary, Paproth 1980).

However, an initial reference point for the Devonian–Carboniferous boundary, while awaiting international definition, has been provisionally described in Ireland at a coastal section at the base of the Castle Slate Member of the Kinsale Formation (Irish National Grid 16242 04069; George *et al.* 1976, pp.6 and 7). George *et al.* accordingly used the Irish name *Courceyan* for the earliest Carboniferous division. It corresponds rather closely to Tournaisian and as the name Tournaisian has priority and is very widely used it would seem that Courceyan could only have regional value.

The Tournaisian Epoch is divided into two ages based in Belgium, namely ***Hastarian*** and ***Ivorian.*** These correspond indeed to the British zonal scheme K + Z and γ respectively and will be used by the British Institute of Geological Sciences as well as internationally by the IUGS Subcommission on Carboniferous Stratigraphy (Ramsbottom 1981). So it seems the Belgian names for these divisions are likely to be used. It does not, of course, follow that the ages will be defined at boundary stratotypes in the eponymous localities. There can only be one such reference point for each boundary and we are not clear whether that will be in Ireland, Belgium or elsewhere. The initial Hastarian boundary has been described at the stratotype at Hastière.

The ***Visean*** Epoch is approximately defined according to the Belgian biostratigraphic sequence (see chart). Attempts at precise definitions of the constituent ages made by George *et al.* (1976) are only briefly indicated here as follows.

The initial ***Chadian*** boundary (after St Chad) lies within the Chatburn Limestone Group at the base of the Bankfield East Beds near Clitheroe, Lancashire, England (National Grid SD 7743 4442). The initial ***Arundian*** boundary lies at the base of the Pen y holt Limestone on the east side of Hobbyhorse Bay in south Pembrokeshire, Wales (*arundo,* Latin for hobby-horse, National Grid SR 8800 9563). The initial *Holkerian* boundary lies at the base of the Park Limestone, in sea cliffs near Holker Hall, Cumbria, England (National Grid SD 3330 7827). The initial *Asbian* boundary lies at the base of the Potts Beck Limestone at Little Asby Scar, Cumbria, England (National Grid NY 6988 0827). The initial *Brigantian* (named for the Celtic tribe of Brigantes) boundary lies at the base of the Peghorn Limestone (lowest unit of Yoredale facies), in the east branch of River Eden, 5 km SSE of Kirkby Stephen, Cumbria, England (National Grid NY 7832 0375). The terminal Brigantian boundary would end the Dinantian regional sub-period.

Serpukhovian is perhaps an unsatisfactory name for the late Mississippian epoch; it approximates to Namurian A (?Namuralian a possible alternative).

The Serpukhovian ages are standardised in the British Isles (Ramsbottom *et al.* 1978, Ramsbottom 1981). The initial *Pendleian* boundary (as well as Serpukhovian, Namurian A, or Silesian boundaries) was proposed by the Heerlen Conference in 1958 to be at the base of strata containing the 'earliest occurrence' of *Cravenoceras leion* Bisat. A stratotype marker-point has been suggested at Little Mearley Clough, Pendle Hill, Lancashire, England or alternatively at Slieve Anierin, Co. Leitrim, Ireland. Pendleian, ***Arnsbergian***, ***Chokierian***, ***Alportian*** (i.e. to end of Namurian A and to the initial Pennsylvanian boundary) are each approximately equivalent to the goniatite zones used in Britain for many years, respectively E1, E2, H1 and H2, initiated by Bisat in 1928 (Ramsbottom 1981). We do not know of more precise definitions.

2.8.2 The Pennsylvanian Sub-period

The ***Bashkirian*** Epoch as defined here begins with the initial Kinderscoutian boundary which should also perhaps now define the initial Pennsylvanian boundary. The whole Epoch is usefully described with evidence for international correlation by Semichatova and others in Wagner *et al.* (1979).

The three ages ***Kinderscoutian***, ***Marsdenian*** and ***Yeadonian*** were similarly defined in the British Isles, i.e. based on goniatite zones of Namurian B and C (Ramsbottom *et al.* 1978). These may be referred to collectively as early Bashkirian. The ages ***Cheremshanskian*** and ***Melekesskian*** are part of the standard Russian sequence and with all the later ages to the end of the Carboniferous Period are based on successions in the Moscow Basin and the Urals and are adopted for the standard scale but, as yet we think, without definition of boundary stratotype points.

The ***Moscovian*** Epoch is divided into four ages which are described in detail with biostratigraphic evidence for international correlation by Ivanova and others in Wagner *et al.* (1979). So far as we know the ages have been defined in biostratigraphic terms but not standardised at reference points in stratotypes.

The initial boundary of the ***Kasimovian*** Epoch (also the base of the Russian Upper Carboniferous) is suggested by Rotai (1979) to correspond to the incoming of *Protriticites pseudomontiparus – Obsoletes obsoletus,* i.e. at the base of Limestone N_2 of the Donetz succession. The three Kasi-

Chart 2.7. Permian chronostratic scale and correlation.

Period	Epoch	Age	Chron	Fusulinid zones	Brachiopods	Ma	N.W. Europe (Germany)	U.S.S.R. Eastern Russian Platform	U.S.S.R. Timan	Japan	Australia (Queensland)	U.S.A. (Delaware Basin)
Tr	Tr$_1$	Griesbachian				248						
Permian (P)	Late (P$_2$)	Tatarian (Tat)		*Yabeina yasubaensis* and *Lepidolina toriyamai*			BUNTSANDSTEIN	VYATSKIY; SEVERODVINSKIY; URZHUMSKIY	RED CLAYS AND MARLS; PYTYR'YUSKIY	KUMAN	? REWAN; BARALABA; TAMAREE	DEWEY LAKE (OCHOAN; UPPER PERMIAN)
		Kazanian (Kaz)				253	ZECHSTEIN: OHRE 5; ALLER 4; LEINE 3	UPPER KAZANSKIY	VESLYANSKIY		U CURRA LST; ?; PELICAN CREEK	RUSTLER
				Verbeekina verbeeki	*Cancrinelloides*		ZECHSTEIN 2: STASSFURT EVAPORITES; HAUPTDOLOMIT-STINKSCHIEFER	LOWER KAZANSKIY	CHEV'YNSKIY	AKASAKAN	SCOTTVILLE	SALADO; CASTILE; CAPITAN (GUADALUPIAN)
		Ufimian (Ufi)		*Neoschwagerina craticulifera*	?; *Lissochonetes*		ZECHSTEIN 1: WERRA; ZECHSTEINKALK; KUPFERSCHIEFER	SHEMSHINSKIY; SOLIKAMSKY	UST'KULOMSKIY; VYCHEGODSKIY		EXMOOR	WORD
	Early (P$_1$)	Kungurian (Kun)	Irenian	*Neoschwagerina simplex*	*Pseudosyrinx*	258	WEISSLIEGENDES	IREN'SKIY	IREN' SKIY	NABEYAMAN	GEBBIE	LEONARDIAN (LOWER PERMIAN)
			Filippovian					FILIPPOVSKIY	VYL' SKIY			
		Artinskian (Art)	Baigendzinian	*Parafusulina kaerimizensis*	*Sowerbina*; *Antiquatonia*	263	ROTLIEGENDES	IKSKIY	KOMICHANSKIY			
			Aktastinian		*Jakutoproductus*				NERMINSKIY		SIRIUS SHALE	
		Sakmarian (Sak)	Sterlitamakian	*Pseudofusulina vulgaris*	*Tornquistia*; *Attenuatella*	268		STERLITAMAK-SKIY	PEL' SKIY	SAKAMOTOZAWAN	TIVERTON	WOLFCAMPIAN
			Tastubian		*Yakovlevia*			TASTUBSKIY	ILIBEYSKIY		LIZZIE CREEK	
		Asselian (Ass)	Krumaian	*Pseudoschwagerina morikawai*	*Tomiopsis*			KOKHANSKIY	NENETSKIY		BURNETT	
			Uskalikian		*Orthotichia*							
			Surenan		*Kochiproductus*			SOKOL'YEGORSKIY	INDIGSKIY		JOE JOE	
C	Gze	Noginskian		*Triticites*		286				HIKAWAN		

movian ages originated by reference to three foraminiferal zones.

The ***Gzelian*** Epoch as defined by Rotai (1979) corresponds to three fusulinid zones and is divided into two ages. It terminates with the initial Asselian (Permian) boundary. As already indicated, the latest Carboniferous and earliest Permian stratigraphy has been confused, in part by difficulties arising from previous correlations. Until recently the Upper Carboniferous in Soviet usage comprised an upper Orenburgian and a lower Gzelian stage, so it may seem odd that Chart 2.6 shows an earlier Kasimovian and a later Gzelian age to terminate the Carboniferous Period. But this seems now to be well established.

Chart 2.6 is constructed from the papers cited above. The biostratigraphic scheme has been taken from Rotai (1979, p.245), except that the biostratigraphic correlation of the Hastarian, Ivorian and Chadian ages has been modified according to a scheme proposed by the Carboniferous Subcommission in 1979, and kindly made available to us while in press by W. H. C. Ramsbottom, currently President of the Subcommission. The principal change was to show *Gattendorfia* rather than *Wocklumeria* as the initial Carboniferous zone.

2.9 The Permian Period

In 1841, after a tour of Imperial Russia, R. I. Murchison named the Permian System to take in the 'vast series of beds of marls, schists, limestones, sandstones and conglomerates' that surmounted the Carboniferous System throughout a great arc stretching from the Volga eastwards to the Urals and from the Sea of Archangel to the southern steppes of Orenburg. He named it from the ancient kingdom of Permia in the centre of that territory, and the city of Perm which lies in the flanks of the Urals. In 1845 he included rocks now known as Kungurian to Tatarian in age and for a time the underlying strata (Artinskian etc.) were known as Permo–Carboniferous (i.e. intermediate between Carboniferous and Permian).

Already by 1822 (e.g. Conybeare & Phillips) the Magnesian Limestone and New Red Sandstone of England were well known as were the equivalent German Rotliegendes and Zechstein (a traditional miner's name), with its valuable Kupferschiefer. However, they lacked richly fossiliferous strata, were difficult to correlate and inadequate to justify the erection of a new system in western Europe. The lack of fossils had been noted by d'Omalius d'Halloy in 1808 who referred to the Kupferschiefer as Terrain Penéen including at first a part of the Triassic Bunter. Permian in due course displaced Penéen in general use.

J. Mancou in 1853 recognised Permian rocks in a large area from the Mississippi to the Rio Colorado and noted two divisions analogous to those in western Europe. He accordingly suggested the name Dyassic as more suitable than Permian and proposed a combined Dyas and Trias as a major period (Zittel 1901). It is thus appropriate to preserve the division of the Permian Period in two epochs (or subperiods) – Early and Late – and the names for these, Rotliegendes and Zechstein, may be appropriate as recognising the earliest detailed stratigraphic researches in these rocks.

Karpinskiy in 1874 extended the Permian System in Russia downwards to include Artinskian and Sakmarian sediments.

The name Thuringian was introduced in 1874 as equivalent to Zechstein and in 1893 de Lapparent used a three-fold division with Thuringian as Upper Permian, Saxonian (= Upper Rotliegendes) as Middle and Autunian (from Autun, in France) for Lower Permian. As recently as 1978 a three-fold division has been suggested by Waterhouse, his middle epoch spanning Kungurian (Filipovian) to Djulfian or Dzhulfian (named after Dzhul'fa on the River Araks, Caucasus and ?part Tatarian, ?part Early Triassic) times; his Late Permian included our (Tozer's 1967) Early Triassic Griesbachian. After much debate the scheme of stages (or ages used here) was proposed by Likharev and others in 1966 and accepted by the GSL Working Group (Smith *et al.* 1974). However there appear to be many gaps in the Soviet successions, so that while our scheme of ages represents a likely standard for nomenclature and classification, little progress appears yet to have been made with regard to standardisation in stratotypes (Smith *et al.* 1974). To attempt in the space available to characterise the standard divisions by fossil zones for the Russian sequence would lead to excessive simplification, but indications of the origins of the names and classifications of the proposed standard ages are listed below from Kotlyar (1977).

2.9.1 The Early Permian Epoch

The earliest ***Asselian*** rocks in the Urals correlate with the earliest formed rocks containing *Triticites californicus* in North America, at the base of the Wolfcampian; the initial Permian boundary would therefore be defined between the zones of latest Pennsylvanian *Triticites coronadoensis* and the earliest *T. californicus.* It will be noted that the long uncertainty in both classification and correlation may well account for some rocks traditionally regarded as Carboniferous (e.g. later Stephanian C) being indeed Asselian. The boundary problem with respect to the USSR is reviewed by Rauser-Chernousova & Shchgolev (1979).

Rocks of ***Sakmarian*** age (from Sakmara, a tributary of the River Ural, in the southern Urals) originally included all deposits from the top of the Upper Carboniferous to the base of the Artinskian. In 1950 Ruzhentsev distinguished two sub-stages – Asselian (lower) and Sakmarian (upper) and the latter was then raised in rank.

Rocks of ***Artinskian*** age (named after Arti in the central Urals by Karpinskiy in 1874) are widespread in east European USSR and in central Asia and contain *Pseudofusulina* and primitive *Parafusulina.* The name has been used for the upper division of Lower Permian rocks and it has been divided into two sub-stages.

Rocks of ***Kungurian*** age (named after the town Kungur near Perm in the Urals, in 1890) have been included in the Artinskian stage and/or combined with Ufimian.

2.9.2 The Late Permian Epoch

Rocks of ***Ufimian*** age (from Ufa, Russia, in 1915, and also referred to as Ufian) were previously known as the 'lower red unit' or the 'lower division of the Permian System (P_1)' and 'lower

Chart 2.8. Triassic chronostratic scale and correlation.

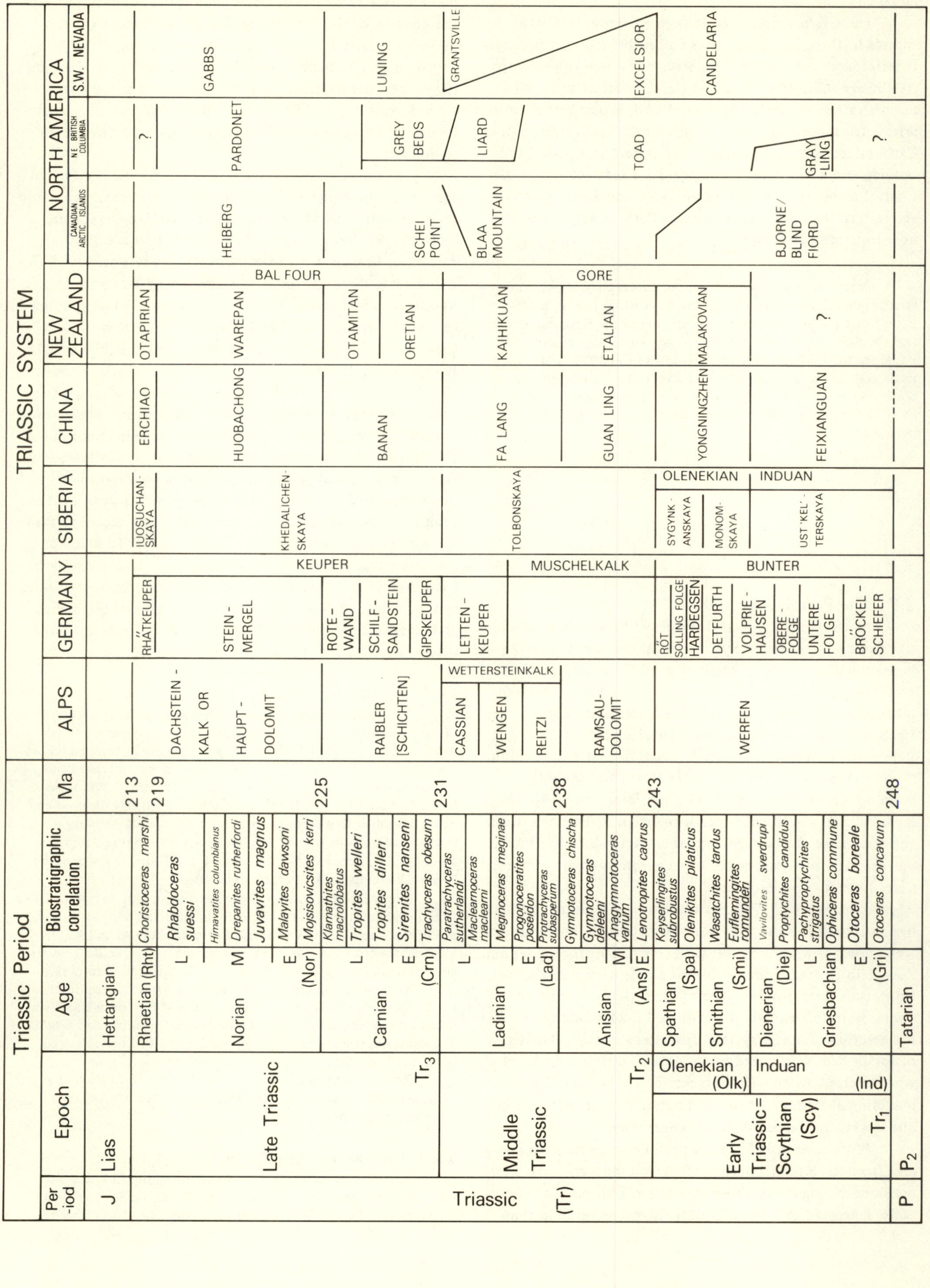

Per-iod	Epoch	Age		Biostratigraphic correlation	Ma	ALPS	GERMANY	SIBERIA	CHINA	NEW ZEALAND	NORTH AMERICA: CANADIAN ARCTIC ISLANDS	NORTH AMERICA: NE BRITISH COLUMBIA	NORTH AMERICA: S.W. NEVADA
J	Lias	Hettangian			213								
Triassic (Tr)	Late Triassic Tr_3	Rhaetian (Rht)		*Choristoceras marshi*	219	DACHSTEIN-KALK OR HAUPT-DOLOMIT	RHÄTKEUPER (KEUPER)	IUOSUCHANSKAYA	ERCHIAO	OTAPIRIAN (BAL FOUR)	HEIBERG	?	GABBS
		Norian	L	*Rhabdoceras suessi*			STEIN-MERGEL	KHEDALICHENSKAYA	HUOBACHONG	WAREPAN		PARDONET	
			M	*Himavatites columbianus*									
			M	*Drepanites rutherfordi*									
			M	*Juvavites magnus*									
			E	*Malayites dawsoni*									
		(Nor)	E	*Mojsisovicsites kerri*	225								
		Carnian	L	*Klamathites macrolobatus*		RAIBLER [SCHICHTEN]	ROTE-WAND		BANAN	OTAMITAN	SCHEI POINT	GREY BEDS	LUNING
			L	*Tropites welleri*									
			L	*Tropites dilleri*			SCHILF-SANDSTEIN						
			E	*Sirenites nanseni*						ORETIAN			
		(Crn)	E	*Trachyceras obesum*	231		GIPSKEUPER						
	Middle Triassic Tr_2	Ladinian	L	*Paratrachyceras sutherlandi*		CASSIAN (WETTERSTEINKALK)	LETTEN-KEUPER	TOLBONSKAYA	FA LANG	KAIHIKUAN	BLAA MOUNTAIN	LIARD	GRANTSVILLE
			L	*Maclearnoceras maclearni*									
				Meginoceras meginae		WENGEN	MUSCHELKALK						
			E	*Progonoceratites poseidon*		REITZI							
		(Lad)	E	*Protrachyceras subasperum*	238								
		Anisian	L	*Gymnotoceras chischa*		RAMSAU-DOLOMIT			GUAN LING	ETALIAN (GORE)		TOAD	EXCELSIOR
			L	*Gymnotoceras deleeni*									
			M	*Anagymnotoceras varium*									
		(Ans)	E	*Lenotropites caurus*	243								
	Early Triassic = Scythian (Scy) Tr_1; Olenekian (Olk)	Spathian		*Keyserlingites subrobustus*		WERFEN	RÖT; SOLLING FOLGE; HARDEGSEN (BUNTER)	SYGYNKANSKAYA (OLENEKIAN)	YONGNINGZHEN	MALAKOVIAN			CANDELARIA
		(Spa)		*Olenikites pilaticus*									
		Smithian		*Wasatchites tardus*			DETFURTH	MONOMSKAYA					
		(Smi)		*Euflemingites romunderi*									
	Induan (Ind)	Dienerian		*Vavilovites sverdrupi*			VOLPRIE-HAUSEN	UST'KEL'TERSKAYA (INDUAN)	FEIXIANGUAN	?	BJORNE/BLIND FIORD	GRAY-LING	
		(Die)		*Proptychites candidus*			OBERE-FOLGE						
		Griesbachian	L	*Pachyproptychites strigatus*			UNTERE FOLGE						
			L	*Ophiceras commune*									
			E	*Otoceras boreale*			BRÖCKEL-SCHIEFER					?	
		(Gri)	E	*Otoceras concavum*	248								
P	P_2	Tatarian											

Permian Red Group'. Ufimian was adopted as the initial age of the upper epoch in 1960.

The ***Kazanian*** age (after the town of Kazan on the Middle Volga) was recently divided into seven (chrons). Some authors have combined it with Ufimian as the Kama age. Indeed Ufimian and Kazanian ages together comprise the earlier of the two divisions of the Late Permian Epoch.

Rocks of ***Tatarian*** age (named in 1887 after the Tartar people) were known as 'upper mottled units' or 'mottled marl stage' and were originally included within the Triassic System. Their exact equivalence is not easy to establish on a global scale; this difficulty reflects the regressive or other distinctive environments towards the end of the Paleozoic Era.

Stevens, Wagner & Sumsion (1979) described and listed twelve fusuline zones for the Early Permian succession in central Cordilleran America. Because they are not correlated with the Russian ages we have not attempted to fit them into Chart 2.7, but list them below, in order to give some indication of the sequence of fusuline faunas.

Leonardian
Parafusulina spiculata
Parafusulina communis
Parafusulina allisonensis
Parafusulina leonardensis

Wolfcampian
Schwagerina aculeata
Pseudoschwagerina convexa
Schwagerina cf. *S. crebrisepta*
Eoparafusulina linearis
Pseudofusulina hueconensis
Schwagerina bellula
Pseudofusulina attenuata
Triticites californicus

The fusulinid zones on Chart 2.7 are taken from a list from Japan, which is rich in fossils of the period (Takai, Matsumoto & Toriyama 1963). Brachiopods in the adjacent column are from the Canadian Arctic, which is not so remote from the USSR; the list and correlation are from Smith *et al.* (1974).

The other columns on Chart 2.7 are derived as follows: NW Europe and USA from Smith *et al.* (1974); Japan from Takai *et al.* (1963); Australia from Waterhouse (1978); the USSR columns from the unnumbered table that plots a comparative scheme of Permian sections abstracted from the whole work by Likharev (1966). Details of these two USSR columns are summarised in that work in Tables 1 and 2 and 4 and 5 respectively.

2.10 The Triassic Period

The Triassic System was established in Germany from the three-fold division into Bunter, Muschelkalk and Keuper by Alberti in 1834. The traditional stages (Scythian, Anisian, Ladinian, Carnian, Norian) were established in the marine Northern Calcareous Alps of Austria where, however, the ammonite zones first used have proved to be incomplete or not always in chronological sequence. For this reason, and because of the excellent Arctic successions and those from the Western Cordillera of North America, Tozer (1967) proposed 'a standard for Triassic time' in which ammonite zones were used to characterise the standard stages more precisely. New stages were proposed to divide Scythian or Early Triassic, for which epoch the name is useful and is retained. In that work Tozer went so far as to define boundary reference points in type sections; and because this is the essential method for defining the chronostratic scale we adopt it here almost in its entirety. Silberling & Tozer (1968) elaborated this scheme for North America as a whole. Therefore the chronostratic scale of the Triassic Period is largely from Table II of Tozer (1967) and the definitions are selected from that work as follows.

2.10.1 The Early Triassic (Scythian) Epoch

The Permian–Triassic transition is nowhere known to be fully represented by fossiliferous strata. A hiatus is generally evident and this is usually regarded as belonging to Permian time followed by the initial Triassic (Griesbachian) age that corresponds in Arctic Canada with the *Otoceras woodwardi* zone of the Himalayas. This is a controversial question involving as it does the difficult and classic problem of the Paleozoic–Mesozoic boundary so that other points of view abound. The boundary is thus proposed (perhaps unsatisfactorily) at the base of the Blind Fjord Formation of north-west Axel Heiberg Island.

Some scientists may doubt the value of dividing the Scythian Epoch into ages, it having been regarded as an age itself (e.g. Kummel in 1957). However, Spath in 1935 divided Early Triassic into Early and Late 'Eo-Trias' each with three subdivisions. His Early and Late divisions corresponded to the Induan and Olenekian ages of northern Siberia introduced by Kiparisova and Popov in 1956 (but changed in 1964). Tozer's four ages (1967) are accepted here. His Dienerian–Smithian boundary divides Scythian as above.

The ***Griesbachian*** Age (from Griesbach Creek, Axel Heiberg Island) was proposed by Tozer in 1965 (1967) and divided into two sub-ages each with two ammonite zones.

The initial ***Dienerian*** (from Diener Creek, Ellesmere Island) boundary is in the Blind Fjord Formation of north-western Ellesmere Island (Tozer 1967, note 16) and is generally recognised by the appearance of Gyronitidae. It is correlated in the Himalayas at the boundary between *Otoceras* and *Meekoceras* beds (Diener in 1912, see Tozer (1967)), and in the Salt Range between the *Ophiceras connectens* bed and the Lower Ceratite Limestone (Kummel & Teichert 1966).

The initial ***Smithian*** (from Smith Creek, Ellesmere Island) is also in the Blind Fjord Formation of north-western Ellesmere Island, corresponding to the boundary between the Induan and Olenekian stages of Kiparisova & Popov (1956, 1964).

The initial ***Spathian*** boundary (from Spath Creek, Ellesmere Island) is in the lower shale member of the Blaa Mountain Formation, the type locality being at Spath Creek, Ellesmere Island (see Tozer 1967; note 11) with a useful additional sequence in Axel Heiberg Island.

2.10.2 The Middle Triassic Epoch

The type locality for the *Anisian* Age is in Austria but it lacks ammonoids near its base. Tozer (1967) defined its initial boundary by the Caurus zone in the type locality (east limb of anticline west of Mile Post 375, Alaska Highway, north-east British Columbia, note 28).

The initial *Ladinian* (from Ladini, people of Tyrol) boundary is (for the same reason) not defined in the original Ladinian rocks of Italy but in the Humboldt Range, Nevada, USA (Tozer 1967).

2.10.3 The Late Triassic Epoch

Carnian is from the Carnic Alps (the alternative spelling Karnian as used by Tozer (1967) is from German rather than from the Latin form). The stage was introduced by Mojsisovics in 1869. The initial boundary of this age is based in the type locality at Ewe Mountain, 4 miles ENE of Triangulation Station 6536, Toad River area, north-east British Columbia (Tozer 1967, note 23).

The *Norian* (or Juvavic) stage of the Eastern Alps (Mojsisovics in 1895, and Diener in 1926) was divided into three parts but doubt has been thrown on their sequential arrangement.

Rocks of *Rhaetian* (from the Rhaetic Alps) age are not rich in pelagic fossils. They have been included within the Jurassic System (e.g. Arkell 1933) and there remains some question of correlation of the younger rocks with *Rhaetavicula contorta.*

The type locality of the initial Rhaetian boundary is at Brown Hill, Peace River, north-east British Columbia (Tozer 1967, note 20). The type locality for the original stage is at Kendelbachgraben, St Wolfgang, Austria, and the Rhaetian division (older than Hettangian with *Psiloceras planorbis*) is now unambiguously Triassic. However, according to our chart it spans only one ammonite zone in contrast to the six zones of Norian Age. Tozer (1979) suggested abandoning the Rhaetian division as an age and incorporating it in the Late Norian Sevatian sub-stage so as to replace the *Rhabdoceras suessi* and *Choristoceras marshi* zones by three zones of a revised Sevatian sub-stage – namely Cordilleranus zone, Amoenum zone (both approximately equivalent to the original Suessi zone), and a third Crickmayi zone to replace the Marshi zone. For the present we retain Rhaetian as a useful age.

A further point concerns the time interval immediately preceding the Jurassic Period as defined approximately by the arrival of *Psiloceras planorbis.* The Blue Lias begins with pre-Planorbis beds which are therefore Triassic. Hallam (1981) has suggested that such a conclusion would amount to stratigraphic pedantry because 'common sense dictates that the horizon of most notable facies change over a wide area, coinciding as it does almost exactly with the appearance of the first Jurassic ammonites, should be taken as marking the system boundary'. In spite of these strictures we adhere to the recommendation of the Stratigraphy Committee of the GSL (George *et al.* 1969, p.53) and reaffirmed in Section 2.11.1 of this chapter (following Cope *et al.* 1980b) for the definition of the initial Hettangian (Jurassic) boundary. Lithostratigraphic boundaries are defined at facies changes but chronostratigraphic boundaries should be standardised within a sequence of uniform facies so the Blue Lias would appear to serve this purpose well and its basal pre-Planorbis beds are therefore Triassic.

2.10.4 Triassic Time Scale

As explained in Chapter 1, Chapters 3 and 4 were completed before Chapter 2 and its charts. The time scale adopted the principle of distributing ages equally between tie-points. It will be seen that the calculations were made on the basis that Scythian was one age unit and Rhaetian another. We now argue that on a quantitative basis Scythian corresponds to more than a division of unit duration and, from what appears above, Rhaetian would appear to represent about a sixth or so of Norian time. Unfortunately we cannot just apply the Rhaetian surplus of time to augment our Scythian interval because a tie-point falls between them. To re-adjust the ages through a much longer time span is of course necessary to make the best time scale, and we leave it to the next attempt at a time scale to do this. We merely point the way to a needed revision, partially satisfied that our main contribution may be to display the reasoning behind the construction of a time scale.

The columns to the right of Chart 2.8 have been based as follows: Alps on Sherlock (1948); Germany on Warrington *et al.* (1980); Siberia on Kiparisova, Radchenko & Gorskiy (1973); China on Chen (1974); New Zealand on Suggate, Stevens & Te Punga (1978); Canadian Arctic Islands and NE British Columbia on Tozer (1967); and SW Nevada of Kummel (1961, p.574).

2.11 The Jurassic Period

Between 1797 and 1815 William Smith published successions and geologic maps of England and Wales in which detailed stratigraphy of successive strata of Jurassic age played a key part, the sequence in England being especially well displayed. Many of these were grouped as the Oolite Formation by Buckland in 1818 or Oolitic Series (divided into Lower, Middle and Upper Oolites) overlying the Lias by Conybeare & Phillips (1822). They were equated with the Jura-Kalkstein of Alexander von Humboldt who in 1795 so referred to the Calcaire de Jura (thinking they were older than the Muschelkalk). From this Alexander Brongniart in 1829 first used the name Terrains Jurassiques but only for the Lower Oolitic Series of Conybeare and Phillips. In Britain until recently the name Jurassic coexisted with the earlier named constituent parts: Lias and Oolites. The above history is surveyed by Zittel (1901), Wilmarth (1925), Arkell (1933, 1956) and Torrens in Cope *et al.* (1980a, b).

Because of the immense wealth of fossils, particularly ammonites, in the Jurassic rocks of Britain, biostratigraphic zonation was generally further advanced than for rocks of all other periods and many distinguished scientists (e.g. Buckman and Oppel) contributed to this. By 1933 Arkell was able to review the Jurassic rocks of Britain, and in 1956 of the world. The scheme here is the same as that developed by Arkell as a European standard from 1946 to 1956 except dividing the Bajocian and we have used Tithonian in place of Arkell's Portlandian and Purbeckian; the latter with freshwater facies had no ammonite zones to characterise it and is in any case partly Cretaceous in age.

For the three standard Jurassic epochs it may be objected that we have used the old rock names Lias, Dogger and Malm in preference to Early, Middle and Late. We think they are of more use in this rôle than as their diverse rock terms but time will tell. Their context will distinguish their intended use.

Perhaps because of the success and relative precision of ammonite zonation, which may approximate more closely to isochronous horizons than biostratigraphic criteria can attain for other periods, the difference between biozones and chronozones has been confused by many writers. But the principle is simple and is the same for all strata. Time divisions of the chronostratic scale are for general use; they must be *defined* by reference points in boundary stratotypes and *correlation* with these reference sections will be correspondingly good because of the advantages that Jurassic ammonites confer. Biostratigraphic units on the other hand are defined only by the presence of their stated fossil content. Chart 2.10 illustrates biostratigraphic correlation with the reference scale.

In the following list of ages the definition of the initial boundaries follows the UK recommendations to the Colloque Jurassique in Luxembourg (July 1967) and the UK Contribution to the International Geological Correlation Programme submitted in Prague in 1968 to the IUGS by the Royal Society (later reprinted in George *et al.* 1969). The boundary definitions will only be indicated here and in any case generally await a firm decision. These ages are summarised and approximately equivalent rock sequences in some other parts of the world are given in Chart 2.9. The classic 74 or so zones are listed in Chart 2.10 (Cope *et al.* 1980a, b). It may be noted that each age on average spans about 5 biozones and 10–15 subzones and these are potential chrons for standardisation in the future. However, as now used they are conceived as standard time divisions or chrons. To make this clear chrons and subchrons, although named from the original zonal name, will in due course be defined by reference points in boundary stratotypes and they should be printed in roman rather than in italic type.

2.11.1 The Early Jurassic (Lias) Epoch

The ***Hettangian*** initial boundary (name from Hettange, France) is recognised by the first appearance of the genus *Psiloceras*. The initial boundaries of the Planorbis Chron and of the Jurassic Period coincide with it. Oppel (in 1856, pp.24–8) described sections at Lyme Regis and in quarries near Uplyme, Dorset, England, as being characteristic of the Planorbis Zone; he also referred to the coastal section at Watchet, Somerset, England. Morton (1971, p.84) recommended that the coastal section between Blue Anchor and Quantock's Head, in the Watchet area, be regarded as the type area of the Planorbis Zone. Cope *et al.* (1980b) regard the Planorbis Subzone as being 'clearly and unequivocally acceptable as the basal subzone of the basal Jurassic Planorbis Zone' (p.22). The portion of the Blue Lias Formation below the base of the Planorbis Subzone, together with the 'Watchet Beds' and the Penarth Group, are thus of Triassic age.

The ***Sinemurian*** initial boundary (name from Sémur, France) is also that of the Bucklandi Chron and the Conybeari Subchron. No type area for the Bucklandi Zone was given by Oppel in 1856. The Conybeari Subzone was founded in the Keynsham area, Somerset, England, but lack of permanent sections renders it unsuitable as a type area. Morton (1971, p.85) recommended that the type area be designated 50 miles SSW of Keynsham, on the Dorset coast, SW of Lyme Regis. Here the base of the Conybeari Subzone is placed at the base of Lang's bed 21, which outcrops at Seven Rock Point and at Devonshire Head. An exactly designated type locality remains to be decided.

The ***Pliensbachian*** initial boundary is also that of the Taylori Subchron and the Jamesoni Chron. According to Morton (1971, p.85) 'there is no explicit type section for the Taylori Subzone, but it was first used with reference to the Dorset coast section' (south England). There it is seen in Lang's (1928) bed 105, the base of the Belemnite Marls, separated from the underlying bed 104 by a non-sequence (Spath 1956, p.148). Bed 105 outcrops near Charmouth. At Pliensbach in south-west Germany, the upper two subzones of the Sinemurian Raricostatum Zone are absent (Geyer, 1964, p.165). Morton (1971, p.85) regarded the Pliensbach section as being 'acceptable for defining the Pliensbachian Stage in terms of its basal subzone. . . Here, the Taylori subzone rests nonsequentially on the lower part of the Raricostatum Zone.'

The ***Toarcian*** initial boundary (name from Thouars, France) is also that of the Tenuicostatum Chron. Morton (1971, p.85) designated the outcrop west of Kettleness, on the north Yorkshire coast, England, as the type section of the Tenuicostatum Zone, and therefore for the basal Toarcian. The marker-point for the base of the Zone is between beds 28 and 29 of Howarth (1955).

2.11.2 The Middle Jurassic (Dogger) Epoch

The name ***Áalenian*** was proposed by Mayer-Eymar in 1864 for the lowest part of the 'Braunjura' in the vicinity of Aalen, Germany at the northern edge of the Swabian Alps. The initial Aalenian (and so Middle Jurassic) boundary is recognised at the base of the Opalinum Zone. However only later Aalenian is represented near the present-day village of Aalen-Attenhofer so an initial boundary stratotype elsewhere is needed. Some workers have regarded Aalenian as Early Bajocian in a three-fold division (e.g. Morton and others at the Colloque Jurassique at Luxembourg, 1967). We have accepted Aalenian as a distinct age and so divide Bajocian into Early and Late only. Arkell's Scissum Zone inserted between Opalinum and Murchisonae zones is not so useful in Europe where the index fossil is found in earlier and later beds.

The ***Bajocian*** was introduced by d'Orbigny in 1852 for strata outcropping near Bayeux, France (hence the name). Its initial boundary would be taken at the base of the Discites Zone (since the Aalenian is being treated here as a separate age). On the north-western periphery of the Anglo–Paris Basin, the Bajocian is represented by the Middle and Upper Inferior Oolites (and the Aalenian by the Lower Inferior Oolite).

The ***Bathonian*** initial boundary (name from Bath, England) is that of the Zigzag Chron (and the Convergens Subchron) at the base of bed 23 of Sturani (1967) at the Bas Auran section, 4 km W of Barrême, Basses-Alpes, south-east France.

The initial boundary of the ***Callovian*** (name from Kellaway, England; Kellaways Rock) and that of the initial Macrocephalus Subchron was proposed with the Chippenham–Trowbridge area, Wiltshire, England as the type area (of the Macrocephalus Subzone). No exact type section has yet been designated.

2.11.3 The Late Jurassic (Malm) Epoch

The initial boundary of the ***Oxfordian*** (name from Oxford, England; Oxford Clay Formation) is that of the Mariae Chron and Scarburgense Subchron, and seems to be better defined in the cliff of Cornelian Bay, 3 km SE of Scarborough, Yorkshire, England, than in the earlier standard on the shore at Auberville, Normandy, France. Fortunately, conflict is avoided, since 'the Oxford Clay of the

Chart 2.9. Jurassic chronostratic scale and correlation.

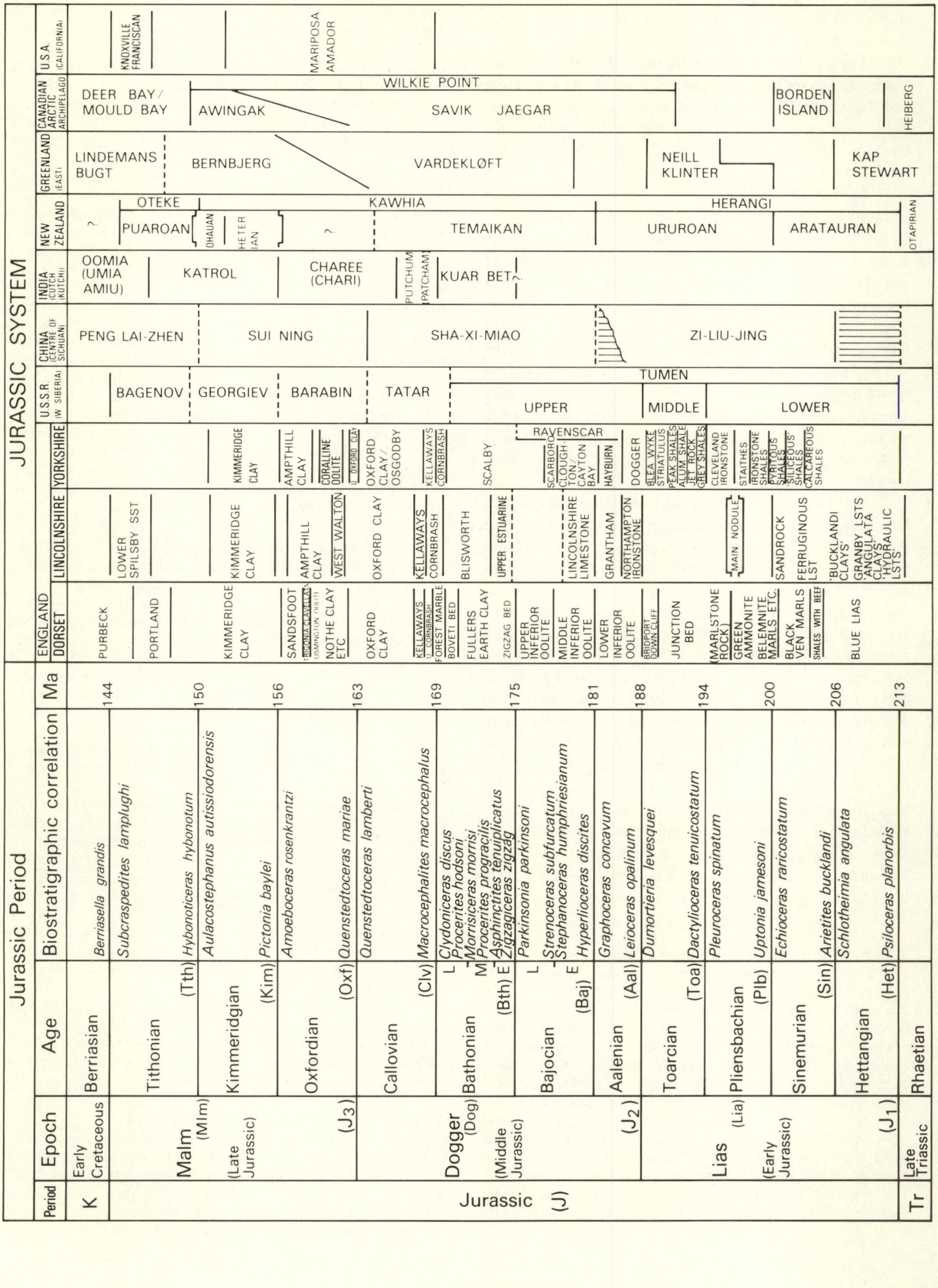

Jurassic Period

Period	Epoch	Age	Biostratigraphic correlation	Ma
K	Early Cretaceous	Berriasian	*Berriasella grandis*	144
Jurassic (J)	Malm (Mlm) (Late Jurassic) (J3)	Tithonian (Tth)	*Subcraspedites lamplughi* *Hybonoticeras hybonotum*	150
		Kimmeridgian (Kim)	*Aulacostephanus autissiodorensis* *Pictonia baylei*	156
		Oxfordian (Oxf)	*Amoeboceras rosenkrantzi* *Quenstedtoceras mariae*	163
	Dogger (Dog) (Middle Jurassic) (J2)	Callovian (Clv)	*Quenstedtoceras lamberti* *Macrocephalites macrocephalus*	169
		Bathonian (Bth)	L *Clydoniceras discus* *Procerites hodsoni* M *Morrisiceras morrisi* *Procerites progracilis* E *Asphinctites tenuiplicatus* *Zigzagiceras zigzag*	175
		Bajocian (Baj)	L *Parkinsonia parkinsoni* E *Strenoceras subfurcatum* *Stephanoceras humphriesianum* *Hyperlioceras discites*	181
		Aalenian (Aal)	*Graphoceras concavum* *Leioceras opalinum*	188
	Lias (Lia) (Early Jurassic) (J1)	Toarcian (Toa)	*Dumortieria levesquei* *Dactylioceras tenuicostatum*	194
		Pliensbachian (Plb)	*Pleuroceras spinatum* *Uptonia jamesoni*	200
		Sinemurian (Sin)	*Echioceras raricostatum* *Arietites bucklandi*	206
		Hettangian (Het)	*Schlotheimia angulata* *Psiloceras planorbis*	213
Tr	Late Triassic	Rhaetian		

JURASSIC SYSTEM

Region	Units (youngest to oldest)
ENGLAND – DORSET	PURBECK; PORTLAND; KIMMERIDGE CLAY; SANDSFOOT; TRIGONIA CLAVELLATA; OSMINGTON OOLITE; NOTHE CLAY ETC; OXFORD CLAY; KELLAWAYS; U. CORNBRASH; FOREST MARBLE; BOVETI BED; FULLERS EARTH CLAY; ZIGZAG BED; UPPER INFERIOR OOLITE; MIDDLE INFERIOR OOLITE; LOWER INFERIOR OOLITE; BRIDPORT DOWN CLIFF; JUNCTION BED; (MARLSTONE ROCK); GREEN AMMONITE; BELEMNITE MARLS ETC.; BLACK VEN MARLS; SHALES WITH BEEF; BLUE LIAS
ENGLAND – LINCOLNSHIRE	LOWER SPILSBY SST; KIMMERIDGE CLAY; AMPTHILL CLAY; WEST WALTON; OXFORD CLAY; KELLAWAYS; CORNBRASH; BLISWORTH; UPPER ESTUARINE; LINCOLNSHIRE LIMESTONE; GRANTHAM; NORTHAMPTON IRONSTONE; MAIN NODULE; SANDROCK; FERRUGINOUS LST; 'BUCKLANDI CLAYS'; GRANBY LSTS; 'ANGULATA CLAYS'; 'HYDRAULIC LSTS'
ENGLAND – YORKSHIRE	KIMMERIDGE CLAY; AMPTHILL CLAY; CORALLINE OOLITE; U. OXFORD CLAY; OXFORD CLAY/OSGODBY; KELLAWAYS; CORNBRASH; SCALBY; RAVENSCAR: SCARBORO, CLOUGHTON/CAYTON BAY, HAYBURN; DOGGER; BLEA WYKE; STRIATULUS; PEAK SHALES; ALUM SHALE; JET ROCK; GREY SHALES; CLEVELAND IRONSTONE; STAITHES; IRONSTONE SHALES; PYRITOUS SHALES; 'SILICEOUS' SHALES; CALCAREOUS SHALES
U.S.S.R. (W. SIBERIA)	BAGENOV; GEORGIEV; BARABIN; TATAR; TUMEN: UPPER, MIDDLE, LOWER
CHINA (CENTRE OF SICHUAN)	PENG LAI-ZHEN; SUI NING; SHA-XI-MIAO; ZI-LIU-JING
INDIA (CUTCH (KUTCH))	OOMIA (UMIA AMIU); KATROL; CHAREE (CHARI); PUTCHUM (PATCHAM); KUAR BET ?
NEW ZEALAND	OTEKE: ?, PUAROAN, OHAUAN, HETERIAN; KAWHIA: ?, TEMAIKAN; HERANGI: URUROAN, ARATAURAN, OTAPIRIAN
GREENLAND (EAST)	LINDEMANS BUGT; BERNBJERG; VARDEKLØFT; NEILL KLINTER; KAP STEWART
CANADIAN ARCTIC ARCHIPELAGO	DEER BAY/MOULD BAY; WILKIE POINT: AWINGAK, SAVIK, JAEGAR; BORDEN ISLAND; HEIBERG
U.S.A. (CALIFORNIA)	KNOXVILLE FRANCISCAN; MARIPOSA AMADOR

Chart 2.10. Jurassic biostratigraphic zonation.

Period	Epoch	Age	Substage	"Epoch"	JURASSIC PLANKTONIC ZONATIONS: AMMONITE ZONES (from Cope, *et al* 1980a&b)	DINOFLAGELLATE ZONES (from Sarjeant, 1979)	NANNOFOSSIL ZONES (from van Hinte, 1978b)
Jurassic (J)	Late	Tithonian (Tth)	[Volgian (Vol)] [Portlandian (Por)]	Malm (Mlm)	*Subcraspedites lamplughi*	*Dichadogonyaulax culmula*	*Nannoconus colomi*
					Subcraspedites preplicomphalus		
					Subcraspedites primitivus		
					? Titanites (Paracraspedites) oppressus		
					Titanites anguiformis		
					Galbanites (Kerberites) kerberus		
					Galbanites okusensis		
					Glaucolithites glaucolithus		
					Progalbanites albani		
			[Kimmeridgian (Late) (Kim)]		*Virgatopavlovia fittoni*	*Gonyaulacysta perforans*	*Parhabdolithus embergeri*
					Pavlovia rotunda		
					Pavlovia pallasioides		
					Pectinatites (Pectinatites) pectinatus		
					Pectinatites (Arkellites) hudlestoni	*Gonyaulacysta longicornis*	
					Pect.(Virgatosphinctoides) wheatleyensis		
					Pectinatites (Virgato.) scitulus		*Watznaueria communis*
					Pectinatites (Virgato.) elegans		
		Kimmeridgian (Early) (Kim)			*Aulacostephanus autissiodorensis*		
					Aulacostephanus eudoxus	*Epiplosphaera bireticulata*	
					Aulacostephanoides mutabilis		
					Rasenia cymodoce		
					Pictonia baylei	*Stephanelytron redcliffense*	
		Oxfordian (Oxf)	Late		*Amoeboceras rosenkrantzi*		*Vekshinella stradneri*
					Amoeboceras regulare		
					Amoeboceras serratum		
					Amoeboceras glosense		
			Middle		*Cardioceras tenuiserratum*		
					Cardioceras densiplicatum		
			Early		*Cardioceras cordatum*	*Wanaea fimbriata*	*Actinozygus geometricus*
				(Mlm)	*Quenstedtoceras mariae*		*Diadozgus dorsetense*
	Middle	Callovian (Clv)		Dogger (Dog)	*Quenstedtoceras (Lamberticeras) lamberti*		*Discorhabdus jungi*
					Peltoceras athleta		
					Erymnoceras coronatum	*Polystephanophorus paracalathus*	*Podorhabdus rahla*
					Kosmoceras (Gulielmites) jason		*Podorhabdus escaigi*
					Sigaloceras calloviense		*Stephanolithion bigoti*
					Macrocephalites (M.) macrocephalus	*Meiourogonyaulax callomonii*	*Stephanolithion hexum*
		Bathonian (Bth)	Late		*Clydoniceras (Clydoniceras) discus*		*Stephanolithion speciosum var. octum*
					Oppelia (Oxycerites) aspidoides		
					Procerites hodsoni	*Dichadogonyaulax sellwoodii*	*Diazomatolithus lehmani*
			Middle		*Morrisiceras (Morrisiceras) morrisi*		
					Tulites (Tulites) subcontractus	*Ctenidadinium continuum*	
					Procerites progracilis		
			Early		*Asphinctites tenuiplicatus*		
					Zigzagiceras (Zigzagiceras) zigzag		
		Bajocian (Baj)	Late		*Parkinsonia parkinsoni*	*Nannoceratopsis spiculata*	*Stephanolithion speciosum s.s.*
					Strenoceras (Garantiana) garantiana		
					Strenoceras subfurcatum		
			Early		*Stephanoceras humphriesianum*		
					Emileia (Otoites) sauzei		
					Witchellia laeviuscula		
					Hyperlioceras discites		
		Aalenian (Aal)			*Graphoceras concavum*	*Polysphaeridium? deflandrei*	
					Ludwigia murchisonae		
				(Dog)	*Leioceras opalinum*		*Discorhabdus tubus*
	Early	Toarcian (Toa)		Lias (Lia)	*Dumortieria levesquei*	No zonation	*Podorhabdus cylindratus*
					Grammoceras thouarsense		
					Haugia variabilis		
					Hildoceras bifrons		
					Harpoceras falciferum		
					Dactylioceras tenuicostatum		
		Pliensbachian (Plb)			*Pleuroceras spinatum*		
					Amaltheus margaritatus		
					Prodactylioceras davoei		
					Tragophylloceras ibex		*Crepidolithus crassus*
					Uptonia jamesoni		
		Sinemurian (Sin)	Late		*Echioceras raricostatum*		
					Oxynoticeras oxynotum		*Palaeopontosphaera dubia*
					Asteroceras obtusum		
			Early		*Caenisites turneri*		*Parhabdolithus liasicus*
					Arnioceras semicostatum		*Parhabdolithus marthae*
					Arietites bucklandi		*Crucirhabdus primulus*
		Hettangian (Het)			*Schlotheimia angulata*		*Annulithus arkelli*
					Alsatites liasicus		
				(Lia)	*Psiloceras planorbis*		

Yorkshire coast begins as far as one can tell with beds of the same age as those at the base of the Mariae Zone in its type-locality, Normandy; and the Mariae Zone is divided into Subzones, the lowest of which has its type section on the Yorkshire coast' (Morton 1971, p.89), i.e. the Scarburgense Subzone.

The initial boundary of the *Kimmeridgian* (name from Kimmeridge, originally Kimeridge, Dorset, England; Kimmeridge Clay Formation; = Kimeridge, of Arkell) and that of the Baylei Chron, appears to be better defined between Osmington (Black Head) and Ringstead Bay, on the coast of Dorset, England, than on the coast of Normandy, France. Morton (1971, p.90) specified that the base of the Baylei Zone is to be defined at Ringstead, Dorset. There is some difference of opinion as to the actual span of this age. We have opted for Tithonian to follow it (see below) so that the 'Kimmeridgian' as used here is restricted to Early Kimmeridgian of others, with the initial Tithonian boundary corresponding to that of the Hybonotum Chron. The middle and upper divisions of a tripartite Kimmeridgian stage are correlated as of Tithonian age.

The ***Tithonian***, from Tithon, spouse of Eos (Aurora) Goddess of Dawn (of Cretaceous Period), was defined by Oppel in 1863, in the Mediterranean area, to include all deposits which lie between a restricted Kimmeridgian (as used here) and the 'Valanginian', with its lower boundary coinciding with the base of the Gravesiana Zone (corresponding to Elegans–Hybonotum Zone of the zonation used here). No stratotype was designated. We prefer Tithonian to Volgian (e.g. of Cope *et al.* 1980a) because it is based on Tethyan, rather than Boreal, faunas and so serves better as a standard for correlating northern and southern hemispheres. It is now regarded as being followed naturally by the earliest Cretaceous age (Berriasian). The two versions of the initial boundary of the Volgian, shown on Chart 2.10, are from Gerasimov *et al.* (1975), with the earlier boundary being that used by Russian workers in the northern Urals, and the later one being that which has been used by Casey (1963, 1967) for the English succession. Arkell's (1933 and 1956) scheme of Jurassic stages used the full Kimmeridgian followed by Portlandian and Purbeckian. His Portlandian is thus contained entirely within the Tithonian age span while his Purbeckian strata in England are not a satisfactory basis for international correlation, being largely of freshwater facies.

The columns in Chart 2.9 were obtained as follows: 'Jurassic Period' on the left tabulates the conclusions of the text above. The three British columns are based on Cope *et al.* (1980a, b); the Siberian column is from Krymgol'ts (1972); Sichuan, China from Wang & Liu (1980); Cutch (India) from Arkell (1956, p.386); New Zealand from Suggate *et al.* (1978); East Greenland from Surlyk (1977); Canadian Arctic Archipelago from Johnson & Hills (1973); and California from Arkell (1956, p.553).

The succession of ammonite zones shown in Chart 2.10 is that of Cope *et al.* (1980a, b); this is a more detailed zonation than that compiled by Van Hinte (1978b) and is taxonomically more up-to-date. A (deliberate) discrepancy arises if Charts 2.9 and 2.10 are compared. For Chart 2.10 it was thought best to retain the zonation of Cope *et al.* (1980a, b) as a coherent scheme, rather than attempt to combine it with any other scheme. Since we have elected to use the term Tithonian, however, we feel it reasonable, in Chart 2.9, to use the traditional *Hybonoticeras hybonotum* as the initial Tithonian ammonite zone, even though this form is not mentioned in Chart 2.10. The nannofossil zonation is taken from Van Hinte (1978b) and is calibrated to the ammonite zonal scheme as much as possible; this nannofossil zonation is essentially that used by Barnard & Hay in 1974.

2.12 The Cretaceous Period

Chalk characterises the Anglo–Paris–Belgian area and was the basis for the Cretaceous System as one of five major terrains set up by J. J. d'Omalius d'Halloy in 1822. In 1823 he defined as Cretaceous terrains those corresponding 'to the formation of the chalk, with its tufas, its sands and its clays'.

Already William Smith had mapped four strata between the 'lower clay' and the 'Portland Stone' namely: 'White Chalk, brown or grey chalk, Greensand and Micaceous clay or brick earth' (the last later referred to as Blue Marl and in 1788 as Gault). In 1822 Conybeare and Phillips listed these in two groups, Chalk and the formations below, so a two-fold division, adopted in England and France at an early stage, has persisted in the two epochs commonly used.

In California the Shasta Series was recognised in 1869 (by W. M. Gibb) as early Cretaceous, and in 1887 P. T. Hill showed that the Comanche Series in Texas was of similar age and older than the late Cretaceous Gulf Series. Chamberlin and Salisbury in 1906 proposed a Comanchean System, but the USGS uses the name as a provincial series otherwise. Comanche and Gulf have been suggested as suitable international names for the two major time divisions of the Cretaceous Period (Wilmarth 1925) and Gulf might yet serve this purpose for the second epoch.

Alternative three-fold divisions have also been proposed for example by Leymerie in 1841 who introduced Neocomian for the lower division and the lower white chalk as its upper division. D'Orbigny developed this into five stages: Neocomian, Aptian, Albian, Turonian and Senonian and later added Urgonian (approximately = Barremian) and Cenomanian. The current complement of twelve ages into which this developed, and which is generally accepted internationally, is too elaborate for some. There may indeed be a case for epochs and sub-epochs using the names Neocomian and Senonian which would need to be clearly defined anew if so used, as for example by Haug who included Albian to Turonian as Middle Cretaceous (Zittel 1901). Because of this tendency these divisions are included in Charts 1.2 and 2.11.

Neocomian Sub-epoch. Ever since Thurmann, in 1836, coined the name Neocomian for strata in the vicinity of Neuchâtel, Switzerland, there has been confusion as to how much of the early Cretaceous Period should be embraced by the name. Some authors have used it to signify earliest Cretaceous up to and including the Valanginian Age; others have extended it to include the Barremian and even Aptian. We have followed the recommendation of Barbier, Debelmas and Thieuloy in the Colloque sur le Crétacé inférieur (Barbier & Thieuloy 1965, Debelmas & Thieuloy 1965) that the Neocomian be regarded as an informal term embracing the lowest three stages of the Cretaceous, i.e. Berriasian, Valanginian and Hauterivian.

Senonian Sub-epoch. The Senonian, like the Neocomian, is an informal name which means different things to different authors, the question here being whether or not the Maastrichtian is included. According to the *Lexique stratigraphique international* (Vol. 1, Europe, fasc. 4a, Crétacé, p.318) 'dans le sens de d'Orbigny (1842), créateur du terme, le Sénonien correspond à l'ensemble des couches comprises entre le Turonien et le Danien, c'est-à-dire qu'il débute avec la craie de Villedieu et se termine avec la craie de Maestricht'. This is the sense in which the Senonian is regarded in this work.

Leaving further discussion of the complex history of Cretaceous stages we set out the sequence of time divisions that has almost wholly come to be accepted internationally.

2.12.1 Early Cretaceous

Six ages are well established.

Berriasian was proposed by H. Coquand in 1871. The stratotype is near the village of Berrias, Ardeche, south-east France. This is the initial Cretaceous age and is adopted here in preference to the Boreal Ryazanian division because it was defined in the Tethyan province, i.e. for the same reason that Tithonian was preferred to Portlandian or Volgian. Originally conceived as a subdivision of the Valanginian, it was subsequently often referred to as 'Infra-Valanginian' until the term 'Berriasian' was eventually brought back into use. The initial Berriasian boundary thus equals the initial Cretaceous boundary; it would be placed at, or near, the introduction of *Berriasella grandis*, but we do not know of a defined marker-point in the stratotype.

The ***Valanginian*** was proposed by E. Desor in 1853, with a type locality at the Seyon Gorge, near Valangin, Neuchâtel, Switzerland (Valendis in German). No standard definition for its base is known, but it would approximate to the introduction of *Kilianella pertransiens*.

The ammonite zonation was worked out in the Vocontian facies (to the south), since the stratotype exhibits shallow, sublittoral sediments only, which are poor in ammonites. Barbier & Thieuloy (1965) thus recommend the designation of a parastratotype section in the Vocontian trough, preferably in the region of the Hautes-Alpes.

Hauterivian was proposed by Renevier in 1874, from Hauterive, near Neuchâtel, Switzerland. Its initial boundary would be at about the introduction of *Endemoceras* in north-west Europe (*Lyticoceras* in the Tethyan realm). Debelmas & Thieuloy (1965) recommended that a parastratotype section be designated in the region of the Vocontian trough, as in the case of the Valanginian (and for the same reasons).

Barremian was named by Coquand in 1861 who mentioned the localities at Barrême, near Digne and Angles (Basses-Alpes, south-east France). Busnardo (1965) subsequently designated the Angles roadside section as the stratotype. There is probably scope for an initial boundary point in that section, at or near the introduction of *Pulchella pulchella*.

Aptian was proposed by d'Orbigny in 1840 for strata containing an 'Upper Neocomian' fauna, and named after the village of Apt, Basses-Alpes, south-east France. Three chrons are based on the sub-ages, Bedoulian (Toucas 1888), Gargasian (Kilian 1887) and Clansayesian (Breistroffer 1947). The Aptian initial boundary would be near the arrival of *Prodeshayesites fissicostatus* (in England), or *Deshayesites deshayesi* (in France).

Albian was proposed by d'Orbigny in 1844 for the interval between Aptian and what is now called Cenomanian. Its name was derived from Alba, the Roman name for Aube, France. It appears to be an age of considerable duration, so a three-fold subdivision into Early, Middle and Late is reasonable. Following Breistroffer (1947) its initial boundary is regarded as commencing with *Leymeriella tardefurcata*. A composite stratotype has been proposed by Larcher, Rat & Malapris (1965) based on several exposures in the Aube area. The latest part of the Late Albian chron was referred to as the 'Vraconian' by Renevier in 1867; this term is now regarded as being superfluous, as it is merely another name for the time interval represented by Spath's (1923) *Stoliczkaia dispar* zone.

2.12.2 Late Cretaceous

Six ages are well established.

Cenomanian was proposed by d'Orbigny in 1847 with a type locality in the environs of Le Mans (Roman Cenomanum), in the Sarthe, France; however, he did not designate a type section. More recently Marks (1967) proposed a composite section in the area of St Ulphace–Théligny–Moulin de l'Aunay as the stratotype for the Cenomanian stage. The Cenomanian initial boundary would be defined near the arrival of *Mantelliceras mantelli*. The GSL (Rawson *et al*. 1978) equated the termination of the Cenomanian with the base of a broad ammonite zone of *Mammites nodosoides*, while Marks (1977) proposed the initial occurrence of the planktonic Foraminifer *Praeglobotruncana helvetica* as indicative of this same boundary.

Turonian was introduced by d'Orbigny in 1842 but in 1847 he separated the lower part off as the Cenomanian. The name Turonian is derived from Tours – Roman Turones, or Touraine – Roman Turonia. No stratotype was proposed but in 1852 d'Orbigny designated the type area as lying between Saumur (on the river Loire) and Montrichard (on the river Cher), France. The lowest formation of the type Turonian, within d'Orbigny's geographical limitations, contains *Mammites nodosoides*. The initial appearance of *M. nodosoides* is taken as the initial boundary of the Turonian; the *M. nodosoides* zone is recognisable over much of the world. Unfortunately, *Praeglobotruncana helvetica*, widely regarded as an index form for the Turonian in Tethyan regions, has not yet been found in Touraine.

In 1856, Coquand divided the 'Upper Chalk' of the Saintonge (Charente-maritime, France) into three 'stages' with the lowest being subdivided into three 'sub-stages'. In 1857, he raised the lowest two 'sub-stages' to stage rank, as the ***Coniacian***, with the stratotype being the section described by Coquand in 1857 at Cognac, Charente (hence the name), in the northern part of the Aquitaine Basin, France. The stratotype appears to contain no foraminifera or nannofossils, and the basal sands at Cognac contain no fossils of correlative value. However, the base of the Coniacian is by definition that of the Senonian, the standard formation for which is the Craie de Villedieu, of Touraine, whose lowest member contains an assemblage containing *Barroisiceras haberfellneri*. The initial appearance of this *Barroisiceras* species is thus regarded as indicating the initial boundary of the Coniacian.

Santonian, the uppermost 'sub-stage' of Coquand (in 1856) was raised to stage rank in 1857 (and named after the village of Saintes, France). In 1858, Coquand designated a section on the road between Javresac and Saintes as the stratotype. At present it is not possible to determine the faunal base of the Santonian in its stratotype area, or its upper limit.

The ***Campanian*** is the second 'stage' of Coquand who introduced the term 'Campanian' as a stage in 1857. In 1858 Coquand made it clear that the stratotype was the hillside section at Aubeterre-sur-Dronne (La Grande Champagne – hence 'Campanian'). De Grossouvre, between 1895 and 1901 worked out possible ammonite zonations leading to the present one. The initial Campanian boundary approximates to the base of the *Placenticeras bidorsatum* zone but the age itself does not seem to be well founded, since recent work on the microfauna of the sections at Aubeterre has shown that the bulk of the type Campanian is actually Maastrichtian as now understood (Rawson *et al.* 1978). Séronie-Vivien designated parastratotype sections near Né in the Charente valley, but since the macrofossils there are scarce and undistinctive, they do not provide a working standard for the base of the Campanian (Rawson *et al.* 1978).

The ***Maastrichtian*** was introduced in 1849 by Dumont who separated the 'calcaire de Maastricht' from the 'craie sénonienne'.

Chart 2.11. Cretaceous chronostratic scale and correlation.

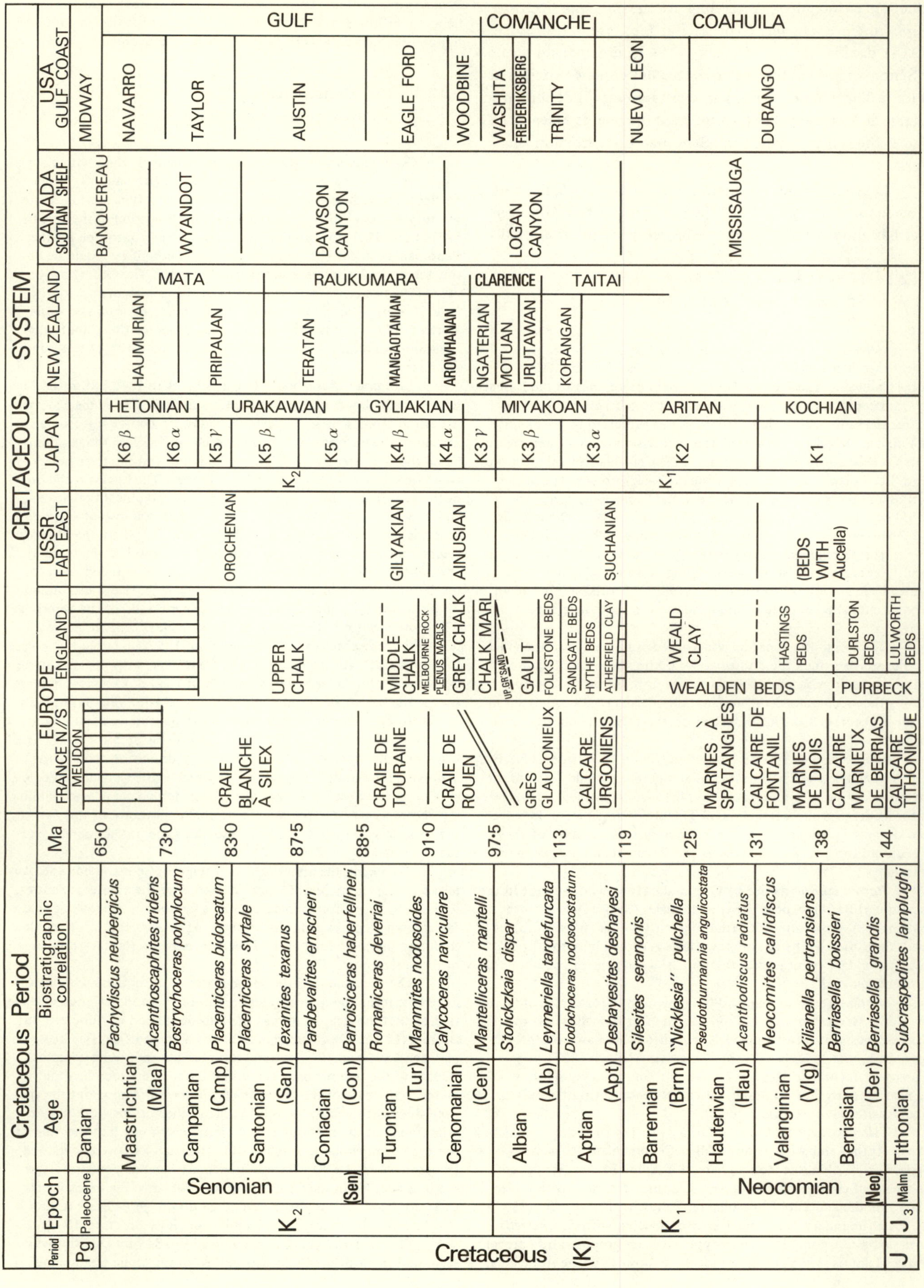

Period	Epoch	Age	Biostratigraphic correlation	Ma	EUROPE: FRANCE N//S	EUROPE: ENGLAND	USSR FAR EAST	JAPAN	NEW ZEALAND	CANADA SCOTIAN SHELF	USA GULF COAST
Pg	Paleocene	Danian			MEUDON					BANQUEREAU	MIDWAY
Cretaceous (K)	K_2; Senonian	Maastrichtian (Maa)	*Pachydiscus neubergicus*; *Acanthoscaphites tridens*	65·0				K6β (HETONIAN)	HAUMURIAN (MATA)	BANQUEREAU	NAVARRO (GULF)
	K_2; Senonian	Campanian (Cmp)	*Bostrychoceras polyplocum*; *Placenticeras bidorsatum*	73·0				K6α; K5γ (HETONIAN / URAKAWAN)	PIRIPAUAN (MATA)	WYANDOT	TAYLOR
	K_2; Senonian	Santonian (San)	*Placenticeras syrtale*; *Texanites texanus*	83·0	CRAIE BLANCHE À SILEX	UPPER CHALK	OROCHENIAN	K5β (URAKAWAN)	TERATAN (RAUKUMARA)	DAWSON CANYON	AUSTIN
	K_2; Senonian (Sen)	Coniacian (Con)	*Parabevalites emscheri*; *Barroisiceras haberfellneri*	87·5				K5α (URAKAWAN)			
	K_2	Turonian (Tur)	*Romaniceras deveriai*; *Mammites nodosoides*	88·5	CRAIE DE TOURAINE	MIDDLE CHALK; MELBOURNE ROCK	GILYAKIAN	K4β (GYLIAKIAN)	MANGAOTANIAN (RAUKUMARA)		EAGLE FORD
	K_2	Cenomanian (Cen)	*Calycoceras naviculare*; *Mantelliceras mantelli*	91·0	CRAIE DE ROUEN	PLENUS MARLS; GREY CHALK; CHALK MARL	AINUSIAN	K4α (GYLIAKIAN); K3γ (MIYAKOAN)	AROWHANAN (RAUKUMARA); NGATERIAN (CLARENCE)	LOGAN CANYON	WOODBINE
	K_1	Albian (Alb)	*Stolickzkaia dispar*; *Leymeriella tardefurcata*	97·5	GRÈS GLAUCONIEUX	UP. GR'SAND; GAULT; FOLKSTONE BEDS		K3β (MIYAKOAN)	MOTUAN; URUTAWAN (CLARENCE); KORANGAN (TAITAI)		WASHITA; FREDERIKSBERG; TRINITY (COMANCHE)
	K_1	Aptian (Apt)	*Diodochoceras nodosocostatum*; *Deshayesites deshayesi*	113	CALCARE URGONIENS	SANDGATE BEDS; HYTHE BEDS; ATHERFIELD CLAY	SUCHANIAN	K3α (MIYAKOAN)			
	K_1	Barremian (Brm)	*Silesites seranonis*; "*Nicklesia*" *pulchella*	119		WEALD CLAY (WEALDEN BEDS)		K2 (ARITAN)		MISSISAUGA	NUEVO LEON
	K_1; Neocomian	Hauterivian (Hau)	*Pseudothurmannia angulicostata*; *Acanthodiscus radiatus*	125	MARNES À SPATANGUES						DURANGO (COAHUILA)
	K_1; Neocomian	Valanginian (Vlg)	*Neocomites callidiscus*; *Kilianella pertransiens*	131	CALCAIRE DE FONTANIL; MARNES DE DIOIS	HASTINGS BEDS (WEALDEN BEDS)	(BEDS WITH Aucella)	K1 (KOCHIAN)			
	K_1; Neocomian (Neo)	Berriasian (Ber)	*Berriasella boissieri*; *Berriasella grandis*	138	CALCAIRE MARNEUX DE BERRIAS	DURLSTON BEDS (PURBECK)					
J	J_3; Malm	Tithonian	*Subcraspedites lamplughi*	144	CALCAIRE TITHONIQUE	LULWORTH BEDS (PURBECK)					

Chart 2.12. Cretaceous biostratigraphic zonation.
The key to abbreviations of generic names in Sigal's foraminiferal zonation is as follows: *Glt.* – *Globotruncana*, *Wh.* – *Whiteinella*, *Rtl.* – *Rotaliporal*, *Pl.* – *Planomalina*, *Hed.* – *Hedbergella*, *Tic.* – *Ticinella*, *Gld.* – *Globigerinelloides*, *Schk.* – *Schackoina*, *Ctes.* – *Comorotalites*, *Clh.* – *Clavihedbergella*, *Gav.* – *Gavelinella*, *Cauc.* – *Caucasella*, *Doroth* – *Dorothia*, *Lent.* – *Lenticulina*.

CRETACEOUS BIOSTRATIGRAPHICAL ZONATIONS

EPOCH	CHRONOSTRATI - GRAPHIC AGE		TETHYAN PELAGIC MACROFOSSIL ZONES (FROM VAN HINTE, 1978a)	FORAMINIFERAL ZONES (SIGAL, 1977)	AGE IN Ma	CALCAREOUS NANNO - PLANKTON ZONES (SISSINGH,1977)
Late	Maastrichtian (Maa)	L	*Pachydiscus neubergicus*	*Glt. mayaroensis*	65·0	26 *Nephrolithus frequens*
						25 *Arkhangelskiella cymbiformis*
						24 *Reinhardtites levis*
				Glt. gansseri		23 *Tranolithus phacelosus*
		E	*Acanthoscaphites tridens*	*Glt. stuarti/ Glt. falsostuarti*		
	Campanian (Cmp)	L	*Bostrychoceras polyplocum*	*Glt. calcarata*	73·0	22 *Tetralithus trifidus*
			Hoplitoplacenticeras vari	*Glt. elevata/ Glt. stuartiformis*		21 *Tetralithus nitidus*
						20 *Ceratolithoides aculeus*
		E	*Delawarella delawarensis*			19 *Calculites ovalis*
			Placenticeras bidorsatum			18 *Aspidolithus parcus*
						17 *Calculites obscurus*
	Santonian (San)	L	*Placenticeras syrtale* w. *Eupachydiscus isculensis*	*Glt. concavata carinata*	83·0	16 *Lucianorhabdus cayeuxii*
						15 *Reinhardtites anthophorus*
		E	*Texanites texanus*			14 *Micula staurophora*
	Coniacian (Con)	L	*Parabevalites emscheri* (*Protexanites, Paratexanites. T. pseudotexanus*)	*Glt. concavata*	87·5	13 *Marthasterites furcatus*
		E	*Barroisiceras haberfellneri*	*Glt. sigali / Glt. schneegansi*		12 *Lucianorhabdus maleformis*
	Turonian (Tur)	L	*Romaniceras deveriai*		88·5	
		M	*Romaniceras ornatissimum*	*Glt. helvetica*		11 *Tetralithus pyramidus*
			Romaniceras bizeti			
		E	*Mammites nodosoides* (*Kanabiceras septemseriatum, Metoicoceras whitei, Inoceramus labiatus*)	*Wh. archaeocretacea*		
	Cenomanian (Cen)	L	*Calycoceras naviculare*	*Rtl. cushmani*	91·0	10 *Microrhabdulus decoratus*
		M	*Acanthoceras rhotomagense*			
		E	*Mantelliceras mantelli*	*Rtl. globotruncanoides* *Rtl. brotzeni*		9 *Eiffellithus turriseiffeli*
Early	Albian (Alb)	L	*Stoliczkaia dispar*	*Rot. appenninica / Pl. buxtorfi*	97·5	
			Mortoniceras inflatum	*Tic. breggiensis*		8 *Prediscosphaera cretacea*
			Diploceras cristatum	*Hed. rischi / Tic. primula*		
		M	*Hoplites lautus/H. nitidus*			
			Hoplites dentatus	*Hed. planispira*		
		E	*Douvilleiceras mammilatum*			7 *Chiastozygus litterarius*
			Leymeriella tardefurcata	*Tic. bejaouaensis*		
	Aptian (Apt)	L	*Diodochoceras nodosocostatum*		113	
			Cheloniceras subnodosocostatum	*Hed trochoidea.*		
				Gld. algeriana.		
				Gld ferreolensis		
		E	*Aconoceras nisus*	*Schk. cabri*		
			Deshayesites deshayesi	*Gld. maridalensis / Gld. blowi*		
				Gld. gottisi / Gld. duboisi		
				Hed. similis		
	Barremian (Brm)	L	*Silesites seranonis* / *Pulchella provincialis*	*Ctes. aptiensis*	119	6 *Micrantholithus obtusus / hoschulzii*
				(? *Ctes. intercedens*)		
		E	"*Nicklesia*" *pulchella* / *Pulchella caicedi*	*Clh. eocretacea*		
			Pulchella didayi	*Hed. sigali*		5 *Lithraphidites bollii*
			Pulchella pulchella			
	Hauterivian (Hau)	L	*Pseudothurmannia angulicostata*	*Ctes bartensteini* *Gav. gr. djaffaensis – sigmoicosta*	125	
			Subsaynella sayni	*Cauc.gr. hauterivica*		4 *Cretarhabdus loriei*
		E	*Crioceras duvali*	*Doroth. ouachensis*		
			Acanthodiscus radiatus	*Hapl. vocontianus*		
	Valanginian (Vlg)	L	*Neocomites callidiscus*	*Lent ouachensis var bartensteini/Doroth hauteriviana*	131	3 *Calcicalathina oblongata*
			Himantoceras trinodosum	*Lent. gr. eichenbergi-meridiana*		
			S. verrucosum	*Lent. busnardoi*		
		E	*K. campylotoxa*			2 *Cretarhabdus crenulatus*
			K. roubaudi	*Lent. guttata*		
			K. pertransiens			
	Berriasian (Ber)	L	*Berriasella boissieri* — Ryazanian		138	1 *Nannoconus steinmanni*
		E	*Berriasella grandis* — Volgian		144	

The stratotype has now been fixed by the Comité d'etude du Maastrichtian, i.e. units Ma–Md of Uhlenbroek (1911) – exposed in the ENCI quarry at St Pietersburg on the outskirts of Maastricht (in South Holland). The type Maastrichtian corresponds only to the later Maastrichtian of current usage. A revised concept, based on the belemnite succession, regards the Maastrichtian as being coextensive with the range of *Scaphites* (*Hoploscaphites*) *constrictus*, with its initial boundary marked by the entry of *Belemnitella lanceolata* (Rawson *et al.* 1978).

The Cretaceous ammonite zones shown in Chart 2.12 are taken from the compilation of Van Hinte (1978a, with sources cited). Of the many Cretaceous zonal schemes that have been erected on the basis of planktonic foraminifera, that of Sigal (1977) has been selected here for inclusion in Chart 2.12; this zonation is particularly applicable to the Mediterranean area in general. The calcareous nannoplankton zonation in Chart 2.12 is that of Sissingh (1977) and is calibrated, as far as is possible, with the planktonic foraminiferal zonation.

Chart 2.11 was compiled with the assistance of N. F. Hughes, for the European columns. The USSR column is from Nalivkin (1973, tables 55 and 56); the Japan column is from Takai *et al.* (1963); the New Zealand column is from Stevens (1980); the Canada column from Ascoli (1976); and the USA column from Murray (1961) and Postuma (1971).

2.13 The Tertiary Sub-era

The name 'Tertiary' is a survivor from an early attempt at stratigraphical classification, i.e. Primary, Secondary, etc. (now obsolete). It was applied first in Italy by Arduino in 1760 to rocks that are still recognised as Tertiary; Brogniart, in 1810, applied the term 'tertiare' to strata which overlay the Cretaceous chalks in the Paris Basin. In its present usage the name is useful and unambiguous, referring to the post-Mesozoic – pre-Quaternary interval.

The word 'Neozoic' was used by Lyell and his contemporaries to embrace all systems from Triassic onwards. The intention at that time was to complement Paleozoic (see Section 2.3). Then J. Phillips, in 1841, divided these systems between his Mesozoic and Cainozoic groups (Cenozoic in this work), and the term 'Neozoic' has fallen into disuse, although at various times it has been used as a synonym of 'Tertiary' and also of 'Cenozoic' (Tertiary + Quaternary).

Lyell, in 1833, divided the European Tertiary into Newer Pliocene, Older Pliocene, Miocene and Eocene, based on the percentage of living amongst living plus extinct mollusc species. Further studies in the middle of the nineteenth century established the presence of extensive marine, brackish, freshwater and continental sediments in northern Europe, in between Lyell's Eocene and Miocene; for these Beyrich, in 1854, proposed the name 'Oligocene'. The Tertiary classification attained its modern aspect in 1874 when Schimper distinguished the lower part of the 'Eocene' as 'Paleocene'. This was based on younger Paleocene strata in western Europe and was distinguished more by facies than by an evolutionary sequence; so separation from Eocene strata was hardly justified by the initial descriptions. But when the American Paleocene, with richly fossiliferous Early Paleocene strata, was considered the division was well justified.

The two-fold classification of Tertiary time derives from European stratigraphy where it is divided by the climax of the Alpine Orogeny. The older division was indeed Lyell's original Eocene and the younger was his Miocene and Pliocene. To combine the two latter, Hornes in 1853 introduced the name Neogene. He also proposed Paleogene which was at first synonymous with Eocene but had more use when Oligocene was distinguished, and it came into its own when Paleogene could usefully combine Paleocene, Eocene and Oligocene after 1874. These divisions became established in Europe: for example E. Suess established his five Mediterranean stages which divided the Neogene within Alpine Europe, and in France in 1902 E. Haug used the name Nummulitique (for Paleogene) and this usage continued there for fifty years. It was only after the IUGS was set up around 1960 that there was a proper authority to settle these matters. So, in response to what seemed to be a prevailing opinion in Europe, George and others in 1968, in a subcommission to the Commission on Stratigraphy of the IUGS, recommended 'that the Cenozoic Era should be divided informally into Tertiary and Quaternary sub-Eras, and that the former unit should include the Paleogene and Neogene Periods/Systems. The Paleogene was divided into Paleocene, Eocene and Oligocene Epochs/Series and the Neogene into Miocene and Pliocene Epochs/Series' (Curry *et al.* 1978, p.2). There are now IUGS subcommissions (of the Commission on Stratigraphy) on both Paleogene and Neogene stratigraphy in parallel with subcommissions on the stratigraphy of other geologic periods.

A comprehensive review of the historical development of the numerous Tertiary 'stages' was provided by Berggren (1971).

2.14 The Paleogene Period

The recommendation by the International Subcommission on Stratigraphic Classification (ISSC) to define stages by boundary stratotypes is not easily carried out in the Paleogene basins of western Europe. One major reason is the lack of continuous sections; instead there are many relatively small outcrops scattered throughout these basins, mainly in France, England, Belgium, Germany and Denmark. The intercalation of marine, marginal-marine and nonmarine sediments, the lateral facies changes and the nature of the fossil assemblages themselves, all combine to render inter-basin correlation difficult, and the possibility of worldwide correlation even more remote. These factors have hindered the correlation and relative positioning of the various proposed stratotypes and have given rise to a bewildering number of 'stage' names; 'most of the European Tertiary stage names in common use today were originally defined as lithostratigraphical (facies) units within cycles of sedimentation' (Berggren 1971, p.696). In Appendix 2 the origins of fifty Paleogene stage names are listed.

The development, over the last three decades or so, of a biostratigraphic zonation based upon planktonic foraminifera has allowed detailed and long-ranging correlations. Such a zonation was initially restricted to low-to-middle latitudes, but with the addition of a calcareous nannoplankton zonation and, latterly, of a radiolarian zonation, the biostratigraphic scheme can be extended into higher latitudes and into the deep ocean (DSDP cores). The integration of these zonal schemes forms a biostratigraphic framework into which the various European Paleogene stratotypes may be placed in their most plausible relative positions. An example of this is shown in Hardenbol & Berggren (1978, fig. 4). Until more precise standardisation has been achieved we also redefine some stages in terms of the biostratigraphic zonal schemes as tabulated in Chart 2.13. Chart 2.14 lists the nominate taxa upon which the P-zone scheme of Blow (1969) is based, together with the calcareous nannofossil taxa typifying the NP-zonal scheme (after Martini 1971). The zonation of Blow (1969, 1979) is based on planktonic foraminifera. In addition to referring to the Cenozoic zones by their nominate taxon (or taxa) Blow numbered them in 'ascending' order in two groups. The prefix 'P' is applied to the Paleogene zones (P1 to P22) and the prefix 'N' to Neogene and Quaternary zones (N4 to N25). Martini (1971) similarly numbered his nannoplankton zones with the prefix 'NP' for the Paleogene zones (NP1 to NP25), and 'NN' for the Neogene (and Quaternary) zones (NN1 to NN25).

The three epochs of the Paleogene Period are considered in time sequence.

2.14.1 The Paleocene Epoch

In Europe, this epoch is represented by two main lithostratigraphic units: (1) a lower, carbonate unit, on which the Danian Age is based, and (2) an upper, clastic unit on which the Thanetian Age is based.

Danian, named by Desor in 1847, has type localities at Stevns Klint and Faxse (= Faxe = Faxoe), Denmark. Danian (with typical chalk facies) was commonly considered as the latest Cretaceous division to be succeeded by Montian (Belgium) as the earliest Tertiary stage. When it was shown that Montian was a different facies of Danian age the debate as to whether to assign it to the Mesozoic or Cenozoic era was probably settled in favour of the latter (about 1960) because already the equivalent Midway rocks (with rich faunas) in the Gulf region of the USA were well established as basal Tertiary. The Danian has few equivalent stages in Europe (Montian in Belgium is in part equivalent to the type Danian) although sediments which have been correlated with it are widespread, especially in central and eastern Europe.

Hansen (1970, p.25) defined the Danian as 'the time interval between the rocks found above the Maastrichtian White Chalk exposed on Stevns Klint up to and including the time of deposition of the rocks found below the Selandian basal conglomerate exposed at the locality Hvallose, in Jyland'. This definition would include part of zone P1a to P2. Martini (1971) and Perch-Nielsen (1972) have recognised zones NP1 to NP3 from outcrops in the type area.

Hardenbol & Berggren (1978) extended the definition of Danian to include all of P1 and P2 (= NP1 to NP3 (part)).

Thanetian proposed by Renevier in 1873 is named after the Thanet Sands, Isle of Thanet, Kent, England. Equivalents in Europe are the type Selandian (Denmark), type Landenian and type Heersian (Belgium) and the 'Sable de Bracheux' (Paris Basin). Martini (1971) reported zone NP8 in the Thanetian, which can be correlated with zone P4 (Bramlette & Sullivan 1961). El-Naggar (1966a, b) reported zone P3 in the type Heersian of Belgium.

Hardenbol & Berggren (1978) extended the Thanetian to include zones P3 to lower P6, i.e. latest NP3 to NP9.

2.14.2 The Eocene Epoch

In western Europe this consists, traditionally, of three chronostratigraphic divisions, in which scheme we list four ages.

(i) ***Early Eocene***, ***Ypresian*** (Dumont in 1849) is named from the Ypres Clay, south of Ypres (Ieper), Belgium. In the type area, shallow to moderately deep-water sands and clays are developed. Equivalents in other basins are – London Clay (England), Rosnaes Clay (Denmark) and Eocene 3 (Germany). The Cuisian stage of the Paris Basin is represented by sandstones believed to be equivalent to the upper part of the Ypresian.

Zones NP11 and 12 have been reported from the type area of the Ypresian (Hay & Mohler 1967, Martini 1971), and also from the Rosnaes Clay and the Eocene 3. Zone NP12 has been correlated with foraminiferal zone P8 by Hay and others in 1967.

Hardenbol & Berggren (1978) extended the Ypresian chronostratigraphic unit to include the biostratigraphically identified zone P6 (upper part) to P9, i.e. zones NP10 to NP13 (Berggren 1971, 1972).

(ii) ***Middle Eocene***,

(a) ***Lutetian*** (de Lapparent in 1883) has no precise stratotype designated but it was proposed to accommodate the 'Calcaire Grossier' of the Paris Basin. This is represented in the Paris Basin mainly by shallow-water carbonates; equivalents in other basins are – part of the Bracklesham Beds (England), and Brussels sands (Belgium). The Lutetian chronostratigraphic division was traditionally regarded as the Middle Eocene stage and so is often taken to include all foraminiferal zones P10 to P14. Bouche (1962) and Hay and others in 1967 identified NP14 in the Lutetian type area, and Bukry & Kennedy (1969) made a tentative correlation of this with foraminiferal zone P10.

(b) *Bartonian* (Mayer-Eymar in 1858) used to be thought to be the northern equivalent of the Mediterranean Priabonian, but the latter contains *Isthmolithus recurvus*, i.e. zone NP19, while the Bartonian does not and is, therefore, presumably older. However, zone NP17 has been reported from the Barton Beds, Hampshire, England (Martini 1971) and this has been correlated with foraminiferal zone P14 (Berggren 1972) or even with P13 (Roth, Baumann & Bertolino 1971) These datings would place the Bartonian within the traditional Lutetian chronostratigraphic unit. Since there seems to be no suggestion of an overlap of the Lutetian *sensu stricto* and the Bartonian *sensu stricto*, Hardenbol & Berggren (1978) considered that confusion in the world-wide correlation framework would be minimised by restricting the Lutetian to foraminiferal zones P10 to P12 (= NP14 to lower NP16) and regarding the Bartonian as P13 and P14 (= upper NP16 to NP17), all in the Middle Eocene.

(iii) ***Late Eocene***, ***Priabonian*** (Munier-Chalmas and de Lapparent in 1893) is named from Priabona, Vicenza province, northern Italy, where the stratotype was restricted by Roveda (1961) to the section from Boro to Grenella. The type section has been found to contain zone P16 and P17 (part) (Hardenbol 1968); also the nannofossil zones NP19 (Martini 1971) and NP19 and 20 (Roth *et al.* 1971). At the Eocene Colloquium of 1968, the Priabonian was considered to include foraminiferal zones P15 to P17 (part), i.e. nannofossil zones NP18 to NP21 (part).

Chart 2.13. Paleogene chronostratic scale and correlation.

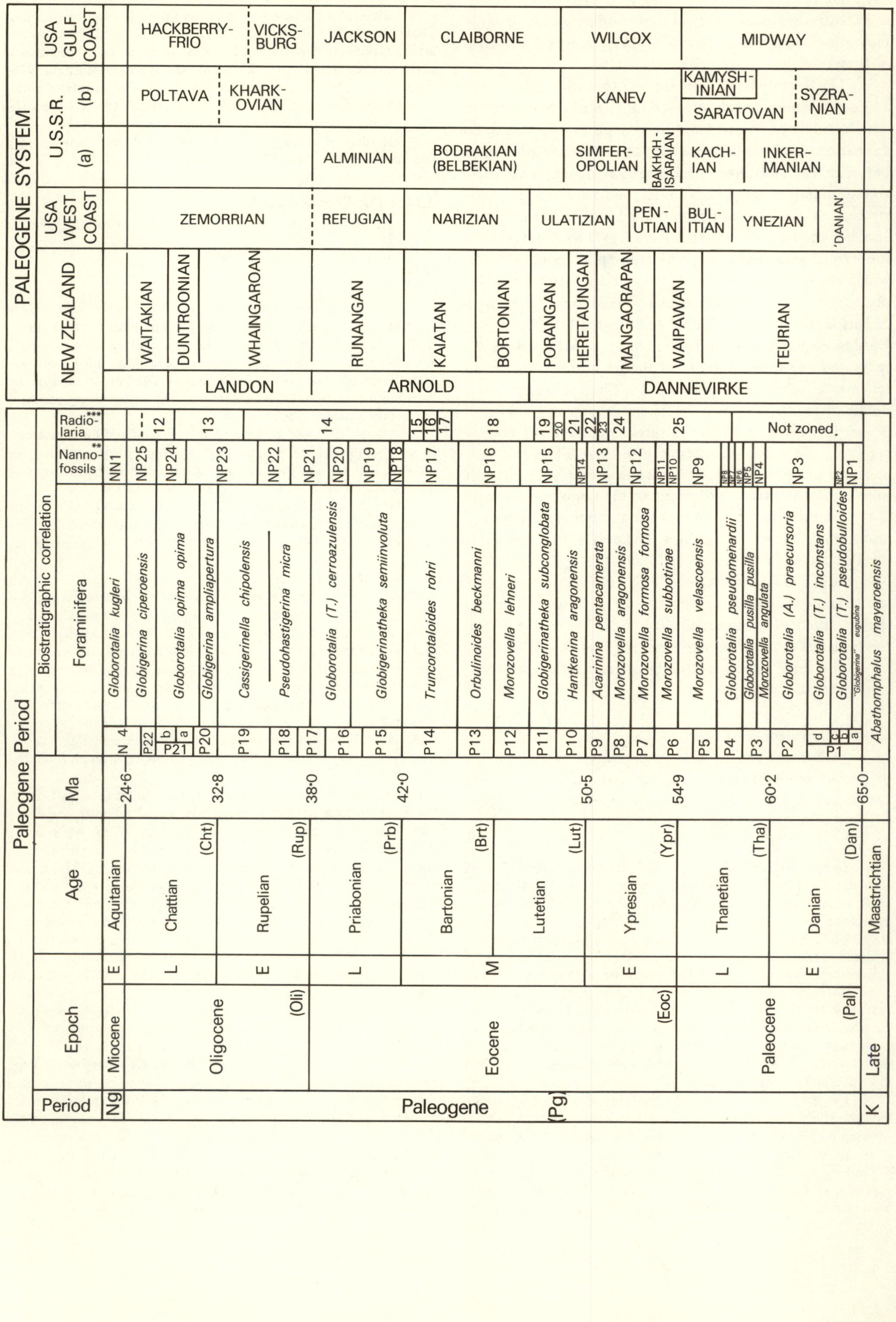

Paleogene Period

Period	Epoch		Age	Ma	Foraminifera zone	Foraminifera	Nannofossils**	Radiolaria***
Ng	Miocene	E	Aquitanian		N 4	*Globorotalia kugleri*	NN1	
				24·6				
Paleogene (Pg)	Oligocene (Oli)	L	Chattian (Cht)		P22	*Globigerina ciperoensis*	NP25	- - - / 12
					P21 b, a	*Globorotalia opima opima*	NP24	12
					P20	*Globigerina ampliapertura*	NP23	13
				32·8				
		E	Rupelian (Rup)		P19	*Cassigerinella chipolensis*	NP23	13
					P18	*Pseudohastigerina micra*	NP22	14
				38·0				
	Eocene (Eoc)	L	Priabonian (Prb)		P17	*Globorotalia (T.) cerroazulensis*	NP21	14
					P16		NP20	14
					P15	*Globigerinatheka semiinvoluta*	NP19, NP18	14
				42·0				
		M	Bartonian (Brt)		P14	*Truncorotaloides rohri*	NP17	15, 16, 17
					P13	*Orbulinoides beckmanni*	NP16	18
			Lutetian (Lut)		P12	*Morozovella lehneri*	NP16	18
					P11	*Globigerinatheka subconglobata*	NP15	19, 20
					P10	*Hantkenina aragonensis*	NP15, NP14	21
				50·5				
		E	Ypresian (Ypr)		P9	*Acarinina pentacamerata*	NP13	22, 23
					P8	*Morozovella aragonensis*	NP13, NP12	24
					P7	*Morozovella formosa formosa*	NP12	25
					P6	*Morozovella subbotinae*	NP11, NP10	25
				54·9				
	Paleocene (Pal)	L	Thanetian (Tha)		P5	*Morozovella velascoensis*	NP9	25
					P4	*Globorotalia pseudomenardii*	NP8, NP7, NP6	25 / Not zoned
					P3	*Globorotalia pusilla pusilla* / *Morozovella angulata*	NP5, NP4	Not zoned
				60·2				
		E	Danian (Dan)		P2	*Globorotalia (A.) praecursoria*	NP3	Not zoned
					P1 d	*Globorotalia (T.) inconstans*	NP3	Not zoned
					P1 c, b	*Globorotalia (T.) pseudobulloides*	NP2	Not zoned
					P1 a	*"Globigerina" eugubina*	NP1	Not zoned
				65·0				
K	Late		Maastrichtian			*Abathomphalus mayaroensis*		

PALEOGENE SYSTEM

New Zealand series	New Zealand stage	USA West Coast	U.S.S.R. (a)	U.S.S.R. (b)	USA Gulf Coast
Landon	Waitakian	Zemorrian		Poltava	Hackberry-Frio
Landon	Duntroonian	Zemorrian		Poltava	Hackberry-Frio
Landon	Whaingaroan	Zemorrian		Poltava / Kharkovian	Hackberry-Frio / Vicksburg
Arnold	Runangan	Refugian	Alminian		Jackson
Arnold	Kaiatan	Narizian	Bodrakian (Belbekian)		Claiborne
Arnold	Bortonian	Narizian	Bodrakian (Belbekian)		Claiborne
Dannevirke	Porangan	Ulatizian	Bodrakian (Belbekian)		Claiborne
Dannevirke	Heretaungan	Ulatizian	Simferopolian	Kanev	Wilcox
Dannevirke	Mangaorapan	Ulatizian / Penutian	Simferopolian	Kanev	Wilcox
Dannevirke	Waipawan	Penutian	Bakhchisaraian	Kanev	Wilcox
Dannevirke	Teurian	Bulitian	Kachian	Saratovan / Kamyshinian	Midway
Dannevirke	Teurian	Ynezian	Inkermanian	Saratovan / Syzranian	Midway
Dannevirke	Teurian	'Danian'		Syzranian	Midway

2.14.3 The Oligocene Epoch

The complex nomenclature history of the Oligocene Epoch has been reviewed by Berggren (1971). The scheme followed here is that of Hardenbol & Berggren (1978) who recognised two distinct lithostratigraphic units in north-west Europe: (i) a lower, moderately deep-marine, clayey unit, which includes the typical Rupelian rocks, and (ii) an upper, predominantly shallow-marine, sandy unit, which contains the Chattian type section.

Rupelian (Dumont in 1849) was named from the 'argile de Rupelmonde', which is a strict junior synonym of 'Argile de Boom', (Koninck in 1837), at Boom (Anvers), Belgium. According to Banner & Blow (1965) the type Rupelian exhibits foraminiferal zones from the upper part of P18 to the upper part of P19. The nannofossil zone NP23 has been recognised in Boom Clay samples (Martini 1971). Hardenbol & Berggren (1978, p.21) 'arbitrarily placed the boundary between the Chattian and Rupelian Stages at the top of P19, which corresponds approximately to the middle of NP23'. The boundary position was thought to have the best chance of falling between the Rupelian and Chattian stages as defined in their respective type areas.

Chattian (Fuchs in 1894) was named from the 'Kasseler Meeressand', near Kassel, north-west Germany where the holostratotype was designated in 1958. Nannofossil zones NP24 and NP25 have been recognised tentatively in the Chattian (Neochattian type) section at Doberg, near Bunde, Germany (Martini 1971, Benedek & Müller 1974, Martini & Müller 1975). Zone NP25 correlates with the foraminiferal zone P22 and the lower part of N4 (of Blow 1969). The *Globigerinoides* Datum (first evolutionary appearance of *G. primordius*), which defines the base of zone N4, has been shown to occur in Upper Oligocene, pre-Aquitanian, levels (Anglada 1971a & b, Scott 1972, Alvinerie *et al.* 1973, Theyer & Hammond 1974; Lamb & Stainforth 1976, Van Couvering & Berggren 1977). The Oligocene–Miocene boundary therefore occurs within zone N4, i.e. top of zone NP25.

The left-hand columns in Chart 2.13 tabulate the above conclusions. The named foraminiferal zones are from Stainforth *et al.* (1975). The New Zealand column is from Stevens (1980); USA (West Coast) from Berggren & Van Couvering (1974, fig. 1); USA (Gulf Coast) from Murray (1961, fig. 6.3); USSR (a) from Berggren (1971, table 52.19) and USSR (b) from Nalivkin (1973, table 15).

Chart 2.14. Paleogene biostratigraphic zonation.

KEY TO PALEOGENE PLANKTONIC ZONATIONS

FORAMINIFERA

N 3 P 22		*Globigerina angulisuturalis*	P-R-Z
N 2 P 21		*Globigerina angulisuturalis / Globorotalia (T.) opima opima*	Conc.-R-Z
N 1 P 19/ P 20		*Globigerina sellii / Globigerina ampliapertura*	P-R-Z
P 18		*Globigerina tapuriensis*	P-R-Z
P 17		*Globigerina gortanii gorlanii / Globorotalia (T.) centralis*	P-R-Z
P 16		*Cribrohantkenina inflata*	T-R-Z
P 15		*Porticulasphaera semiinvoluta*	P-R-Z
P 14		*Globorotalia (M.) spinulosa spinulosa*	P-R-Z
P 13		*Globigerapsis beckmanni*	T-R-Z
P 12		*Globorotalia (M.) lehneri*	P-R-Z
P 11		*Globigerapsis kugleri / Subbotina frontosa boweri*	Conc.-R-Z
P 10		*Subbotina frontosa frontosa / Globorotalia (T.) pseudomayeri*	Conc.-R-Z
P 9		*Globorotalia (A.) aspensis / Globigerina lozanoi prolata*	Conc.-R-Z
P 8	8b	*Globorotalia aragonensis / Globorotalia (M.) formosa*	Conc.-R-Z
	8a	*Globorotalia (M.) formosa / Globorotalia (M.) lensiformis*	P-R-Z
P 7		*Globorotalia (A.) wilcoxensis berggreni*	P-R-Z
P 6		*Globorotalia (M.) subbotinae subbotinae Globorotalia (M.) velascoensis acuta*	P-R-Z
P 5		*Muricoglobigerina soldadoensis soldadoensis/ Globorotalia (M) velascoensis pasionensis*	Conc-R-Z
P 4		*Globorotalia (G.) pseudomenardii*	P-R-Z
P 3		*Globorotalia (M.) angulata angulata*	P-R-Z
P 2		*Globorotalia (A.) praecursoria praecursoria*	P-R-Z
P 1	1b	*Globorotalia (T.) compressa compressa Eoglobigerina eobulloides simplicissima*	Conc-R-Z
	1a	*Globorotalia (T.) pseudobulloides Globorotalia (T.) archeocompressa*	Conc-R-Z
Pα		*Globorotalia (T.) longiapertura*	P-R-Z

CALCAREOUS NANNOPLANKTON

NP 25	*Sphenolithus ciperoensis*
NP 24	*Sphenolithus distentus*
NP 23	*Sphenolithus predistentus*
NP 22	*Helicopontosphaera reticulata*
NP 21	*Ericsonia ? subdisticha*
NP 20	*Sphenolithus pseudoradians*
NP 19	*Isthmolithus recurvus*
NP 18	*Chiasmolithus oamaruensis*
NP 17	*Discoaster saipanensis*
NP 16	*Discoaster tani nodifer*
NP 15	*Chiphragmalithus alatus*
NP 14	*Discoaster sublodoensis*
NP 13	*Discoaster lodoensis*
NP 12	*Marthasterites tribrachiatus*
NP 11	*Discoaster binodosus*
NP 10	*Marthasterites contortus*
NP 9	*Discoaster multiradiatus*
NP 8	*Heliolithus riedeli*
NP 7	*Discoaster gemmeus*
NP 6	*Heliolithus kleinpelli*
NP 5	*Fasciculithus tympaniformis*
NP 4	*Ellipsolithus macellus*
NP 3	*Chiasmolithus danicus*
NP 2	*Cruciplacolithus tenuis*
NP 1	*Markalius inversus*

RADIOLARIA

Lyc.b.	*Lychnocanium bipes*
Dor.at.	*Dorcadospyris ateuchus*
The.t.	*Theocyrtis tuberosa*
Thy.b.	*Theocyrtis bromia*
Thy.t.	*Thyrsocyrtis tetracantha*
Pod.g.	*Podocyrtis goetheana*
Pod.c.	*Podocyrtis chalara*
Pod.m.	*Podocyrtis mitra*
Pod.a.	*Podocyrtis ampla*
Thy.tr.	*Thyrsocyrtis triacantha*
The.m.	*Theocampe mongolfieri*
The.c.	*Theocotyle cryptocephala*
Pho.s.	*Phormocyrtis striata striata*
Bur.c.	*Buryella clinata*
Bek. b.	*Bekoma bidartensis*

FORAMINIFERA ; Zonation of Blow (1979)
P-R-Z = Partial Range Zone
Conc-R-Z = Concurrent Range Zone
T-R-Z = Total Range Zone

CALCAREOUS NANNOPLANKTON ; Zonation of Martini (1971)

RADIOLARIA ; Zonation from Hardenbol & Berggren (1978) (In: Cohee *et al.*, 1978)

Chart 2.15. Neogene chronostratic scale and correlation.

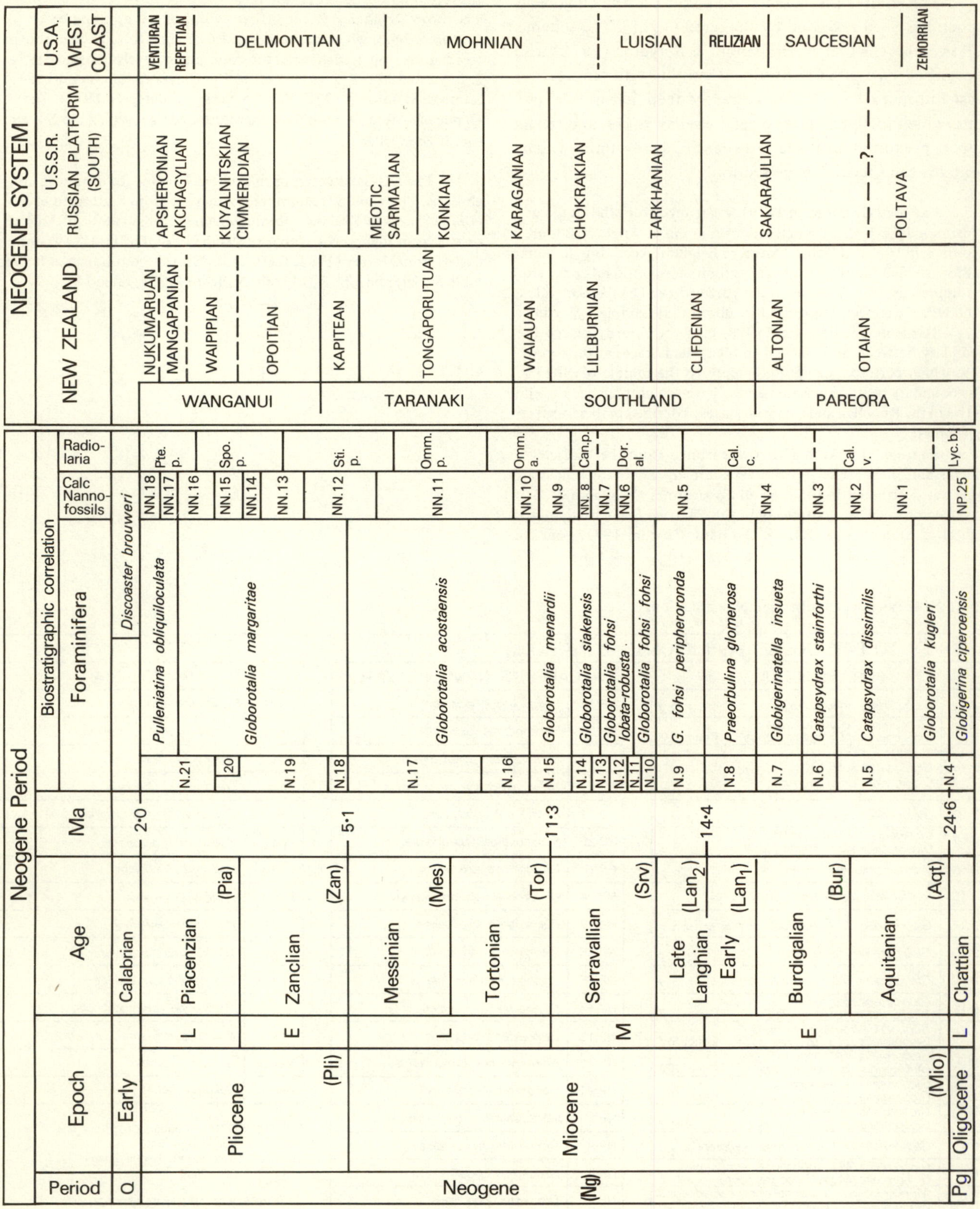

Neogene Period

Period	Epoch		Age	Ma	Biostratigraphic correlation: Foraminifera		Calc Nannofossils	Radiolaria
Q	Early		Calabrian			*Discoaster brouweri*		
				2·0				
Neogene (Ng)	Pliocene (Pli)	L	Piacenzian (Pia)		N.21	*Pulleniatina obliquiloculata*	NN.18 / NN.17	Pte. p.
						Globorotalia margaritae	NN.16	
					20		NN.15 / NN.14	Spo. p.
		E	Zanclian (Zan)		N.19		NN.13	
					N.18		NN.12	Sti. p.
				5·1				
	Miocene	L	Messinian (Mes)		N.17	*Globorotalia acostaensis*	NN.11	Omm. p.
			Tortonian (Tor)		N.16		NN.10	Omm. a.
				11·3	N.15	*Globorotalia menardii*	NN.9	
		M	Serravallian (Srv)		N.14	*Globorotalia siakensis*	NN. 8	Can.p.
					N.13 / N.12	*Globorotalia fohsi lobata-robusta*	NN.7 / NN.6	Dor. al.
					N.11 / N.10	*Globorotalia fohsi fohsi*		
			Langhian Late (Lan_2)		N.9	*G. fohsi peripheroronda*	NN.5	
				14·4				
			Langhian Early (Lan_1)		N.8	*Praeorbulina glomerosa*		Cal. c.
		E	Burdigalian (Bur)		N.7	*Globigerinatella insueta*	NN.4	
					N.6	*Catapsydrax stainforthi*	NN.3	
			Aquitanian (Aqt)		N.5	*Catapsydrax dissimilis*	NN.2	Cal. v.
	(Mio)					*Globorotalia kugleri*	NN.1	
				24·6	N.4			
Pg	Oligocene	L	Chattian			*Globigerina ciperoensis*	NP.25	Lyc. b.

NEOGENE SYSTEM

NEW ZEALAND		U.S.S.R. RUSSIAN PLATFORM (SOUTH)	U.S.A. WEST COAST
WANGANUI	NUKUMARUAN	APSHERONIAN	VENTURAN
	MANGAPANIAN	AKCHAGYLIAN	REPETTIAN
	WAIPIPIAN	KUYALNITSKIAN	DELMONTIAN
		CIMMERIDIAN	
	OPOITIAN		
TARANAKI	KAPITEAN	MEOTIC	
		SARMATIAN	
	TONGAPORUTUAN	KONKIAN	MOHNIAN
SOUTHLAND	WAIAUAN	KARAGANIAN	
	LILLBURNIAN	CHOKRAKIAN	
			LUISIAN
	CLIFDENIAN	TARKHANIAN	
			RELIZIAN
PAREORA	ALTONIAN	SAKARAULIAN	SAUCESIAN
	OTAIAN	?	
		POLTAVA	ZEMORRIAN

Chart 2.16. Neogene biostratigraphic zonation.

KEY TO NEOGENE PLANKTONIC ZONATIONS

FORAMINIFERA		
N.21	*Globorotalia (T.) tosaensis tenuitheca*	Cons-R-Z
N. 20	*Globorotalia (G.) mutticamerata/ Pulleniatina obliquiloculata obliquiloculata*	P-R-Z
N.19	*Sphaeroidinella dehiscens dehiscens/ Globoquadrina altispira altispira*	P-R-Z
N.18	*Sphaeroidinellopsis subdehiscens paenedehiscens / Globorotalia (G) tumida tumida*	P-R-Z
N.17	*Globorotalia (G) tumida plesiotumida*	Cons-R-Z
N.16	*Globorotalia (T) acostaensis / Globorotalia (G) merotumida*	P-R-Z
N.15	*Globorotalia (T) continuosa*	Cons-R-Z
N.14	*Globigerina nepenthes / Globorotalia (T) siakensis*	Conc.-R-Z
N.13	*Sphaeroidinellopsis subdehiscens subdehiscens / Globigerina druryi*	P-R-Z
N.12	*Globorotalia (G) fohsi*	P-R-Z
N.11	*Globorotalia (G) praefohsi*	Cons-R-Z
N.10	*Globorotalia (T) peripheroacuta*	Cons-R-Z
N.9	*Orbulina suturalis / Globorotalia (T) peripheroronda*	P-R-Z
N.8	*Globigerinoides sicanus / Globigerinatella insueta*	P-R-Z
N.7	*Globigerinoides trilobus / Globigerinatella insueta*	P-R-Z
N.6	*Globigerinatella insueta / Globigerinita dissimilis*	Conc.-R-Z
N.5	*Globoquadrina dehiscens praedehiscens / Globoquadrina dehiscens dehiscens*	P-R-Z
N.4	*Globigerinoides primordius / Globorotalia (T) kugleri*	Conc.-R-Z

CALCAREOUS NANNOPLANKTON	
NN.18	*Discoaster brouweri*
NN.17	*Discoaster pentaradiatus*
NN.16	*Discoaster surculus*
NN.15	*Reticulofenestra pseudoumbilica*
NN.14	*Discoaster asymmetricus*
NN.13	*Ceratolithus rugosus*
NN.12	*Ceratolithus tricorniculatus*
NN.11	*Discoaster quinqueramus*
NN.10	*Discoaster calcaris*
NN.9	*Discoaster hamatus*
NN.8	*Catinaster coalitus*
NN.7	*Discoaster kugleri*
NN.6	*Discoaster exilis*
NN.5	*Sphenolithus heteromorphus*
NN.4	*Helicopontosphaera ampliaperta*
NN.3	*Sphenolithus belemnos*
NN.2	*Discoaster druggi*
NN.1	*Triquetrorhabdulus carinatus*

RADIOLARIA	
Pte. p.	*Pterocanium prismatium*
Spo. p.	*Spongaster pentas*
Sti. p.	*Stichocorys peregrina*
Omm. p.	*Ommatartus penultimus*
Omm.a.	*Ommatartus antepenultimus*
Can. p.	*Cannartus (?) petterssoni*
Dor. al.	*Dorcadospyris alata*
Cal. c.	*Calocycletta costata*
Cal. v.	*Calocycletta virginis*
Lyc. b.	*Lychnocanium bipes*

FORAMINIFERA ; Zonation of Blow (1969,1979)
P-R-Z = Partial Range Zone
Cons-R-Z = Consecutive Range Zone
Conc-R-Z = Concurrent Range

CALCAREOUS NANNOPLANKTON ;
Zonation of Martini (1971)

RADIOLARIA ; Zonation of
Riedel & Sanfilippo , 1971

2.15 The Neogene Period

In 1853, the Austrian geologist Hörnes proposed the name Neogene as follows (translated): 'The common occurrence of the Vienna Mollusca not only in the type Miocene but in the type Pliocene induces me to join the nomenclature of these suites in a single category, the two suites to be known under the name Neogene. . .'

During the nineteenth century many Neogene 'stage' names were proposed for Europe; these were mainly lithostratigraphic and were merely a reflection of facies variations. Most of them are no longer in common use, but an interesting historical review of their development was given by Berggren (1971). Stratigraphic nomenclature has now stabil ised to a very large extent, and the Neogene ages mentioned here are now widely accepted as 'standard' ages. In current usage Miocene and Pliocene are both regarded as epochs (series).

2.15.1 The Miocene Epoch

Lyell, in 1833, named the Miocene as the second oldest of his four Tertiary divisions and considered the outcrops in the Touraine (south-west Paris Basin) as representing the type Miocene area.

Aquitanian was introduced by Mayer-Eymar in 1858 for the largely lagoonal, mollusc-rich beds in the Aquitaine Basin, Gironde, France. He noted that the best outcrops occurred in a 10 km section along the stream Saint Jean d'Etampes which runs through Saucats (Moulin de l'Eglise) and La Brede (Moulin Bernachon). This section was designated as the stratotype by Dollfus in 1909 and has been accepted as such by the Committee on Mediterranean Neogene Stratigraphy (1958 meeting at Aix-la-Provence). The Neogene (Aquitanian) initial boundary might be defined near the base of this section (George *et al.* 1969, p.52).

Biostratigraphically, the base of the Aquitanian stage appears to occur within planktonic foraminiferal zone N4 and its upper limit is close to the top of the succeeding zone, N5.

In discussing the stratal sequence in northern Europe Mayer-Eymar in 1858 included in his Aquitanian stage strata which had been explicitly included by Beyrich in 1854 and in 1856 in his (Upper) Oligocene. 'This has led, in no small way, to confusion regarding the affinities of the Aquitanian and assignment of Aquitanian to the Oligocene by some, to the Miocene by others' (Berggren 1971, p.747).

Burdigalian was introduced by Deperet in 1892 for marine strata overlying Aquitanian in the Aquitaine Basin, France (the name is from the richly fossiliferous 'Faluns de Bordeaux'). No single stratotype was designated in 1892, but faluns (crags) of Saucats and Léognan were cited as being Early Burdigalian in age. Dollfus in 1909 designated Le Coquillat, near Léognan, as the type Burdigalian locality. He also referred to Mayer-Eymar's beds 8–10 in the Saint Jean d'Etampes section as Burdigalian. The designation of Le Coquillat was confirmed at the Congrès du Néogène Méditerranéen in Vienna in 1960.

Since the type locality is a small, isolated outcrop (a ditch) and difficult of access, E. Szöts (Berggren 1971, p.752) suggested designating as a 'neostratotype' the section along the Saint Jean d'Etampes stream between Moulin de l'Eglise and Pont-Pourquey, near Saucats; this section has the merit of being continuous with the Aquitanian stratotype. But there can only be one boundary point.

Examination of material from the Aquitanian–Burdigalian type area, in south-west France, led Berggren (1971, p.753) to the conclusion that the Burdigalian appears to be approximately equivalent to the *Catapsydrax dissimilis* and *C. stainforthi* zones of Bolli (N5 and N6 of Blow (1969)) and possibly to the lower part of the *Globigerinatella insueta* zone (N7) of Stainforth *et al.* (1975). Anglada (1971b) reported zone N5 from near the lower boundary of the Burdigalian neostratotype.

Langhian was introduced by Pareto in 1864 for outcrops in the heart of the Langhe, north of Ceva, northern Italy. Cita & Premoli Silva (1960) designated the Bricco della Croce section as the stratotype section, from near Cessole northwards to Case dei Rossi. The stratotype includes most of the 'Pteropod marls'. It appears that Burdigalian is earlier than the *Globigerinoides sicanus* zone (of Bolli), i.e. pre-N8 of Blow, and earlier than the *Praeorbulina glomerosa* zone of Stainforth *et al.* (1975, and Chart 2.16). From the work of Cita & Elter (1960) it appears that 'the base of the Langhian may satisfactorily be drawn at the *base* of the *G. sicanus* zone' (Berggren 1971, p.753), i.e. at the base of N8 and the base of the *P. glomerosa* zone in Chart 2.16. The Langhian straddles the *Orbulina* Datum (base of N9) and so is partly Early Miocene and partly Middle Miocene in age. For this reason we divide Langhian into two ages which for the time being we refer to as Early Langhian and Late Langhian.

Serravallian was proposed by Pareto in 1865 for outcrops in the vicinity of the village of Serravalle Scrivia (Alessandria, Italy), which is the type area (not Serravalle, Firenze, Italy). It was proposed as a stage intermediate between Langhian and Tortonian as then understood. The Serravallian stage has, historically, been 'correlated' with various parts of the 'Helvetian' stage. Drooger (1964) suggested that Serravallian should be used to replace 'Helvetian'. A stratotype has been designated and described by Vervloet (1966), i.e. the Serravalle Formation in the Scrivia valley.

Serravallian embraces more than the average number of Neogene planktonic zones – i.e. zone N10 to within N15 of Blow (1979) (*Globorotalia* (*T.*) *peripheroacuta* to *G.* (*T.*) *continuosa*) – and from within calcareous nannoplankton zone NN5 to NN9 (*Sphenolithus heteromorphus* to *Discoaster hamatus*) of Martini (1971). The relative fineness of these zonations attests to the general diversity 'explosion' of Middle Miocene time, following the relatively low diversities of Oligocene and, to a large extent, of Early Miocene time.

Tortonian was proposed by Mayer-Eymar in 1858 for the 'Blaue Mergel mit *Conus canaliculatus* und *Ancellaria glandiformis* von Tortona'. The stratotype, designated by Gianotti in 1953, is exposed in the valley of Rio Mazzapiedi – Castellania (Tortona, Alessandria Province, Italy); it has been described by Gino *et al.* (1953).

In 1868, Mayer-Eymar restricted his original concept of Tortonian, distinguishing the regressive, increasingly brackish-water strata (Cerithium marls) above marine Tortonian (his lower Messinian) here a part of the Paratethys regressive sequence. However, he did also refer a series of marine units near Messina, Sicily, to his Messinian, thus providing a more objective basis for comparison and correlation. Banner & Blow (1965) recognised late N15 and N16 zones in the type Tortonian. Berggren & Van Couvering (1974) considered that the early part of zone N17 was also represented.

Marine strata from near Messina, Sicily, were referred to by Mayer-Eymar in 1868 as ***Messinian***. Owing to sedimentological and tectonic complications in the vicinity of Messina itself, Selli (1960) selected and described a neostratotype, exposed between Mt Capodarso and Mt Pasquasia, on either side of the Morello River between Caltanisetta and Enna, Sicily. The section is bounded below by Tortonian marls and above by basal Pliocene ('trubi'), and its microfauna has been described by d'Onofrio (1964).

The original concept of Messinian included the diatomaceous marls at the base ('tripoli'), evaporites ('gesso') and the overlying 'trubi' (deep-water lutites with rich planktonic faunas). However Seguenza in 1868 and 1879 restricted the Messinian to the 'tripoli' and the 'gesso', considering the 'trubi' to be basal Pliocene.

Planktonic foraminiferal zone N16 (*Globorotalia* (*G.*) *tumida plesiotumida*) has been recognised in the type section of the Messinian by d'Onofrio (1964), Colalongo (1970) and Blow (1969). Berggren & Van Couvering (1974) concluded that Messinian is equivalent to only the later part of zone N17, i.e. it was thus of particularly short duration.

2.15.2 The Pliocene Epoch

Lyell in 1833 proposed the name Pliocene for the youngest Tertiary deposits which he recognised at the time. He divided his Pliocene (more than 50% of molluscan species still living) into an 'Older Pliocene' which would correspond to the Astian–Piacenzian as subsequently recognised in Italy, and a 'Newer Pliocene' (90–95% of molluscan species still living) for which he subsequently (in 1839) introduced the name Pleistocene. The 'Subapennine' strata of northern Italy were regarded by Lyell in 1833 as typical for his 'Older Pliocene' (= Pliocene *sensu stricto*).

Although some Italian geologists favour a three-fold division of the Pliocene Epoch, we have followed tradition (e.g. Berggren & Van Couvering 1974) in using a two-fold division. The planktonic foraminiferal zones N18 to N21 of Blow (1969, 1979) span the Pliocene interval. Berggren (1973) has constructed an alternative, more detailed, six-fold Pliocene subdivision based on planktonic foraminifera – his zones PL1 to PL6.

Zanclian was proposed by Seguenza in 1868 from Zancla (the pre-Roman name for Messina) for foraminiferal white marls (the 'trubi' of Baldacci in 1886) followed by coral limestones and yellowish sandy marls, in the vicinity of Messina, Sicily. No stratotype was expressly designated but Cita subsequently (in 1974) proposed one at Capo Rosello, near Agrigento, on the southern coast of Sicily. The base of the stratotype section signifies the beginning of permanent open-marine conditions in the Mediterranean following the restricted deposition and desiccation (evaporites) of the latest Miocene (Messinian). The name Zanclian is preferred to the more northerly Tabianian, since the cooler-water fauna of the latter is less favourable for correlation. Banner & Blow (1965) recognised planktonic foraminiferal zone N18 in the type Zanclian, and its base (the Miocene–Pliocene boundary) has been variously placed in the middle of this zone or at its base; the latter course is followed here, to be consistent with Berggren & Van Couvering (1974).

The ***Piacenzian*** was introduced by Mayer-Eymar in 1858 (as the 'Piacenzische Stufe') for the argillaceous facies of the Lower Pliocene with *Nassa semistriata* in northern Italy. It was originally distinguished as a sub-stage of the Astian (de Rouville in 1853) and subsequently erected as a stage by Renevier (1897). (The name Astian now refers to the sandy facies which interfingers with the Piacenzian clays and marls in northern and central Italy). Pareto (1865) adopted the name, using the French equivalent Plaisancian, and specified that the typical development of the stage was to be seen in the hills around Castell' Arquato in northern Italy (Berggren & Van Couvering 1974).

The name derives from the city of Piacenza, which lies approximately mid-way between Parma and Milan in Italy. The name Plaisancian, as stated above, is merely the French translation of Piacenzian, and does not refer to the small village of Plaisance situated between Castell' Arquato and Lugagnana, as erroneously attributed by Movius in 1949.

No specific type section has been designated, but the vicinity of Castell' Arquato appears to meet general approval as a type area, and the Castell' Arquato section was referred to as 'classical' by di Napoli-Alliata in 1954.

The divisions of the Period described above are illustrated in Chart 2.15. The named foraminiferal zones are from Stainforth *et al.* (1975); the New Zealand divisions are from Suggate *et al.* (1978); the USSR South Russian Platform from Nalivkin (1973) and the Californian west coast from Berggren & Van Couvering (1974). The foraminiferal, calcareous nannoplankton and radiolarian zone abbreviations given on Chart 2.15 are expanded on Chart 2.16 where references are given.

2.16 The Quaternary Sub-era (Pleistogene Period)

This latest period in Earth history has been variously named. In the early years of the nineteenth century Alluvium was established and in 1823 Buckland regarded the somewhat older Diluvium as being a product of the Biblical Flood.

In the years leading up to 1840 the widespread erratics in Alpine and northern Europe were interpreted as deposits of former extensive ice. In 1837 the term Drift came to be used for the widespread sands, gravels and boulder clays newly thought to have been deposited from floating ice. Lyell proposed the Pleistocene Period in 1839 to accommodate this ice age and to post-date his already established Pliocene. Pliocene was the last period in the Tertiary Era, thus there was room for a post-Pliocene Quaternary division – so named by Morlot in 1854. It was different from Pleistocene in that it also contained Lyell's original 'Recent', later named Holocene by the IGC in 1885. So the classification determined by historical priority and long usage is:

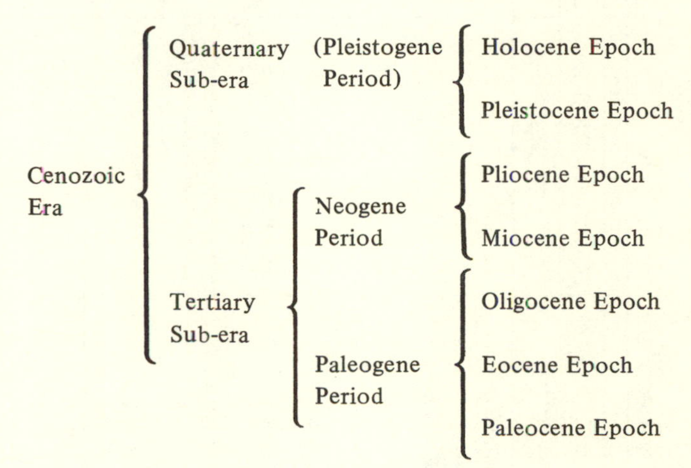

Quaternary is not a satisfactory name in the scheme; Primary and Secondary have been replaced and Tertiary is being replaced by Paleogene and Neogene as formal period names so alternatives Anthropogene (often in use in the USSR) and Pleistogene have been proposed, the latter better fitting the overall nomenclature. However tradition may well prevail and is acknowledged here. Alternative schemes have been proposed (e.g. to include Pleistocene in Neogene). However the scale is not to embody, rather to calibrate, natural events and standardisation is intended to give a convenient and stable chronostratic scale that will not vary with changing opinion.

Chart 2.17. Quaternary (Pleistogene) time scales (the two parts of this chart are independent and correlation across from one to the other is not implied).

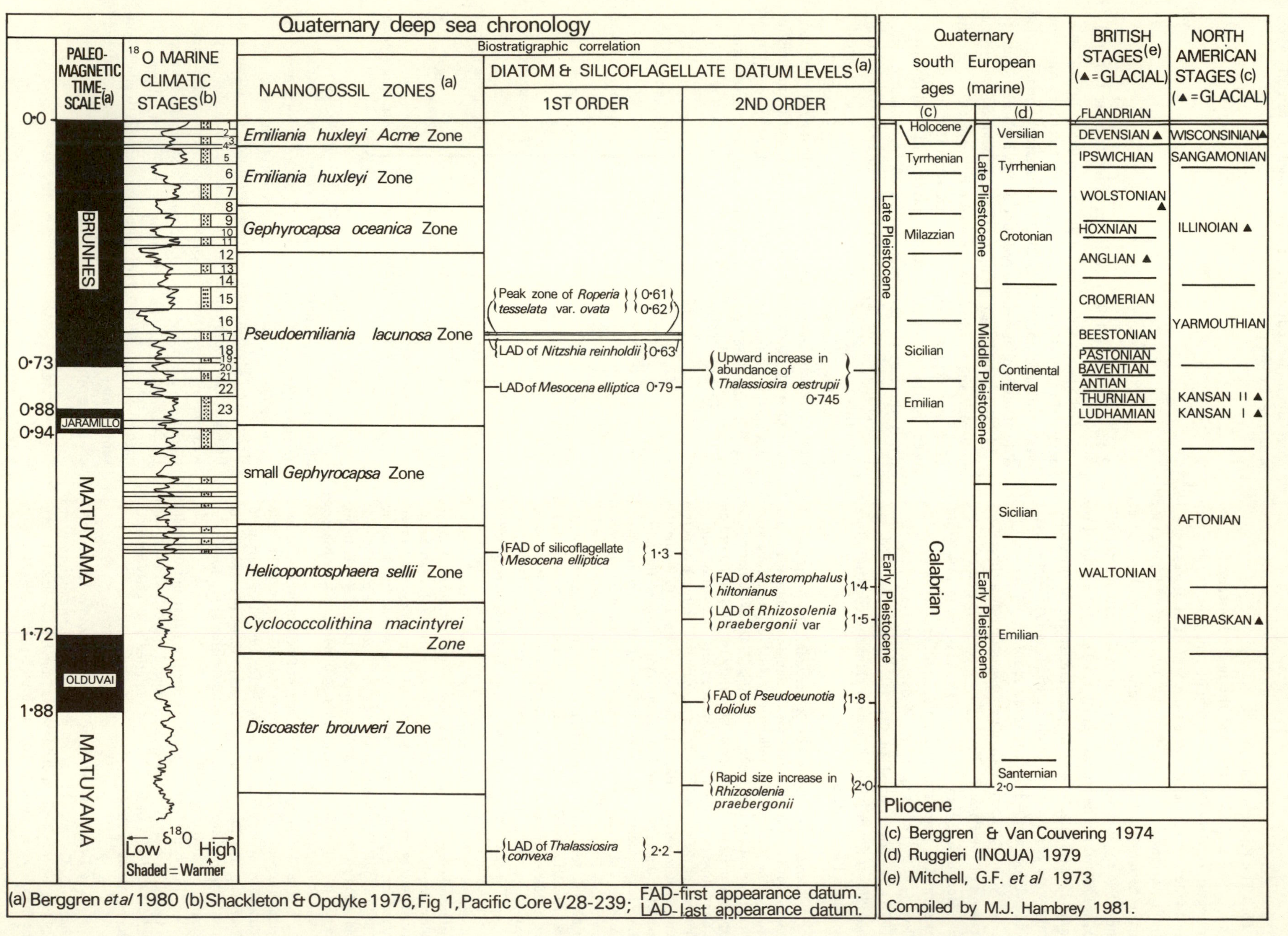

2.16.1 The Pliocene–Pleistocene boundary

An attempt to standardise the initial Pleistocene boundary was made at the IGC in London in 1948 when it was realised that an objective reference in a stratotype near or at the base of Calabrian strata in Italy would be appropriate. This has been pursued for example by the International Geological Correlation Programme but continental correlation has proved difficult, although it is accepted that the boundary would correspond to such a position.

The initial boundary is now placed at the base of the Calabrian Stage, a name originally introduced by Gignoux in 1910. The Pliocene–Pleistocene boundary stratotype should be one of the marine sections preserved on land in southern Italy.

Estimates of its age have ranged from 0.6–4 Ma (Haq, Berggren & Van Couvering 1977). Correlation using paleontological criteria has proved difficult. The initial Calabrian boundary was thought to be marked by the first appearance of both *Arctica islandica* and *Hyalinea baltica* (Šibrava 1978) but Ruggieri (INQUA publication in 1979) showed that the latter appears slightly later. Furthermore, Ruggieri argued for the suppression of Calabrian and replacement by Santernian along with a revision of the rest of the sequence. Various sections in southern Italy have competed for the position of stratotype. On the basis of calcareous plankton at Le Castella and Santa Maria di Catanzaro, as well as of deep-sea sediments, Haq *et al.* (1977) correlated the boundary with the top or slightly above the Olduvai event at 1.6 Ma. Other recent estimates have centred on 1.8 Ma (Šibrava 1978, Nikiforova 1978). More recently a section at Vrica in Calabria has been regarded as more suitable for the stratotype. Here Selli *et al.* (1977) obtained a boundary date of 2.0 ± 0.1 Ma from K–Ar whole-rock analysis in pumice. There can only be one standard for the boundary and as this is now favoured at Vrica and calibrated at about 2.0 Ma then the boundary favoured as terminating the Olduvai event (though important) is a different one and not the Pliocene–Pleistocene boundary.

2.16.2 Pleistocene terrestrial sequences

Attempts to make a stratigraphic sequence out of successive cycles of continental glaciation have proved to be totally inadequate in distinguishing particular fluctuations of climate amongst so many fluctuations. Furthermore, continental classifications have been shown to be oversimplified and incomplete, based to a large extent on erroneous concepts.

In 1973 a GSL Working Group revised the Quaternary stratigraphy of the British Isles (Mitchell *et al.* 1973, see Chart 2.17). The new terminology for the British stages was partly intended to discourage older terrestrial correlations and to face up to the recognition of a large number of glacial stages in the Quaternary Sub-era. However, this scheme is far from satisfactory, being essentially a climatic rather than a rock sequence. Lack of knowledge in key areas has led to the unfounded assignment of ages in places far removed from the stratotypes (Bowen 1978). Continental sequences of glacial tills, terraces, etc. were worked out first in the Alps and northern Europe long before there was any means of correlation between them or an adequate time scale by which to order them. They are still difficult to correlate and so have not been included in the chart. Two such sequences are given in Evans (1971) and Berggren & Van Couvering (1974) (who each fitted the divisions differently on their time scales) and they are combined in the list below. Evans compressed all the glacial and interglacial periods of Europe into 1 Ma, although he recognised that some terrestrial sequences contained a gap representing 1 Ma and that the Pliocene–Pleistocene boundary occurred at 2.1 Ma.

For the Alps the sequence in increasing age is:

- Würm Glacial
  - Riss/Würm Interglacial
- Riss Glacial
  - Mindel/Riss Interglacial
- Mindel Glacial
  - Günz/Mindel Interglacial
- Günz Glacial
- (Berggren & Van Couvering gave two glacial Günz sub-stages)
  - Donau/Günz Interglacial
- Donau Glacial
- (Evans listed four glacials under Donau)
  - ?Biber Glacial
- (Berggren & Van Couvering place Biber glacial at around 3 Ma).

For northern Europe the sequence (increasing age) is:

  - Flandrian (Holocene)
- Weichselian Glacial
  - Eemian Interglacial
- Saalian Glacial
  - Holsteinian Interglacial
- Elsterian Glacial
  - Cromerian Interglacial
- Menapian Glacial
  - Waalian Interglacial
- Eburonian Glacial
  - Tiglian Interglacial
- (Berggren & Van Couvering indicate this as extending back into pre-Quaternary time).

In the USSR the following sequence (increasing age) was given by Flint (1971) (with additions from Sachs & Strelkov (1961)):

- Valdayan/Zyryanka Glacial
  - Mikulino Interglacial
- Moscovian Glacial
  - Odintsovo Interglacial
- Dnepr/Samarovo Glacial
  - Likhvin Interglacial
- Oka/Demyanka Glacial.

The central North American sequence also suffers from major problems of interpretation and the rigid classical approach is unsuitable (Bowen 1978). However, recent radiometric and paleomagnetic data have enabled some time constraints to be placed on the sequence. The Nebraskan Stage, which is poorly known, is often regarded as consisting of any till older than Kansan; but several glacial horizons between 1.2 and 0.7 Ma have been dated and the oldest glacial sediments in this area are dated around 2 Ma; a single glacial stage is clearly inappropriate. Radiometric ages have enabled considerable refinement of the Wisconsinian Stage. The sequence as a whole is presently being re-assessed.

An independent record of Late Pleistocene and Holocene climatic changes has been derived from $^{18}O/^{16}O$ ratios in cores through the Greenland and Antarctic ice sheets (Johnsen *et al.* 1972) and from other areas.

2.16.3 Pleistocene marine sequence

Because the span of Quaternary time is so near our own, a different order of discrimination is possible and different methods are being rapidly developed. The principal development in a new Pleistocene time scale depends on the regularity of the climatic cycle that was discovered around 1875 by Croll and developed especially by Milankovitch. This approach was not taken too seriously by Quaternary geologists until there were quantitative ways of testing it. Zeuner (1945), Emiliani (1965) and Evans (1971) were amongst those to recalculate and relate the astronomical parameters, testing, for example, a 40 000 year cycle against other phenomena, such as the newly established oxygen-isotope curve from the oceans. The first rigorous treatment using wide ranging techniques was by Hays, Imbrie & Shackleton (1976). Isotope studies from the bottom sediments of the Atlantic and Pacific Oceans since then have indicated as many as 21 Quaternary glacial ages and, as with Cenozoic studies, the continental evidence is so incomplete as compared with the oceanic sequences that terrestrial glacial-interglacial stratigraphy in future must depend on the ocean record for interpretation.

Because the Quaternary time scale permits a treatment in many respects different from pre-Quaternary studies it is not discussed in Chapter 2 nor in much detail in Chapter 3. The discussion that follows is a brief review of the present situation.

Radiometric ages within the last 40 000 years or so depend on ^{14}C and, for the main part of Pleistocene time, on K–Ar, ^{230}Th–^{234}U and ^{231}Pb–^{235}U determinations. Berggren *et al.* (1980) outlined the chronologies of different classes of events which need to be calibrated by chronometric methods. The principal scales used are: magnetostratigraphic; the oxygen-isotope scale, which extends from the present-day to the Jaramillo magnetic reversal event (0.88–0.94 Ma), i.e. stages 1 to 23 (Chart 2.17); calcium carbonate content curves for the equatorial Pacific, back to the same reversal event; microfossil-assemblage changes; and planktonic-foraminiferal coiling scales in certain parts of the oceans. Correlation between the various scales is established by simultaneous analysis and comparison with magnetic reversal stratigraphy in individual cores and by many other methods, including comparison of amino acid content and thermoluminescence. The ages of all levels on the chart, except those of magnetic reversals, have been assigned by interpolation and extrapolation and thus are subject to modification.

Datum levels relate to the apparent first and last appearance of taxa in the evolution of oceanic biota (FAD's and LAD's respectively): they are mainly determined from the appearances, acmes and extinctions of calcareous nannoplankton, diatoms, planktonic foraminifera and radiolaria.

The age of Pleistocene marine sediments is based either directly or indirectly on the magnetic polarity time scale, which in turn rests upon K–Ar dating of terrestrial lavas. K–Ar-derived ages are available for the chron boundaries. The ages of events are often interpolated from sea-floor magnetic anomalies and marine sediments, assuming constant rates of sea-floor spreading and deposition respectively.

The oxygen-isotope scale makes use of the fact that, when continental ice builds up as a result of global cooling and sea level is lowered, the ice is depleted in ^{18}O relative to the ocean water, leaving the ocean water enriched in ^{18}O. The oxygen-isotope composition of calcareous foraminifera and coccoliths, and of siliceous diatoms varies in direct proportion to that of the water (see Shackleton & Opdyke (1976) for discussion of the limitations of isotope stratigraphy). The 16 stages of Emiliani (1966) obtained from Caribbean and Atlantic sediment cores were extended to 22 by Shackleton & Opdyke (1973) after analysis of an equatorial Pacific core. These cores did not extend to the Pliocene–Pleistocene boundary, but subsequently another equatorial Pacific core (Shackleton & Opdyke 1976) and an Atlantic core (Van Donk 1976) extended through the boundary to provide a complete record of Quaternary climatic change. Shackleton & Opdyke's core with 23 defined stages is shown in Chart 2.17. The Brunhes–Matuyama boundary, and the Jaramillo and Olduvai events were all recognised, thus enabling an average accumulation rate of 1 cm per 1000 years to be determined. However, Shackleton & Opdyke regarded their earlier core as showing steadier accumulation rates during the Brunhes Chron; thus they regarded their estimates of ages determined by interpolation in this earlier core as being the best available. They have suggested that this core be used as the Late Pleistocene stratotype. Van Donk recognised a total of 21 isotopically determined interglacial stages and an equal number of glacial or near-glacial stages. However, his numbering scheme (which extended beyond 23 stages) has not been formally adopted.

A biostratigraphic framework for the Pleistocene Epoch based on datum levels of calcareous nannoplankton, was established by Gartner from several cores in equatorial and temperate regions (Berggren *et al.* 1980). His revised zonation (Gartner 1977), as given by Berggren *et al.* (1980) is shown on Chart 2.17; it has been tied in to the geomagnetic polarity scale and ages have been assigned to the interpolation after correlation between cores. Eight diatom and two silicoflagellate datum levels have been correlated with the paleomagnetic time scale and indirectly to the oxygen-isotope record. These are ranked into first and second order according to their reliability (Berggren *et al.* 1980). Planktonic foraminiferal datum levels (not shown here) have also been determined.

2.16.4 The Pleistocene–Holocene boundary

Corresponding to a climatic event at around 10 000 radiocarbon years BP this boundary is likely to be standardised in a lacustrine varved sequence in Sweden. It was proposed at the Eighth INQUA Congress in Paris in 1969 and has since been accepted. The change is well established in a variety of sediments, notably in Scandinavia, and corresponds to the following boundaries: European Pollen Zones III/IV, the Younger Dryas/Preboreal and Late Glacial/Postglacial (Mörner 1976).

3

Chronometric calibration of age boundaries

3.1 Introduction

Our reason for deciding to review the chronometric calibration of age boundaries was the need for a new time scale in our own work. The most commonly used time scales, such as that in Harland *et al.* (1964) and in its supplement (Lambert 1971), need revision to take into account the new standardised decay constants (Steiger & Jäger 1977) as well as more recent data.

Given a reasonably well distributed set of stratigraphically controlled radiometric ages, the setting up of a time scale would be straightforward. Standard statistical methods could be applied to the data to give the age of each stratigraphic boundary together with estimates of the errors involved. The statistical problem is very similar to that of finding the ages of magnetic reversals from radiometric data (Cox & Dalrymple 1967, Mankinen & Dalrymple 1979).

Unfortunately, this straightforward procedure cannot be applied to the entire Phanerozoic record for several reasons. The first is that, except for the Cenozoic and the later Cretaceous stratigraphic boundaries, the data controlling the ages of chronostratigraphic boundaries are sparse and unevenly distributed. A second problem is the heterogeneity of the available data set, which includes ages determined by several different methods on a range of minerals from a variety of geologic settings. For example, glauconite ages are used for dating throughout the Phanerozoic, particularly in the younger periods. However, opinion about their reliability varies (e.g. Lanphere & Jones 1978 compared to Odin 1978a, b in Cohee *et al.* 1978). All glauconites used in the time scale presented here are plotted separately on Figs. 3.3–3.7. A third problem is that the biostratigraphic age of a dated mineral may be difficult to establish locally and the correlation of the local biostratigraphic age with the global stages may be imperfect. A final problem is that many of the available ages do not date a stratigraphic division directly but serve only to bracket its age within broad limits. We have used all generally accepted radiometric ages whatever their geologic setting, but have done this in a way that will enable the interested reader to recognise and, should he wish, reject some of the data. The use of a large data base that does not exclude any generally accepted data introduces considerable stability into the time scale, although possibly at the risk of also introducing some bias.

3.2 Data base

Although many isotopic ages are published annually, only a few are of value in improving the numerical age estimates of stratigraphic boundaries. Experience gained during the review of available data convinced us that it was preferable to use previously published age compilations that had been assessed by one or more geochronologist, rather than to attempt to make piecemeal additions to these compilations ourselves.

Cenozoic. We have not discovered any comprehensive, systematically organised set of Cenozoic radiometric ages. However, there are several detailed reviews of the Cenozoic time scale and we did not consider that the methods described below (Section 3.5) would produce results significantly different from these reviews, although they would have provided a useful visual representation of the accuracy within which the ages of individual boundaries are known. We have therefore adopted Berggren & Van Couvering's (1974) and Hardenbol & Berggren's (1978) ages for Cenozoic stratigraphic boundaries, as modified by Ness, Levi & Couch (1980) to take into account the new recommended decay constants (Steiger & Jäger 1977). The ages of several sub-epoch boundaries that were poorly defined on the basis of radiometric dating have been adjusted by using magnetostratigraphy (Chapter 4).

Mesozoic and Paleozoic. The primary source for all Mesozoic and Paleozoic age determinations used here is Cohee *et al.* (1978), particularly Armstrong's pre-Cenozoic data file. Armstrong scrutinised all the items in the two Geological Society of London time-scale publications (Harland *et al.* 1964, Harland *et al.* 1971), rejecting about one third of the pre-Cenozoic items in the first publication and a few in the second, and adding to his own file many ages that had subsequently appeared in the geological literature. Despite the passage of fourteen years, this modified pre-Cenozoic data file contains only about 40 more dates than are in the pre-Cenozoic part of the time scale of Harland *et al.* (1964).

The only changes we have made to Armstrong's compilation are described in Section 3.3. They include adding a few new items from other papers in Cohee *et al.* (1978) and some Paleozoic fission-track ages (McKerrow, Lambert & Chamberlain 1980); interpreting some of the isotopic ages as radiometrically young and modifying the stratigraphic ranges of some items. As one might expect, the results of our review are similar to those of Armstrong (1978). In view of the great interest of Paleozoic stratigraphers in the fission-track data, it is interesting to report that their inclusion in the data file did not in general change the estimates of the ages of the stratigraphic boundaries significantly (Section 3.11).

3.3 Stratigraphic divisions used

The stratigraphic divisions used by Armstrong (1978) differ somewhat from those adopted in this chapter. It should also be pointed out that for reasons already stated (Chapter 2), some of the divisions adopted in this chapter differ from those selected in Chapter 2. The summary time scale (Section 3.11) lists the divisions adopted in this review.

The principal differences between Armstrong's usage and our own are:

Armstrong (1978)	This review
Jurassic	
Bathonian/Bajocian = 1 unit	2 units
Aalenian/Toarcian = 1 unit	2 units
Sinemurian/Hettangian = 1 unit	2 units
Triassic	
Scythian not subdivided	Scythian subdivided into 4 units (Spathian, Smithian, Dienerian, Griesbachian) but these units aggregated in the summary time scale (Section 3.11)
Permian	
Ufimian age not recognised	Ufimian recognised; assumed equivalent to lower part of Armstrong's Kazanian
Asselian age not recognised	Asselian recognised; assumed equivalent to lower part of Armstrong's Sakmarian
Devonian	
Couvinian recognised	Couvinian assumed equivalent to Eifelian
Silurian	
Downtonian recognised	Downtonian assumed equivalent to Pridoli
Cambrian	
Potsdamian recognised	Potsdamian equated with Merioneth
Acadian recognised	Acadian equated with St David's
Georgian recognised	Georgian equated with Caerfai

Because we have used Armstrong's data file as the main source of ages, we have little data on the Bathonian–Bajocian, Aalenian–Toarcian or Sinemurian–Hettangian boundaries. We have no data at all on the ages of the Spathian–Smithian, Smithian–Dienerian or Dienerian–Griesbachian boundaries, but, as discussed below (Section 3.10), such a fine subdivision of the Scythian stage has not yet been realised for global work, though these four subdivisions have been provisionally adopted in Chapter 2.

The age of the base of the Ufimian is taken as the age of Armstrong's Kazanian; there are no data in his file for the Kazanian–Ufimian boundary. Similarly, the base of the Asselian is taken as the base of Armstrong's Sakmarian: there are no data in his file for the Sakmarian–Asselian boundary. The assumed equivalence of the Couvinian/Eifelian; Downtonian/Pridoli; Potsdamian/Merioneth; Acadian/St David's; and Georgian/Caerfai divisions is believed to introduce errors that are small compared with the isotopic dating errors.

Some of the epochs of the Carboniferous finally adopted in Chapter 2 – Serpukhovian, Bashkirian, Kasimovian and Gzelian – also differ from the units used in this chapter – Namurian, Westphalian and Stephanian – because the radiometric determinations had been given in relation to the continental sequence which has been used in most previous time scales.

3.4 Corrections

All ages have been corrected using the recommended values of the constants (λ) and isotopic abundances in Steiger & Jäger (1977). These are:

K–Ar, Ar–Ar

$\lambda(^{40}K_{\beta^-}) = 4.962 \times 10^{-10}\ a^{-1}$

$\lambda(^{40}K_e) = 0.581 \times 10^{-10}\ a^{-1}$

% atomic abundance of ^{40}K as % of all K atoms is 0.01167

Rb–Sr

$\lambda(^{87}Rb) = 1.42 \times 10^{-12}\ a^{-1}$

U–Pb

$\lambda(^{238}U) = 1.55125 \times 10^{-10}\ a^{-1}$

$\lambda(^{235}U) = 9.8485 \times 10^{-10}\ a^{-1}$

$^{238}U/^{235}U = 137.88$
(= ratio of number of ^{238}U atoms to ^{235}U atoms)

All calculations start with the published age, do the backwards calculation to find the original ratio of radioactive parent to daughter products, correct the ratio if necessary, and then use the new constants to standardise the age (Table 3.1).

3.5 Method

The ages of stage boundaries have been estimated by a modification of the same method used to estimate the ages of the magnetic reversals from radiometric data (Cox & Dalrymple 1967).

Case 1. B and C adjacent stages. Let B and C be adjacent stages where B is younger than C. Let B_i and C_i be ages known from geologic evidence to belong to stages B and C respectively. Let each age have an error s_{B_i} and s_{C_i}. Let t_e be an estimate of the age of the B–C boundary. We ignore all ages from B that are younger than the time estimate, t_e, and all ages from C (the older stage) that are older.

For the remainder, we find E^2, where E is taken as our measure of the error,

$$E^2 = (B_i - t_e)^2/s_{B_i}{}^2 + (C_i - t_e)^2/s_{C_i}{}^2 \qquad (3.1)$$

and the sum for all relevant ages from B and C. The estimate t_e is increased in suitable increments and the value of t_e at the minimum value, $E^2{}_{min}$, is taken as the best estimate of the age of the B–C boundary.

If age x from stage B is a minimum age – for example, a suspect glauconite – then this age contributes to all error estimates of the B–C and older boundaries for which $t_e < x$.

The age does not contribute at all to any estimates of the younger boundary of B, or to any other still younger boundaries.

A plot of E^2 against t_e usually has a parabolic form, whose shape provides a visual assessment of the tightness of the control ages (Figs. 3.3–3.7). The ideal curve has a unique minimum close to zero and rises symmetrically and steeply on both sides. In reality, there are two common kinds of non-ideal behaviour. In the first, the curve has a flat base, with zero values repeated several times: this corresponds to a case where there are no data in the time range of the repeated zeros (e.g. Fig. 3.4c). In the second, the curve has a minimum that is significantly different from zero – more than 5 say – indicating that there are a significant number of ages from B that are older than $E^2{}_{min}$ and several from C that are younger (e.g. Fig. 3.3f). Marked asymmetry is a third, less common, type of behaviour generally caused by a scarcity of data on one side of the minimum (e.g. Fig. 3.4i).

Case 2. B and C are adjacent stages, but the age is 'bracketed' by A and D. Let A (youngest) and D (oldest) be adjacent stages forming a sequence $A/B/C/D$.

Here we can regard any age from B or C as lying within the composite stage B/C. Let an age be x. Then the A–B boundary $< x <$ C–D boundary. If t_e is a trial estimate, then x contributes to the error of the A–B (or younger) stage boundaries only if $t_e < x$. It is used as an estimate of the C–D (or older) boundaries if $t_e > x$. The age of the internal stage boundary, B–C, cannot be estimated.

Case 3. Stages E and K 'bracket' an age, where A/B/. . . J/K/L . . . form the sequence of stages. This case is simply an extension of Case 2, where the composite stage is now $F/G/H/I/J$.

The bracketed age contributes to the error of stage boundaries A–B, B–C, C–D, E–F if $t_e < x$ and for stage boundaries J–K, L–M . . . if $t_e > x$.

As in Case 2, a bracketed age does not provide any information about any stage boundary within the composite stage, such as F–G, G–H, H–I or I–J.

3.6 Errors

The method requires an error value for each age determination. Although most determinations quote an error, a significant number do not. Errors for these determinations have been estimated by fitting a linear regression line to the available error/time data. The errors adopted ranged from about ±2.7 Ma at 60 Ma to about ± 12 Ma at 500 Ma (Table 3.1).

3.7 Chronograms

The graph of E^2 against time for a particular boundary, together with the data used in the plot will be referred to as a 'chronogram' (Figs. 3.3–3.7). We have used the convention that time runs numerically from left to right: the older boundary is always on the right of the chronogram. We have also used the convention that data relevant to the younger stage are placed on the left on two higher levels than data relevant to the older stage, which are placed on the right at two lower levels. The younger data are shown as crosses and the older data as open circles. Glauconites relevant to the position of the younger stage at a boundary are plotted on the highest level, those glauconites relevant to the older stage are on the lowest level.

Were all the data from particular stages, then successive chronograms would show ages moving from the older, right-hand level to the younger, left-hand level. Bracketed ages are not relevant to any stage boundary ages between the bracketing stages: they disappear from such chronograms and may be absent from several in succession. Thus there is not a 1:1 correspondence between ages in successive chronograms.

The scanning intervals were 0.5 Ma for less than 100 Ma; 1 Ma for 100–199 Ma; 2 Ma for 200–399 Ma; and 4 Ma for more than 399 Ma. This choice of interval means that an age that has a minimum error at time t could also have its true minimum at $t \pm$ (scanning interval)/2. For example, an age of 212 Ma could also have the value of 211 or 213 Ma; an age of 501 Ma could equally well be 499 or 503 Ma.

The net effect of these choices is that the duration of short stages in the Cretaceous could be in error by at least 50 per cent. For example, Fig. 3.3c shows the Santonian–Coniacian boundary at 87.5 Ma and Fig. 3.3d gives the Coniacian–Turonian boundary at 88.5 Ma, suggesting that the Coniacian has a duration of 1 Ma. But the actual duration of the Coniacian could be as short as 0.5 Ma or as large as 1.5 Ma, irrespective of any errors in the data. In the Paleozoic such short stages would be missed in the scan for the minimum error.

3.8 'Tie-points' and interpolations

The available data are too sparse for rigorous statistical methods to be applied. We have attempted to give a quantitative estimate of the error in the age obtained from a chronogram by taking this error as half the age range for which the error did not exceed its minimum value by more than 1.0 (Figs. 5.4 & 5.6). Its significance is readily seen where only two identical ages determine a boundary, one of them being from the younger stage, the other from the older stage. From equation (3.1), E^2 is zero at the boundary. It rises to 1.0 on both sides of the boundary when the trial age differs from the experimental age by the quoted error. Call this difference t. If there are N identical ages, with identical errors, e, then E^2 rises to 1.0 when t is reduced in the ratio $1/N^{1/2}$. In reality, a statistically valid estimate of the error is complicated by the need to combine errors for several determinations with a wide range of values into a single error estimate.

Figures 3.1 and 3.2 (overleaf) Chronometric age plotted against boundaries of chronostratic divisions (abbreviated, see Appendix 3), spaced equally. Key tie-points are shown by black circles. The centre of each circle represents the adopted age. The chronogram age is generally midway along the horizontal error bars except where shown separately as a tick on the error bar. No error bars are shown for Cenozoic boundaries because chronograms (Figs. 3.3–3.7) have not been determined for this time interval. Note that the 'error range' of these figures and Section 3.11 is the range of the error bars.

Figure 3.1. Holocene to Ladinian. (See caption at foot of page 47.)

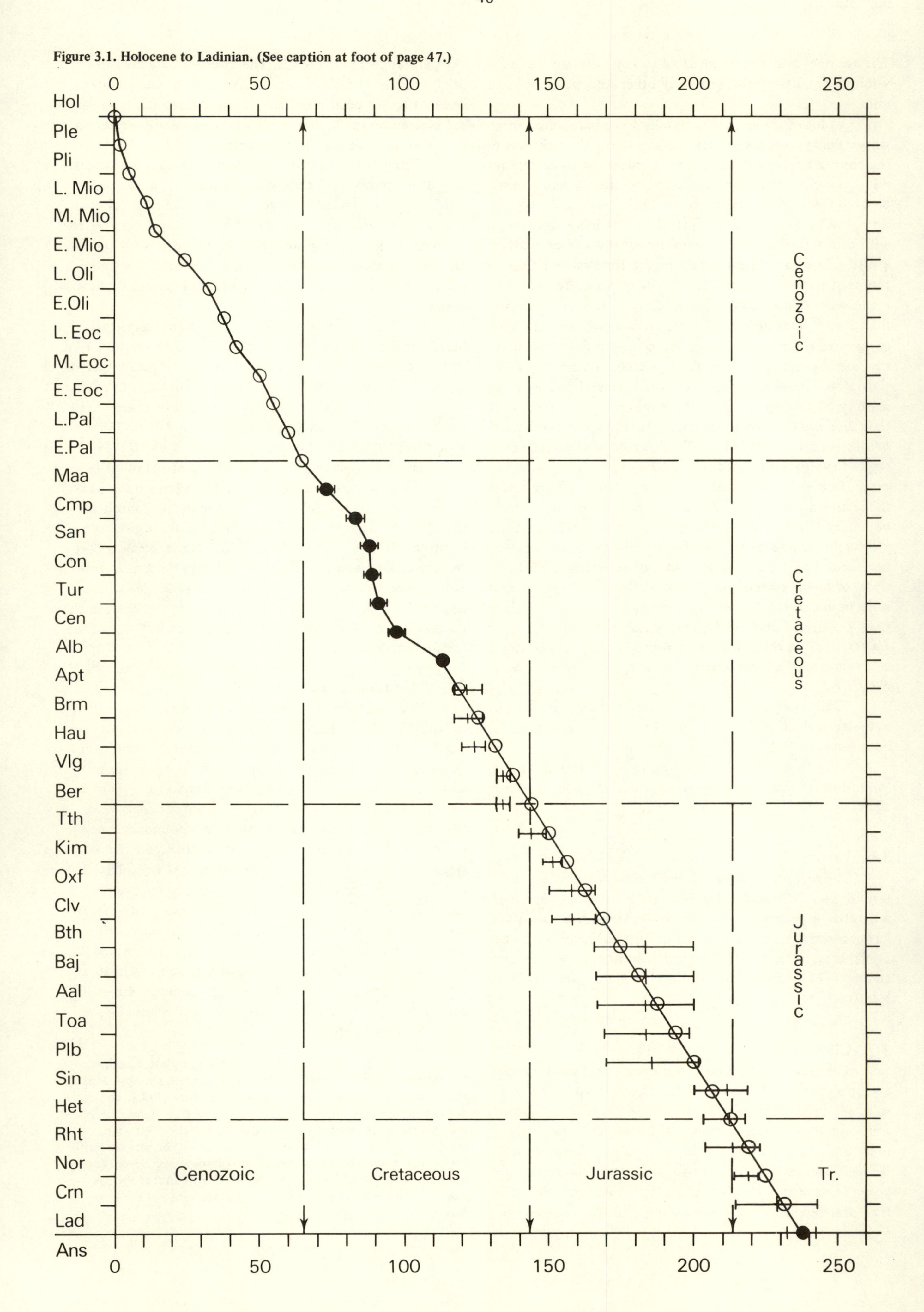

Figure 3.2(a). Rhaetian to Tournaisian.

Figure 3.2(b). Famennian to Caerfai.

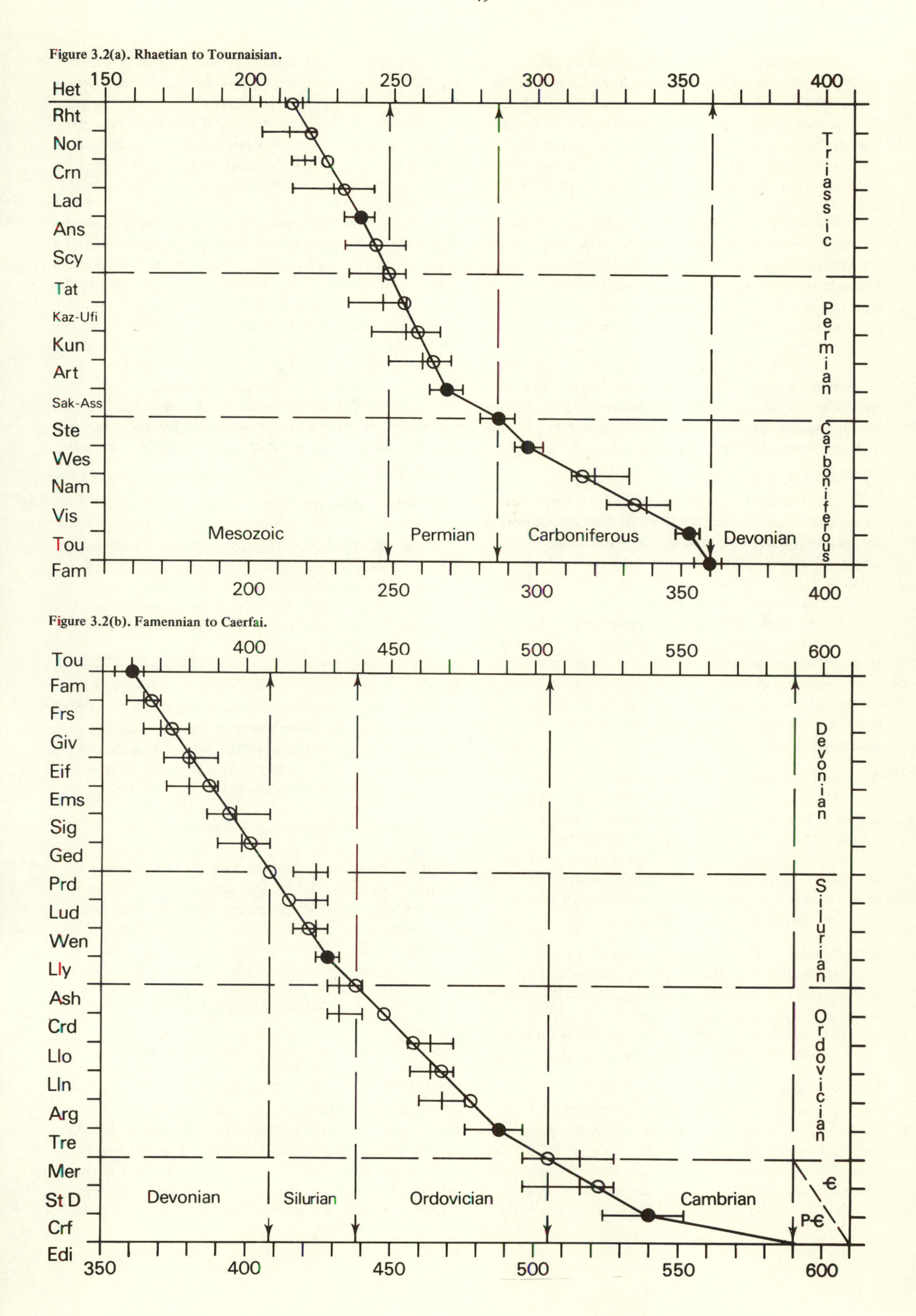

Nevertheless, the estimates form a useful guide to the quality of the data delineating stage boundaries. For the purposes of choosing tie-points, we have accepted only those chronograms with an error range of less than 5 Ma for stage boundaries that are younger than 200 Ma, giving 29 tie-points. For older stage boundaries, we take as tie-points those chronograms where error range is less than 12 Ma, giving 14 tie-points.

The only exceptions to this procedure are: (1) Norian–Carnian has been rejected as a tie-point (error range = 8 Ma), (2) Artinskian–Sakmarian, Sakmarian–Asselian and Arenig–Tremadoc have been accepted as tie-points even though the error ranges are 12, 12 and 20 Ma respectively. The reasons for acceptance or rejection are given in Section 3.9.

The main problem in drawing up a time scale from sparse data is the estimation of the ages of stratigraphic boundaries between tie-points. Four methods are available: relative spacing of ocean-floor magnetic anomalies; relative thicknesses of sediments; relative numbers of zones in each unit; the 'equal-age' hypothesis.

3.8.1 Ocean-floor anomalies

The relative spacing of the ocean-floor anomalies is assumed to represent their relative durations after adjusting for variations in the rate of ocean-floor spreading. The biostratigraphic ages of the anomalies can be found from deep-ocean (e.g. DSDP) cores or land sections. We have used magnetostratigraphy to refine several Cenozoic and Late Cretaceous boundaries by amounts ranging up to 1.2 Ma, all of which are within the uncertainties of the radiometric ages. However, magnetostratigraphy has not proved useful for earlier periods, either because the anomalies are absent (mid-Cretaceous), are poorly defined biostratigraphically (the M-anomalies, Section 4.4) or have been subducted (pre-Jurassic anomalies).

3.8.2 Relative stratigraphic thicknesses

This method has been used for some parts of the time scale (e.g. Boucot 1975, Churkin, Carter & Johnson 1977, Ziegler 1978) and has the greatest potential value for its future overall refinement. However, since there is no systematic stage-by-stage global compilation of relative thicknesses from a variety of depositional environments, we have chosen not to use this method for interpolation.

3.8.3 Relative numbers of zones per age

Although this appears to be a rough guide to the relative durations of stratigraphic ages, variations in the durations of Tertiary ages and in the numbers of zones per age suggest the dangers of applying this approach without detailed consideration. We have not used it for interpolation.

3.8.4 'Equal-age' hypothesis

The later Cretaceous chronograms (Figs. 3.3a–g) clearly show that this hypothesis also is invalid: the ages vary in duration from 1 Ma to more than 15 Ma. Yet it appears to provide a reasonable initial working hypothesis. The hypothesis can be recast as 'average duration of age hypothesis': viewed through a 60 Ma sliding window, the average duration of the ages seen in the window is, except for the Carboniferous, remarkably constant back to the early Paleozoic. In this form it is supported by the available data. For example, the average duration of the seven youngest Cretaceous ages is about 6.6 Ma, whereas the average duration of the twenty next oldest ages is 6.2 Ma. We therefore adopt this average equal duration hypothesis to interpolate between tie-points.

Presumably, the reason for the approximate constancy of average age duration is that, other things being equal, the subjective assessment by paleontologists of what constitutes a stage – essentially the smallest globally recognisable stratigraphic division – represents a certain amount of morphological change of the low-latitude cephalopods used to zone the Mesozoic. This change appears to have proceeded at a steady average rate. We should not expect stages based on other fossil groups to have the same duration.

Radiometric ages interpreted in terms of this hypothesis may offer some guidance as to whether some of the proposed Permo–Triassic subdivisions should be regarded as ages or sub-ages (Section 3.10).

3.9 Selected tie-points

Cenozoic. As noted above (Section 3.2) and later in Section 3.11, we accept the estimates of Ness *et al.* (1980), with minor revisions of sub-epoch boundaries in the Eocene and Paleocene. The Tertiary–Cretaceous boundary is that of Armstrong (1978) and Lanphere & Jones (1978).

Cretaceous. Figs. 3.3a–g are considered to provide reasonable tie-points for the seven youngest Cretaceous ages. The ages of the two youngest boundaries (Maastrichtian–Campanian and Campanian–Santonian) were changed by about 1 Ma from the chronogram minima on the basis of magnetostratigraphy (Chapter 4). The oldest tie-point for the Cretaceous is the Aptian–Albian boundary at 113 Ma (Fig. 3.3g).

Jurassic. There are no Jurassic tie-points.

Triassic. The Ladinian–Anisian boundary (Fig. 3.5b) is accepted as a tie-point (238 Ma). There are no other tie-points. The Norian–Carnian chronogram is rejected as a tie-point because there are few data from Carnian or older rocks compared with Norian and younger rocks (Fig. 3.4i). The asymmetric data distribution causes a marked skewness to the chronogram. The Norian and younger data are also largely from bracketed intrusions such as A447 (Table 3.1).

Permian. Fig. 3.5g (Artinskian–Sakmarian/Asselian) and Fig. 3.5h (Sakmarian/Asselian–Stephanian) are taken as tie-points (268, 286 Ma). The Artinskian–Sakmarian/Asselian chronogram is accepted as a tie-point because the data, though largely from bracketed intrusions, have a higher density and are more evenly distributed than the Norian–Carnian data (Fig. 3.4i). The chronogram is also more symmetric. The Sakmarian/Asselian–Stephanian boundary is accepted as a tie-point because there is a reasonable distribution of data points at the boundary and a fairly symmetric chronogram.

Carboniferous. The Carboniferous tie-points are taken at the Stephanian–Westphalian boundary (Fig. 3.5i) at 296 Ma; at the Visean–Tournaisian boundary (Fig. 3.6c) at 352 Ma and at the Tournaisian–Famennian boundary (Fig. 3.6d) at 360 Ma.

Two items 'pull' the minimum error value for both the Visean–Tournaisian and the Tournaisian–Famennian boundaries towards an old value. These are two whole-rock K–Ar ages with small errors on Scottish basalts (items A4360 and A5360, Table 3.1) at 354 and 366 Ma. Their stratigraphic age is confirmed as probably Visean (E. H. Francis, personal communication 1981). These ages clearly disagree with 342 Ma whole-rock Rb–Sr age for the base of the Tournaisian (item B1) from Bouroz (1978). All three ages have been retained because there are no clear criteria by which to choose amongst them.

Devonian. There are no tie-points.

Silurian. A poor tie-point is given in Fig. 3.7b (Wenlock–Llandovery) at 428 Ma.

Ordovician. A poor tie-point is given in Fig. 3.7f (Arenig–Tremadoc) at 488 Ma. Although three younger age boundaries have lower errors (Caradoc–Llandeilo, Fig. 3.7d; Llandeilo–Llanvirn, Fig. 3.7d and Llanvirn–Arenig, Fig. 3.7e), only Fig. 3.7f is accepted because the values of two other boundaries are identical. The third age boundary is rejected as a tie-point because the first three Arenig or older data points in Fig. 3.7e are an 'illite' (PTS47) and two glauconites (PTS163 and PTSS348), all of which are regarded as probably minimum ages.

Cambrian. A poor tie-point is given in Fig. 3.7i (St David's–Caerfai) at 540 Ma.

3.10 Interpolated values

Cretaceous to mid-Triassic. There are 20 ages and 125 Ma, or 6.25 Ma per age over this interval. Starting at the Albian–Aptian boundary at 113 Ma, the values of successively older ages are found by adding 6 Ma, 6 Ma, 6 Ma and 7 Ma to each age boundary (over five sets each of four ages), giving the results summarised in Section 3.11 and Fig. 3.1. The fluctuations about the line between the two tie-points probably partly reflect real departures from equal duration of ages.

Mid-Triassic to Early Permian. The slope of the line on Fig. 3.2a is very similar to that of the mid-Cretaceous to mid-Triassic line – 6 ages in 30 Ma, or 5 Ma per age. This similarity has been achieved only by regarding the Scythian as a single age and suppressing the Ufimian. Had the Scythian been subdivided or the Ufimian recognised, the slope would have been less and the ages shorter. We regard this as reasonable evidence that the Scythian and Kazanian (from which the Ufimian has been split, Section 3.3) may be sufficiently short that the subdivisions adopted in Chapter 2 may require more detailed investigation before they can be readily recognised globally.

By contrast, the suppression of the Asselian stage (Fig. 3.2a), gives 18 Ma to the Sakmarian. From this evidence we suggest that the Asselian is likely to be a useful unit for global correlation.

Carboniferous. The Stephanian–Westphalian boundary is a tie-point at 296 Ma. The next tie-point at the Visean–Tournaisian boundary is at 352 Ma. There are two intervening age boundaries and 56 Ma. By linear interpolation, the Westphalian–Namurian boundary is placed at 315 Ma; the Namurian–Visean boundary at 333 Ma.

The Carboniferous ages are markedly longer than all other ages (except for the Cambrian). Part of the reason presumably lies in the fact that much of the later Carboniferous has been zoned by non-marine faunas and floras.

The Tournaisian–Famennian tie-point is important in providing the oldest reasonably well-controlled tie-point in the Phanerozoic time scale.

Devonian to Early Silurian. The age boundaries have been found by linear interpolation – 10 ages and 68 Ma, or 6.8 Ma per age. The average age length is remarkably similar to average Mesozoic age length.

Early Silurian to Early Ordovician. These ages use two poor tie-points – 6 ages and 60 Ma, or 10 Ma per age.

Early Ordovician to Early Cambrian. These use two poor tie-points at 488 and 540 Ma: a time interval of 52 Ma to subdivide between 3 ages, by giving 17 Ma, 18 Ma and 17 Ma to each successively older age.

Earliest Cambrian. We have arbitrarily adopted 590 Ma for the initial Cambrian boundary, giving 50 Ma to the Caerfai Epoch.

3.11 Summary time scale (pp. 52–5)

The columns in this table are as follows: in parentheses our preferred abbreviation for the time division; the code number of the time division in columns 8 and 9 of Table 3.1; the name of the time division; its duration in Ma; its initial age in Ma; a figure to denote its uncertainty in Ma as estimated from the chronograms (the age range for which E^2 does not exceed its minimum value by 1.0; the 'error' on the wallchart (Figs. 5.4 and 5.6) is taken as half this value); and finally, references as appropriate.

QUATERNARY (duration = 2.00 ± 0.1 Ma)						
(Hol)	1	Holocene	10 000 years (0.01 Ma)			
(Ple)	2	Pleistocene	2.0 ± 0.1 Ma			Selli *et al.* (1977)
TERTIARY (duration = 63 Ma)						
(Pli)	3	Pliocene	(duration = 3.1 Ma)			
				5.1		Ness *et al.* (1980)
(Mio)	4	Miocene	(duration = 19.5 Ma)			
		Late	6.2			
				11.3		Ness *et al.* (1980)
		Middle	3.1			
				14.4		Ness *et al.* (1980)
		Early	10.2			
				24.6		Ness *et al.* (1980)
(Oli)	5	Oligocene	(duration = 13.4 Ma)			
		Late	8.2			
				32.8		Ness *et al.* (1980)
		Early	5.2			
				38.0		Ness *et al.* (1980)
(Eoc)	6	Eocene	(duration = 16.9 Ma)			
		Late	4.0			
				42.0		Chapter 4
		Middle	8.5			
				50.5		Chapter 4
		Early	4.4			
				54.9		Ness *et al.* (1980)
(Pal)	7	Paleocene	(duration = 10.1 Ma)			
		Late	5.3			
				60.2		Chapter 4
		Early	4.8			
				65.0		Armstrong (1978), Lanphere & Jones (1978), Chapter 4
CRETACEOUS (duration = 79.0 Ma)						
Late						
(Maa)	8	Maastrichtian	8			
				73	4	Chapter 4, Fig. 4.4, tie-point
(Cmp)	9	Campanian	10			
				83	4.5	Chapter 4, Fig. 4.4, tie-point
(San)	10	Santonian	4.5			
				87.5	4.5	Fig. 3.3c, tie-point
(Con)	11	Coniacian	1			
				88.5	2.5	Fig. 3.3d, tie-point
(Tur)	12	Turonian	2.5			
				91	2.5	Fig. 3.3e, tie-point
(Cen)	13	Cenomanian	6.5			
				97.5	2	Fig. 3.3f, tie-point
Early						
(Alb)	14	Albian	15.5			
				113	4	Fig. 3.3g; tie-point for Aptian to Ladinian interval
(Apt)	15	Aptian	6			
				119	9	Fig. 3.3h
(Brm)	16	Barremian	6			
				125	9	Fig. 3.3h
(Hau)	17	Hauterivian	6			
				131	8	Fig. 3.3i
(Vlg)	18	Valanginian	7			
				138	5	Fig. 3.4a

(Ber)	19	Berriasian	6			
				144	5	Fig. 3.4b
JURASSIC (duration = 69 Ma)						
Late						
(Tth)	20	Tithonian	6			
				150	12	Fig. 3.4c
(Kim)	21	Kimmeridgian	6			
				156	6	Fig. 3.4d
(Oxf)	22	Oxfordian	7			
				163	15	Fig. 3.4e
Middle						
(Clv)	23	Callovian	6			
				169	15	Fig. 3.4e
(Bth)	24	Bathonian	6			
				175	34	Fig. 3.4f
(Baj)	25	Bajocian	6			
				181	34	Fig. 3.4f
Early						
(Aal)	26	Aalenian	7			
				188	34	Fig. 3.4f
(Toa)	27	Toarcian	6			
				194	28	Fig. 3.4f
(Plb)	28	Pliensbachian	6			
				200	32	Fig. 3.4f
(Sin)	29	Sinemurian	6			
				206	18	Fig. 3.4f
(Het)	30	Hettangian	7			
				213	14	Fig. 3.4g
TRIASSIC (duration = 35 Ma)						
Late						
(Rht)	31	Rhaetian	6			
				219	18	Fig. 3.4h
(Nor)	32	Norian	6			
				225	8	Fig. 3.4i
(Crn)	33	Carnian	6			
				231	22	Fig. 3.5a
Middle						
(Lad)	34	Ladinian	7			
				238	10	Fig. 3.5b; tie-point for all younger ages to base of Albian; all older ages to base of Artinskian
(Ans)	35	Anisian	5			
				243	22	Fig. 3.5c
Early						
(Spa)	36	Spathian				
				Not estimated		
(Smi)	37	Smithian				
				Not estimated		
(Die)	38	Dienerian				
				Not estimated		
(Gri)	39	Griesbachian				
				Not estimated		
(Scy)	36–39	Scythian	5			
				248	20	Fig. 3.5c

PERMIAN (duration = 38 Ma)						
Late						
(Tat)	40	Tatarian	5			
				253	20	Fig 3.5d
(Kaz)	41	Kazanian				
				Not estimated		
(Ufi)	42	Ufimian	5	(Kazanian/Ufimian)		
				258	24	Fig. 3.5e
Early						
(Kun)	43	Kungurian	5			
				263	22	Fig. 3.5f
(Art)	44	Artinskian	5			
				268	12	Fig. 3.5g; tie-point for all younger ages to base Ladinian.
(Sak)	45	Sakmarian				
				Not estimated		
(Ass)	46	Asselian	18	(Sakmarian/Asselian)		
				286	12	Fig. 3.5h; tie-point
CARBONIFEROUS (duration = 74 Ma)						
(Ste)	47	Stephanian	10			
				296	10	Fig. 3.5i; tie-point for next two ages
(Wes)	48	Westphalian	19			
				315	20	Fig. 3.6a
(Nam)	49	Namurian	18			
				333	22	Fig. 3.6b
(Vis)	50	Visean	19			
				352	8	Fig. 3.6c tie-point for two previous ages
(Tou)	51	Tournaisian	8			
				360	10	Fig. 3.6d; tie-point for all ages to base of Wenlock
DEVONIAN (duration = 48 Ma)						
Late						
(Fam)	52	Famennian	7			
				367	12	Fig. 3.6e
(Frs)	53	Frasnian	7			
				374	18	Fig. 3.6f
Middle						
(Giv)	54	Givetian	6			
				380	18	Fig. 3.6g
(Eif)	55	Eifelian	7			
				387	28	Fig. 3.6g
Early						
(Ems)	56	Emsian	7			
				394	22	Fig. 3.6h
(Sig)	57	Siegenian	7			
				401	18	Fig. 3.6i
(Ged)	58	Gedinnian	7			
				408	12	Fig. 3.7a
SILURIAN (duration = 30 Ma)						
(Prd)	59	Pridoli	6			
				414	12	Fig. 3.7a
(Lud)	60	Ludlow	7			
				421	12	Fig. 3.7a
(Wen)	61	Wenlock	7			
				428	8	Fig. 3.7b; tie-point for younger ages to Tournaisian and all older ages to Arenig initial boundary
(Lly)	62	Llandovery	10			
				438	12	Fig. 3.7c

ORDOVICIAN (duration = 67 Ma)						
(Ash)	63	Ashgill	10			
				448	12	Fig. 3.7c
(Crd)	64	Caradoc	10			
				458	16	Fig. 3.7d
(Llo)	65	Llandeilo	10			
				468	16	Fig. 3.7d
(Lln)	66	Llanvirn	10			
				478	16	Fig. 3.7e
(Arg)	67	Arenig	10			
				488	20	Fig. 3.7f; poor tie-point for all younger ages to initial Wenlock; all older ages to initial St David's boundary
(Tre)	68	Tremadoc	17			
				505	32	Fig. 3.7g
CAMBRIAN (duration = 85 Ma?)						
(Mer)	69	Merioneth	18			
				523	36	Fig. 3.7h
(StD)	70	St David's	17			
				540	28	Fig. 3.7i; poor tie-point for younger ages to initial Arenig boundary; older ages to initial Cambrian boundary
(Crf)	71	Caerfai	50(?)			
				590(?)		Base from Cowie & Cribb (1978); no chronogram available
PROTEROZOIC						
	80	undifferentiated Precambrian				

Figures 3.3–3.7 (see following pages). Chronograms showing chronometric ages, error function and data for each chronostratic boundary (for stratigraphic name abbreviations see Appendix 3). Stars in the upper right indicate tie-points. Age is in Ma, increasing numerically from left to right; the value of an error function is plotted vertically. The function itself is the thick U-shaped curve. Note the scale changes. Glauconites stratigraphically younger than the boundary are plotted as the upper row of crosses; all other younger ages are on the lower row of crosses. Stratigraphically older glauconites are plotted as the lower row of circles; all other older ages are on the upper row of circles. The chronogram age is shown by a full vertical line with the exact value written in the middle of the chronogram. The age adopted is shown by the dashed vertical line ending in a 'T', its value is given in Section 3.11. Where the two coincide only a full line is shown.

Figure 3.3. Chronograms: Maastrichtian–Valanginian. (See caption on page 55.)

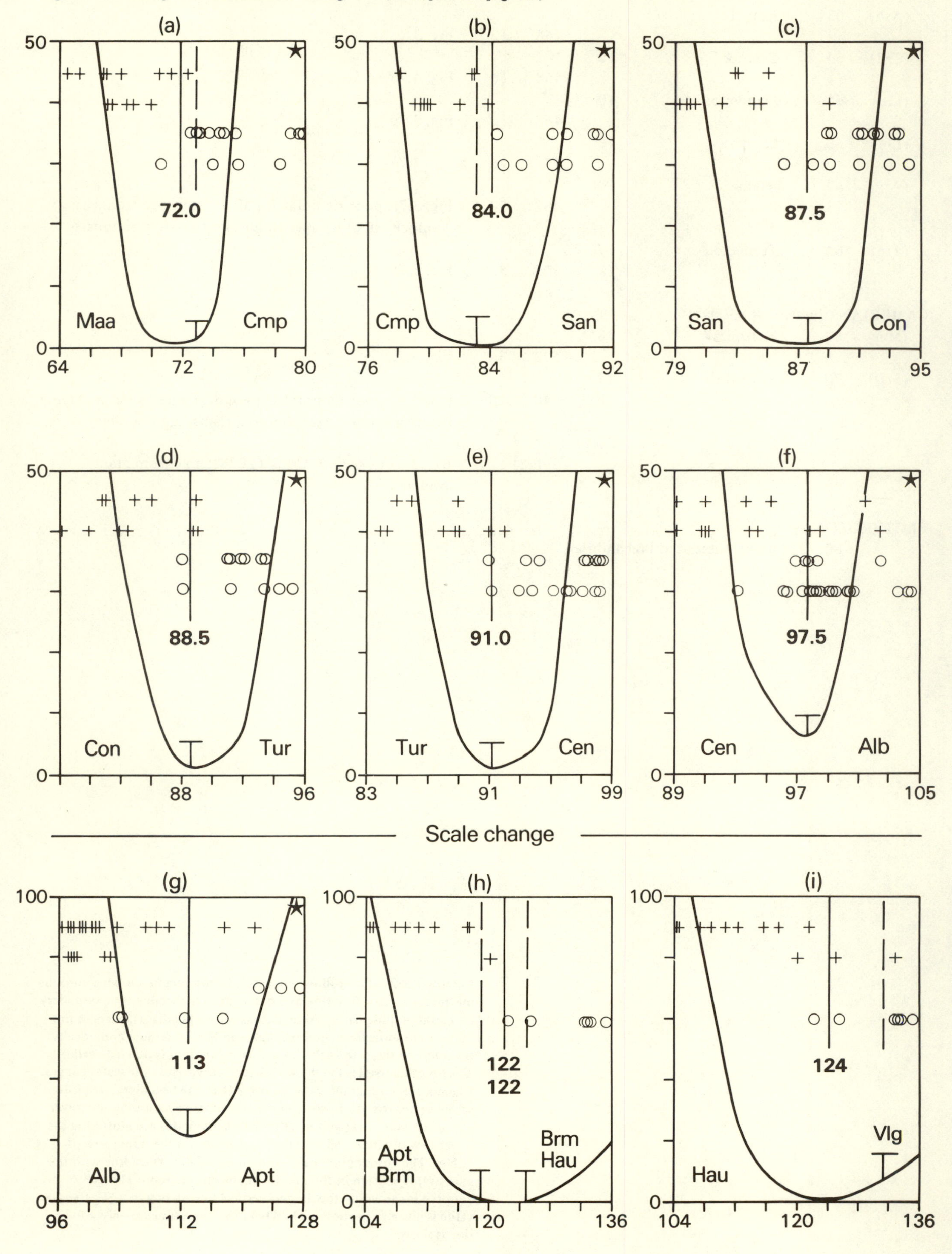

Figure 3.4. Chronograms: Valanginian–Carnian.

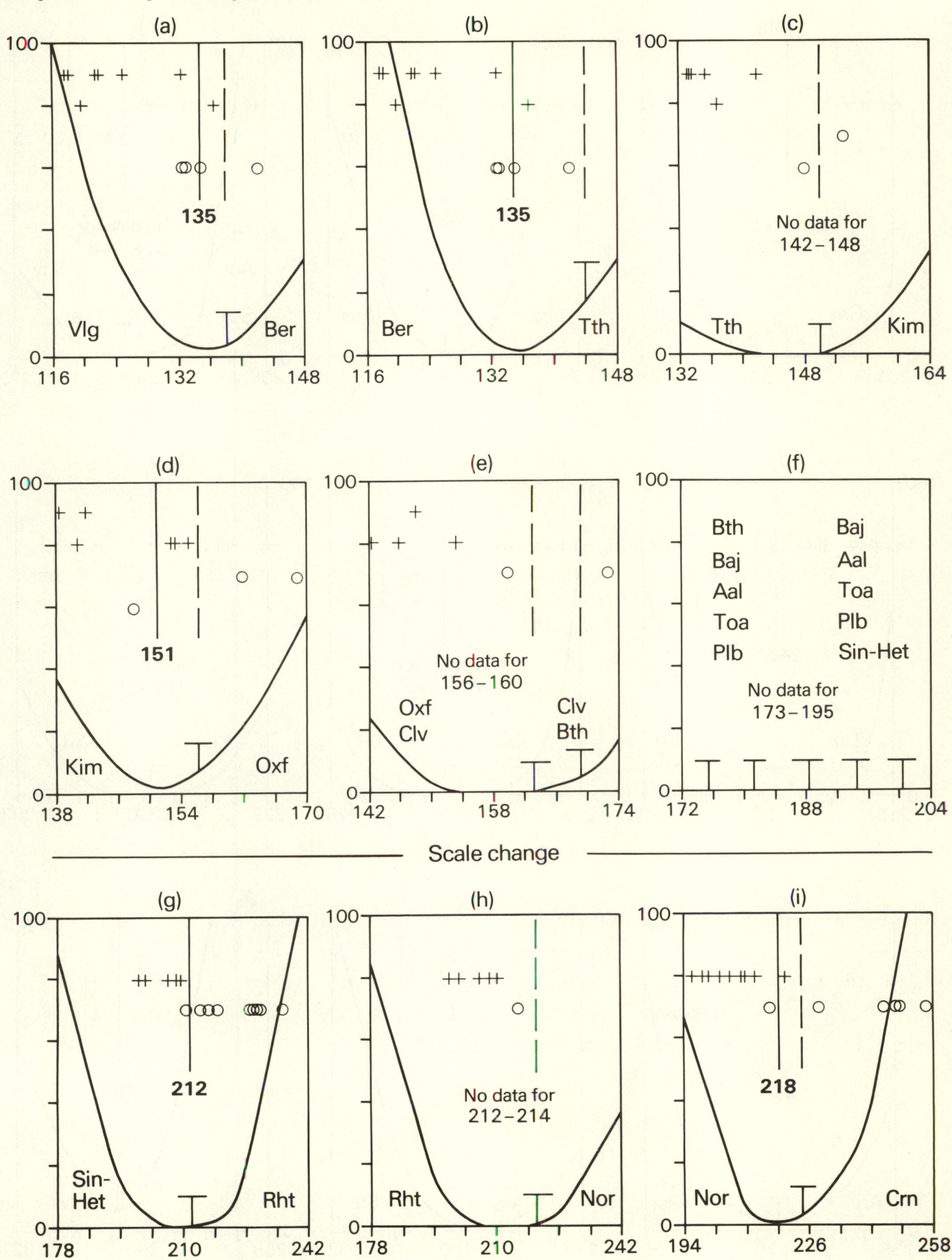

Figure 3.5. Chronograms: Carnian–Westphalian.

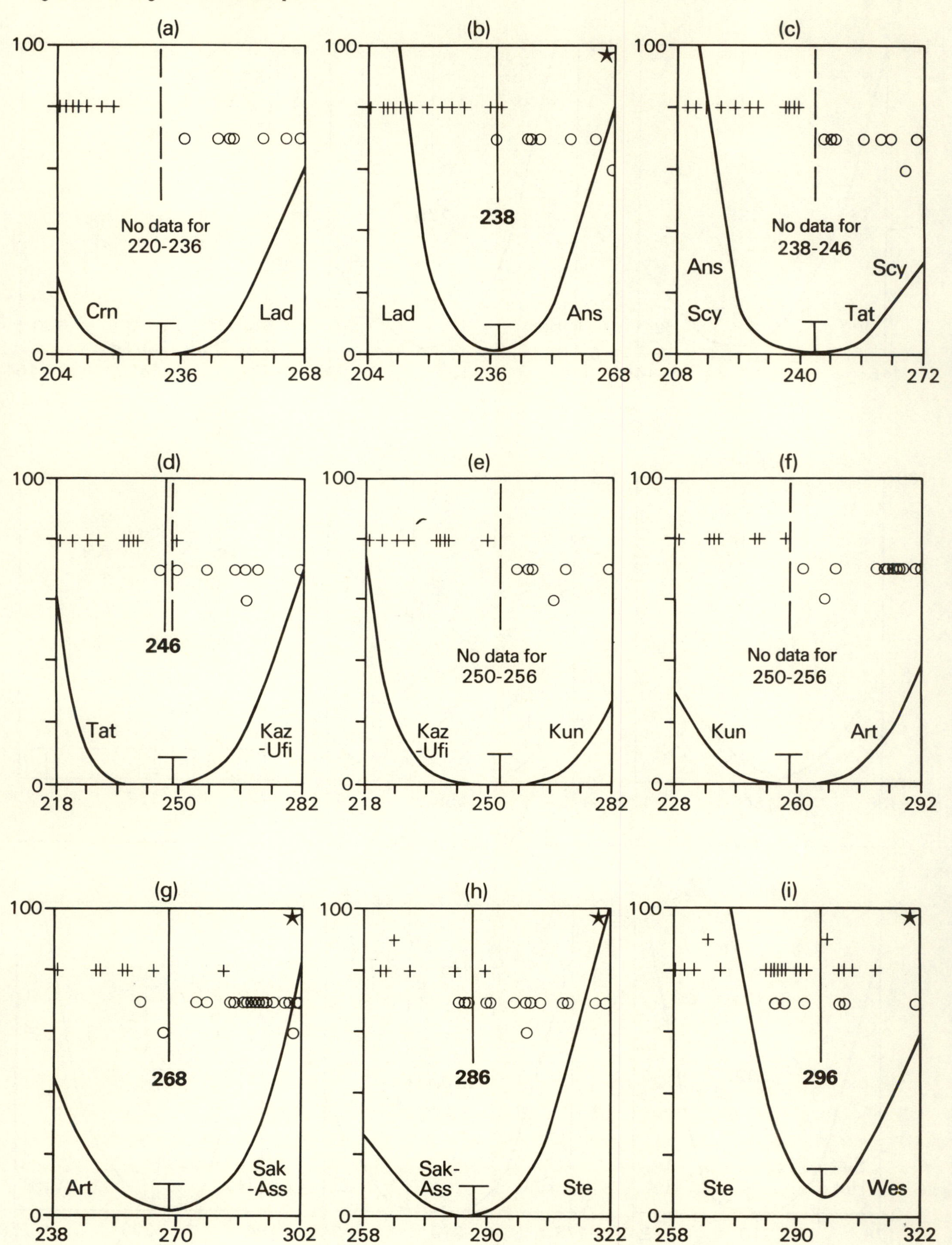

Figure 3.6. Chronograms: Westphalian–Gedinnian.

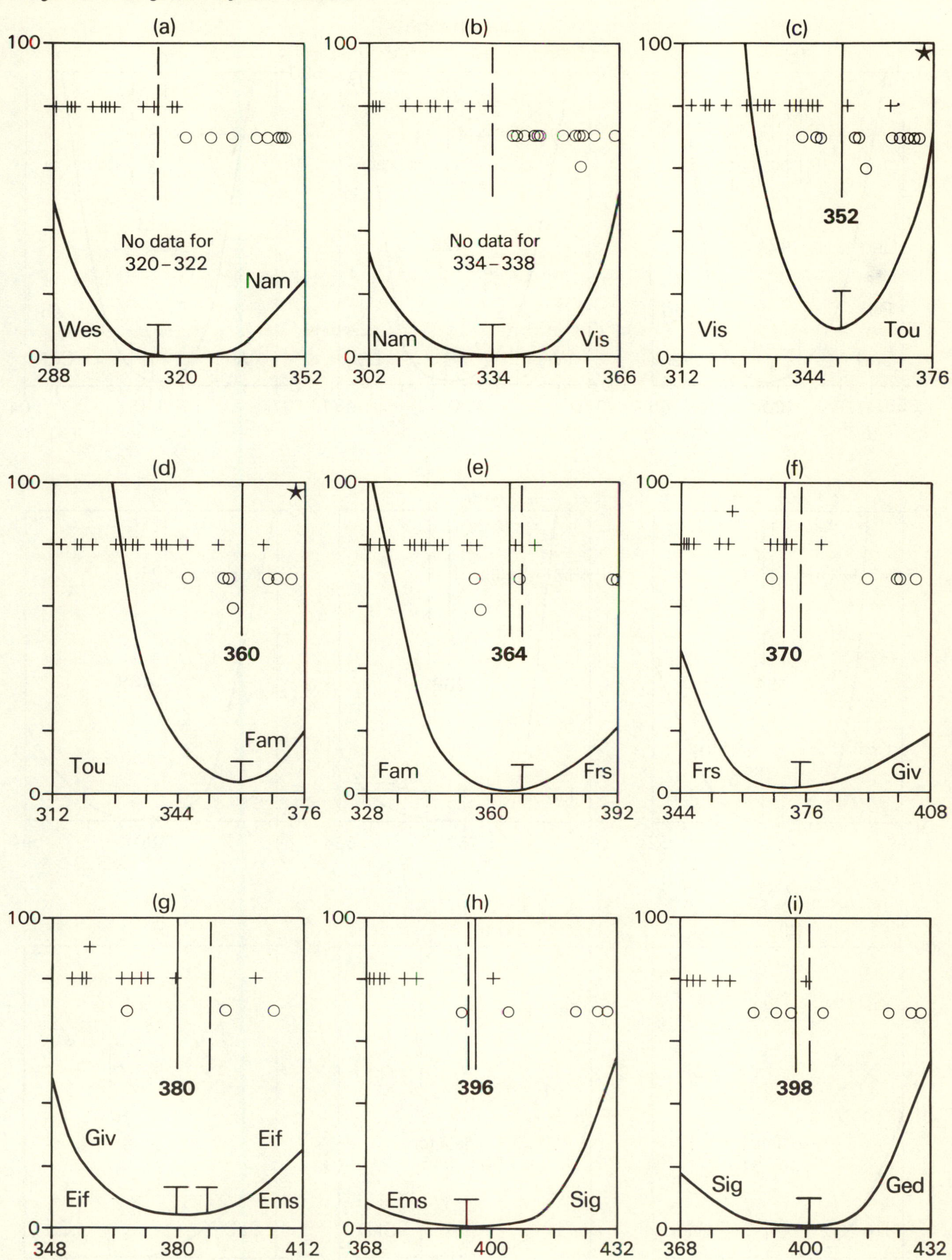

Figure 3.7. Chronograms: Gedinnian–Caerfai.

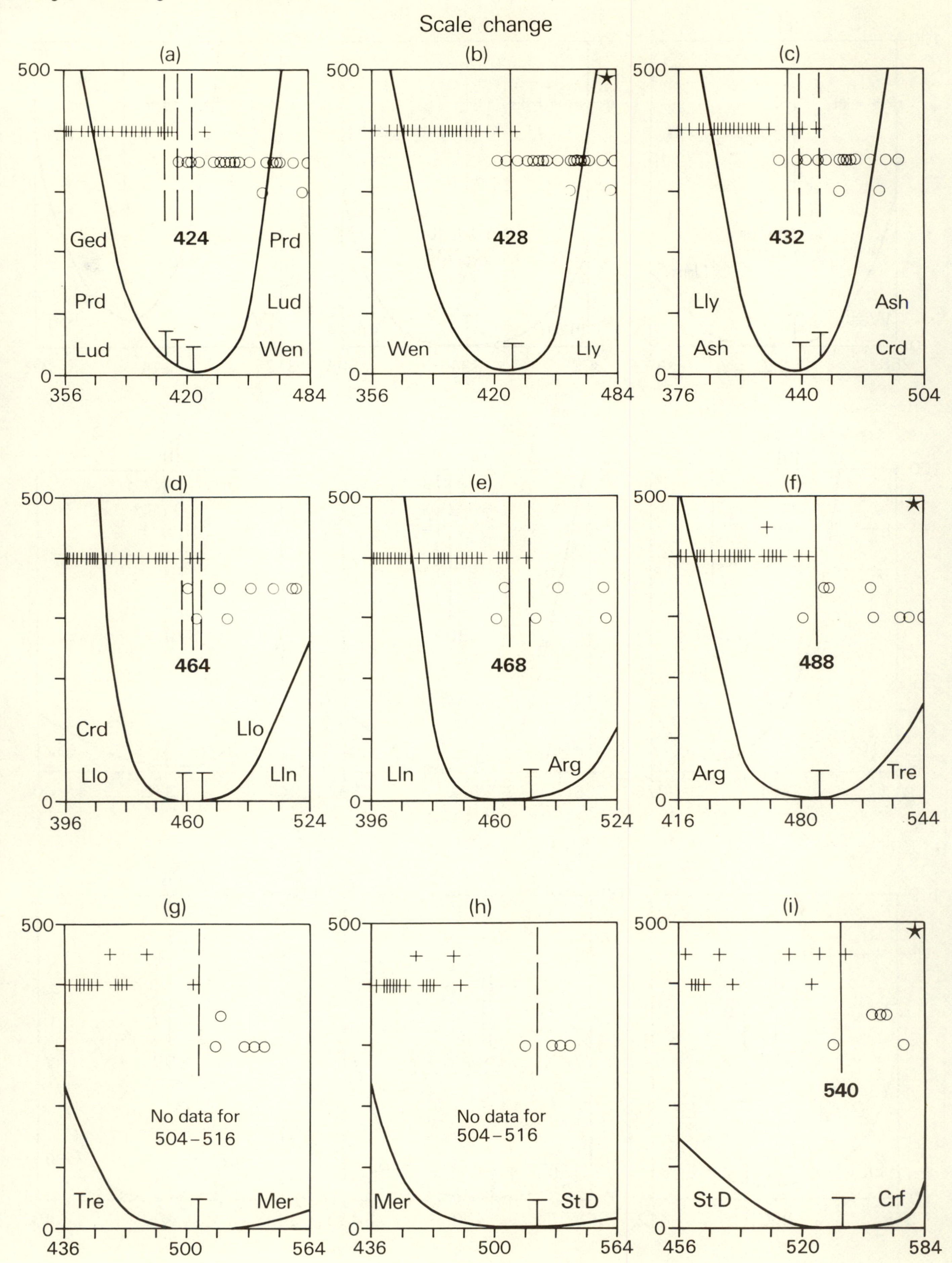

Table 3.1. *Published and standardised ages used in this work*

Column 1 (Item codes)
The letters give the initials of the publication or author; the number gives the number in the publication list or author list.

A = Armstrong, in Cohee *et al.* (1978);
B = Bouroz, in Cohee *et al.* (1978);
LJ = Lanphere & Jones, in Cohee *et al.* (1978);
MLC = McKerrow, Lambert & Chamberlain (1980);
PTS, (PTSS) = Geol. Soc. London, *Phanerozoic time-scale* and *Supplement* (Harland *et al.* 1964, 1971);
SMYH = Shibata, Matsumoto, Yanagi & Hamamoto, in Cohee *et al.* (1978).

Columns 2 and 4 (Dating method and coding for it)
KAr 1, 2, 3 for old Western, old Soviet and new constants;
RbSr 4, 5, 6 for 1.39, 1.47 and new (1.42) constants;
U 7, 8 for old and new ages;
Fi 9, 10 for old and new fission-track ages.
(See Section 3.4 and Appendix 1 for details.)

Column 3 (materials dated)
Bi = biotite; Fel = feldspar; Gl = glauconite; Hb = hornblende; Ign = igneous rocks; Ill = illite; Mis = miscellaneous; San = sanidine; Syl = sylvinite; Ur = uraninite; WR = whole-rock; Zr = zircon.

Column 5 (published age in Ma)

Column 6 (standardised age in Ma)

Column 7 (errors)
Positive values are the published error estimates in Ma. Negative values are estimated from the standardised age and a regression line fitting the published errors to published ages.

Columns 8 and 9 ('age' codes)
The code numbers refer to the 'ages' listed in Section 3.11. If two numbers are given, the age is bracketed; if one number and a zero is given, the value is from within that division. Negative numbers are regarded as minimum values.

(1)	(2)	(3)	(4)	(5)	(6)	(7)	(8)	(9)
PTS198	RbSr	Mis	4	64.00	62.65	5.00	7	8
A467	KAr	Gl	1	61.00	62.45	2.00	7	0
PTS198	U	Mis	7	64.00	63.33	5.00	7	8
PTS198	KAr	Bi	1	67.00	68.59	5.00	7	8
A512	KAr	Gl	1	66.70	68.28	3.00	8	0
A450	KAr	Bi	1	67.40	68.99	0.70	8	0
A451	KAr	Bi	1	68.50	70.12	0.70	8	0
PTS54	KAr	Gl	1	69.00	70.63	-2.70	8	0
PTS13	RbSr	Gl	5	65.00	67.29	-2.70	8	0
A514	KAr	Gl	1	70.80	72.47	3.00	8	0
A468	KAr	Gl	1	63.10	64.60	2.00	8	0
A449	RbSr	Gl	4	68.50	67.05	-2.70	8	0
A513	KAr	Gl	1	69.90	71.55	3.00	8	0
PTS200	KAr	Bi	1	66.00	67.56	-2.70	8	0
PTSS363	KAr	Bi	1	65.50	67.05	1.00	8	0
PTSS364	KAr	Bi	1	64.00	65.52	1.00	8	0
PTS217	U	Zr	7	115.00	113.79	-3.60	8	14
PTS12	KAr	Gl	1	69.00	70.63	4.00	9	0
PTS62	KAr	Gl	1	81.00	82.90	-2.80	9	0
A508	KAr	San	1	80.00	81.87	3.00	9	0
A511	KAr	Gl	1	72.40	74.11	3.00	9	0
A516	KAr	Gl	1	76.50	78.30	3.00	9	0
PTS201	KAr	Bi	1	74.00	75.74	-2.80	9	0
A452	KAr	Bi	1	71.50	73.19	0.70	9	0
A453	KAr	Bi	1	72.60	74.31	0.70	9	0
A454	KAr	Fel	1	71.30	72.98	0.70	9	0
A455	KAr	Bi	1	72.20	73.90	0.70	9	0
A456	KAr	Bi	1	72.80	74.52	0.70	9	0
A457	KAr	Bi	1	77.90	79.73	0.80	9	0
A458	KAr	Bi	1	77.30	79.12	0.80	9	0
A459	KAr	Bi	1	78.20	80.03	0.80	9	0
A469	KAr	Gl	1	63.30	64.80	2.00	-9	0
A470	KAr	Gl	1	81.00	82.90	3.00	9	0
PTSS365	KAr	Bi	1	71.00	72.68	1.00	9	0
A410	KAr	Hb	1	78.00	79.83	2.00	9	0
A417	KAr	Bi	2	86.00	84.02	5.00	9	10
A515	KAr	Gl	1	73.80	75.54	3.00	9	0
PTS229	KAr	Gl	1	83.00	84.94	-3.00	10	0
A460	KAr	Bi	1	82.50	84.43	0.80	10	0
A509	KAr	San	1	87.00	89.03	3.00	10	0
PTS57	KAr	Gl	1	87.00	89.03	-3.15	11	0
PTS208	KAr	Gl	2	78.00	76.21	-3.15	-11	0
A461	KAr	Bi	1	86.80	88.82	0.90	11	0
PTS58	KAr	Gl	1	84.00	85.96	-3.30	12	0
PTS59	KAr	Gl	1	79.00	80.85	-3.30	-12	0
PTS61	KAr	Gl	1	86.00	88.01	-3.30	12	0
A462	KAr	Bi	1	88.90	90.97	0.90	12	0
A510	KAr	San	1	90.00	92.09	3.00	12	0
PTS335	KAr	Ign	2	94.00	91.83	-3.30	12	14
PTS209	KAr	Gl	2	93.00	90.85	-3.45	13	0
PTS211	KAr	Gl	2	100.00	97.69	-3.45	13	0
PTS226	KAr	Bi	1	96.00	98.22	-3.45	13	0
A463	KAr	Bi	1	91.30	93.42	0.90	13	0
A464	KAr	Bi	1	92.10	94.24	0.90	13	0
A495	KAr	Gl	1	91.50	93.62	1.80	13	0
A498	KAr	Gl	1	93.00	95.16	2.30	13	0
A502	KAr	Gl	1	88.90	90.97	2.80	13	0
A517	KAr	Gl	1	99.00	101.28	3.00	13	0
PTS202	KAr	Bi	1	100.00	102.31	3.00	13	0
A418	KAr	Gl	2	90.00	87.93	4.00	13	0
PTS51	KAr	Gl	1	94.00	96.18	-3.60	14	0
PTS56	KAr	Gl	1	96.00	98.22	-3.60	14	0
PTS203	KAr	Bi	1	117.00	119.66	4.00	14	0
PTS204	KAr	Bi	1	96.00	98.22	2.00	14	0
PTS212	KAr	Gl	2	103.00	100.62	-3.60	14	0
PTS219	KAr	Gl	1	108.00	110.47	-3.60	14	0
PTS220	KAr	Gl	2	103.00	100.62	-4.00	14	0
PTS227	KAr	Gl	1	91.00	93.11	-3.60	14	0
PTS228	KAr	Gl	1	97.00	99.24	-3.60	14	0
PTS230	KAr	Gl	1	94.00	96.18	-3.60	14	0
PTS233	KAr	Gl	1	115.00	117.62	-3.60	14	0
PTS237	KAr	Gl	1	119.00	121.70	-3.60	14	0
PTS242	KAr	Gl	1	98.00	100.26	3.00	14	0
PTS336	KAr	Ign	2	105.00	102.57	-3.60	14	0
A428	KAr	Gl	2	106.00	103.55	4.00	14	0
A429	KAr	Gl	2	110.00	107.45	5.00	14	0
A465	KAr	Bi	1	95.30	97.51	1.00	14	0
A466	KAr	Fel	1	95.30	97.51	1.00	14	0
A496	KAr	Gl	1	97.10	99.34	3.30	14	0
A497	KAr	Gl	1	96.00	98.22	4.00	14	0
A499	KAr	Gl	1	95.70	97.91	5.30	14	0
A500	KAr	Gl	1	94.90	97.10	3.70	14	0
A501	KAr	Gl	1	97.20	99.45	6.10	14	0
A518	KAr	Gl	1	96.00	98.22	3.00	14	0
A494	KAr	Gl	1	106.30	108.74	4.20	14	0
PTS49	KAr	Gl	1	110.00	112.52	-3.70	15	0
PTS50	KAr	Gl	1	115.00	117.62	-3.70	15	0
PTS60	KAr	Gl	1	102.00	104.35	-3.70	15	0
PTS213	KAr	Gl	2	107.00	104.52	-3.70	15	0
PTS181	KAr	Gl	2	97.00	94.76	-3.70	-15	0
LJ1	U	Zr	7	115.00	113.79	2.00	1	15
SMYH1	RbSr	WR	5	121.00	125.26	6.00	15	17
SMYH2	RbSr	WR	5	128.00	132.51	12.00	15	17
PTS75	KAr	Bi	2	134.00	130.87	-3.90	17	21
LJ3	U	Zr	8	136.00	136.00	2.00	17	21
PTS215	KAr	Gl	2	136.00	132.82	-4.20	18	0
PTS322	KAr	Gl	2	125.00	122.09	-4.20	18	0
A430	KAr	Gl	2	128.00	125.02	5.00	18	0
LJ4	KAr	Bi	3	136.50	136.50	2.50	18	0
A406	KAr	WR	1	124.00	126.80	-4.20	1	18
PTS328	KAr	Bi	2	134.00	130.87	-4.35	1	19
PTS177	KAr	Gl	1	131.00	133.95	4.00	-19	0
PTS73	KAr	Gl	1	139.00	142.10	-4.50	20	0
A445	KAr	Hb	1	155.00	158.42	-4.50	20	80
A492	KAr	Gl	1	129.60	132.52	6.00	20	0
A1322	KAr	Gl	2	136.00	132.82	-4.50	20	0
PTS178	KAr	Gl	1	132.00	134.97	4.00	20	0
A444	KAr	Bi	1	150.00	153.32	5.00	21	0
A480	KAr	Bi	1	150.00	153.32	-4.60	1	21
A481	RbSr	Bi	5	150.00	155.28	-4.60	1	21
A485	KAr	Gl	1	135.50	138.54	3.00	-21	0
PTS76	KAr	Bi	2	143.00	139.65	-4.60	1	21
PTS77	KAr	Gl	1	136.00	139.05	-4.80	-22	0
A484	KAr	Gl	1	145.00	148.22	3.00	22	0
A479	KAr	Bi	1	168.00	171.66	-4.80	22	27
A431	KAr	Gl	2	145.00	141.61	5.00	-23	0
A419	KAr	WR	2	164.00	160.14	6.00	24	25
PTS89	KAr	Bi	1	169.00	172.69	-5.15	24	28
PTS90	KAr	Bi	2	173.00	168.92	-5.15	1	24
A1358	KAr	Bi	1	218.00	222.56	2.00	26	34
PTSS358	RbSr	WR	5	218.00	225.68	16.00	26	34
A432	KAr	Gl	2	160.00	156.24	6.00	-28	0
A433	KAr	WR	2	175.00	170.87	-5.70	1	28
A475	KAr	Bi	1	200.00	204.25	5.00	28	33
PTSS366	RbSr	Fel	4	200.00	195.77	2.00	28	33
A409	KAr	Bi	1	194.00	198.14	4.00	29	32
A477	KAr	WR	1	202.00	206.29	6.00	29	0
A478	KAr	Hb	1	206.00	210.35	6.00	1	30
A505	KAr	Hb	1	173.00	176.76	8.00	-29	0
A504	KAr	Hb	1	197.00	201.20	6.00	1	29
A506	KAr	Hb	1	195.00	199.16	8.00	1	29
A507	KAr	Hb	1	205.00	209.34	9.00	1	29
A446	RbSr	Mis	4	217.00	212.42	5.00	31	32
A447	KAr	Hb	1	214.00	218.49	4.00	31	32
PTS69	KAr	Bi	1	223.00	227.65	-6.10	31	43
A519	KAr	WR	1	224.00	228.66	5.00	33	34
PTS160	U	Ur	7	218.00	215.71	5.00	33	0
A476	KAr	Bi	5	230.00	238.10	-6.60	1	34
A1361	KAr	Bi	1	231.00	235.78	-6.60	1	34
PTSS361	RbSr	WR	5	230.00	238.10	-6.60	1	34
A520	KAr	WR	1	232.00	236.80	4.00	35	39
PTSS338	KAr	WR	2	255.00	248.85	15.00	40	0

Table 3.1. (*cont.*)

(1)	(2)	(3)	(4)	(5)	(6)	(7)	(8)	(9)
PTSS343	KAr	WR	2	252.00	245.93	-7.60	41	48
A434	KAr	WR	2	255.00	248.85	-7.60	41	0
PTS53	KAr	Syl	2	240.00	234.24	5.00	-43	0
PTS68	KAr	Fel	1	252.00	257.12	-7.90	43	0
A435	KAr	Hb	2	270.00	263.47	-7.90	43	45
PTS46	U	Zr	7	259.00	256.28	-8.10	1	44
PTS45	KAr	Bi	1	259.00	264.23	7.00	1	44
PTSS341	KAr	Bi	2	289.00	281.97	10.00	1	44
A503	RbSr	WR	4	276.00	270.17	7.00	44	45
PTS122	RbSr	Bi	5	282.00	291.93	7.00	44	47
PTS8	KAr	Bi	1	295.00	300.78	6.00	45	47
PTS8	RbSr	Bi	5	275.00	284.68	-8.20	45	47
PTS120	KAr	Gl	2	274.00	267.36	-8.20	45	0
PTS174	KAr	Bi	2	350.00	341.36	-8.20	45	50
PTS176	KAr	WR	1	295.00	300.78	19.00	45	49
PTS192	KAr	Bi	1	284.00	289.62	-8.20	1	45
PTSS344	KAr	Bi	2	288.00	281.00	-8.20	45	80
PTSS345	KAr	Bi	2	268.00	261.52	-8.20	45	48
A420	KAr	Bi	2	293.00	285.87	15.00	45	47
A422	RbSr	Bi	4	291.00	284.85	20.00	45	47
A1038	RbSr	Bi	5	275.00	284.68	-8.40	45	47
A483	KAr	Bi	1	285.00	290.63	5.00	45	47
A421	U	Zr	7	287.00	283.99	10.00	45	47
PTS30	KAr	Bi	1	278.00	283.53	-8.50	47	0
B2	RbSr	WR	5	300.00	310.56	10.00	47	0
PTS31	KAr	Bi	1	327.00	333.23	-8.50	47	50
PTS63	RbSr	Bi	5	288.00	298.14	8.00	47	0
PTS65	KAr	Bi	1	298.00	303.82	-8.50	47	0
PTS171	RbSr	Bi	5	320.00	331.27	-8.50	47	50
PTSS340	KAr	WR	2	310.00	302.42	15.00	47	0
PTS119	RbSr	Bi	5	300.00	310.56	10.00	47	50
PTSS356	RbSr	WR	5	308.00	318.84	7.00	47	48
PTS29	KAr	Ill	1	295.00	300.78	-8.70	48	0
PTS64	KAr	Gl	2	308.00	300.47	-8.70	48	0
PTSS360	KAr	WR	1	308.00	313.96	10.00	1	48
A436	KAr	WR	2	292.00	284.89	-8.70	48	0
A438	KAr	Bi	2	299.00	291.71	-8.70	48	49
A439	RbSr	Bi	4	291.00	284.85	-8.70	48	49
A1360	KAr	WR	1	313.00	319.03	16.00	1	48
A2360	KAr	WR	1	334.00	340.32	17.00	50	0
A3360	KAr	WR	1	338.00	344.37	4.00	50	0
PTS66	KAr	Bi	1	327.00	333.23	-8.80	49	0
PTS191	KAr	WR	1	322.00	328.16	12.00	49	0
PTSS339	KAr	Bi	2	330.00	321.89	10.00	49	0
A423	KAr	Gl	2	308.00	300.47	10.00	-49	0
PTS172	RbSr	Bi	5	328.00	339.55	-9.00	50	0
PTS173	RbSr	Bi	5	334.00	345.76	7.00	50	0
A413	RbSr	Bi	5	347.00	359.22	-9.00	50	80
A4360	KAr	WR	1	347.00	353.49	7.00	50	0
PTS98	KAr	Bi	1	370.00	376.78	-9.00	50	56
A5360	KAr	WR	1	359.00	365.64	6.00	50	0
B1	RbSr	WR	5	330.00	341.62	6.00	51	0
PTS6	KAr	Bi	1	391.00	398.02	7.00	51	58
PTS6	RbSr	Bi	5	397.00	410.98	11.00	51	58
PTSS347	RbSr	WR	4	379.00	370.99	17.00	51	52
A424	KAr	Bi	2	355.00	346.22	10.00	51	53
A1354	RbSr	WR	5	359.00	371.64	15.00	52	53
PTS2	U	Ur	7	350.00	346.33	10.00	52	0
PTS5	KAr	Bi	1	404.00	411.17	8.00	52	58
PTS95	KAr	Bi	1	350.00	356.53	-9.30	52	0
PTSS354	KAr	Bi	1	362.00	368.68	6.00	52	53
A440	KAr	WR	2	345.00	336.49	10.00	-52	0
A2354	RbSr	WR	5	367.00	379.92	22.00	52	53
PTS94	KAr	Bi	1	340.00	346.40	6.00	-53	0
A425	KAr	Gl	2	366.00	356.93	10.00	53	0
A441	KAr	Hb	2	365.00	355.95	20.00	53	54
A442	KAr	Bi	2	355.00	346.22	10.00	1	53
A443	KAr	Bi	2	355.00	346.22	10.00	-53	0
A488	RbSr	WR	4	400.00	391.55	-9.40	53	58
PTS1	KAr	Bi	1	393.00	400.05	12.00	54	0
PTS96	RbSr	WR	4	375.00	367.08	-10.00	56	0
A489	KAr	WR	1	390.00	397.01	-10.00	1	57
A448	RbSr	WR	4	395.00	386.65	16.00	1	57
PTS3	KAr	San	1	385.00	391.96	15.00	58	0
PTSS355	RbSr	WR	4	413.00	404.27	5.00	58	0
A522	RbSr	WR	4	438.00	428.75	4.00	58	0
A471	RbSr	WR	4	394.00	385.68	20.00	1	58
A472	KAr	Hb	1	402.00	409.15	10.00	1	58
PTS97	KAr	Bi	1	394.00	401.06	-10.50	1	60
MLC6	Fi	Zr	10	407.00	407.00	8.00	60	0
PTSS408	RbSr	Bi	4	447.00	437.56	6.00	60	62
MLC7	Fi	Zr	10	416.00	416.00	9.00	61	0
MLC8	Fi	Zr	10	422.00	422.00	10.00	61	0
PTS93	KAr	Bi	1	390.00	397.01	12.00	1	61
A1093	RbSr	Bi	5	399.00	413.05	16.00	1	61
A415	KAr	Hb	3	432.90	432.90	3.30	62	0
MLC9	Fi	Zr	10	437.00	437.00	11.00	62	0
A482	RbSr	WR	4	430.00	420.92	15.00	62	68
A490	KAr	Bi	1	437.00	444.52	13.00	62	64
A491	RbSr	Bi	5	430.00	445.14	20.00	62	64
A521	RbSr	WR	4	455.00	445.39	15.00	62	0
PTS156	U	Zr	7	447.00	442.31	3.00	64	0
PTS157	KAr	Bi	1	440.00	447.55	4.00	64	0
MLC10	Fi	Zr	10	466.00	466.00	11.00	64	0
MLC11	Fi	Zr	10	465.00	465.00	10.00	64	0
A411	RbSr	WR	4	474.00	463.99	-11.10	64	65
PTSS350	KAr	Bi	1	475.00	482.88	-11.10	64	67
PTSS351	KAr	Bi	1	445.00	452.60	-11.10	64	0
A487	RbSr	WR	4	460.00	450.28	10.00	1	64
PTS156	RbSr	Bi	4	473.00	463.01	7.00	64	0
A1156	KAr	Bi	1	420.00	427.34	5.00	64	0
MLC12	Fi	Zr	10	477.00	477.00	11.00	65	66
A416	KAr	Hb	1	460.00	467.75	5.00	66	67
A412	U	Zr	7	510.00	504.65	10.00	66	80
MLC13	Fi	Zr	10	493.00	493.00	11.00	67	0
PTS47	KAr	Ill	1	457.00	464.72	-11.50	67	0
PTS163	KAr	Gl	1	453.00	460.68	-11.50	67	0
A407	RbSr	WR	5	474.00	490.69	5.00	68	80
A414	RbSr	WR	4	515.00	504.12	7.00	1	68
PTSS348	KAr	Gl	2	492.00	479.44	-11.70	68	0
PTS186	U	Zr	7	523.00	517.52	-11.80	69	80
PTS70	KAr	Bi	1	518.00	526.24	-12.00	1	70
A426	KAr	Gl	2	530.00	516.35	20.00	70	0
A473	RbSr	Gl	4	542.50	531.04	20.00	70	0
A474	RbSr	WR	4	555.00	543.27	18.00	70	0
A486	U	?	7	570.00	564.02	25.00	71	80
PTS42	KAr	Bi	1	553.00	561.50	-12.10	71	80
PTS183	RbSr	Gl	4	584.00	571.66	30.00	71	0
PTS185	KAr	Gl	2	550.00	535.77	-12.10	71	0
PTSS352	RbSr	WR	5	574.00	594.21	11.00	71	80
PTSS353	RbSr	WR	4	569.00	556.98	4.00	71	80
PTS116	KAr	Gl	2	573.00	558.10	-12.30	-80	0
PTS117	KAr	Gl	2	595.00	579.46	-12.30	-80	0
PTS118	KAr	Gl	2	615.00	598.87	-12.30	-80	0
A427	KAr	Gl	2	590.00	574.60	15.00	-80	0
PTS55	U	Ur	7	620.00	613.50	20.00	80	0

4
Magnetostratigraphic time scale

4.1 Geomagnetic polarity reversals

4.1.1 Global synchroneity

The basis of magnetostratigraphy is the retention by rocks of a magnetic imprint acquired in the geomagnetic field that existed at the time the rocks formed. The processes that produce the geomagnetic field occur in the Earth's core, where fluid motions driven by convection comprise a dynamo that generates a global magnetic field which is roughly symmetrical about the Earth's rotation axis. For reasons that are still not well understood, at irregular times the currents flowing in the core reverse their direction, producing a reversal in the polarity of the magnetic field. By convention the polarity is *normal* when the field at the Earth's surface is directed towards the north and the plunge or *inclination* of the field is directed downward in the northern hemisphere and upward in the southern hemisphere. When the polarity is *reversed*, the field points to the south and the sign of the inclination is reversed in both hemispheres. These 180 degree changes in the direction of the magnetic field imprinted on rocks forming at the Earth's surface provide the physical basis for magnetic polarity stratigraphy. Because polarity reversals are recorded simultaneously by rocks being formed all over the world, magnetostratigraphic units, unlike lithostratigraphic and biostratigraphic units, are not time-transgressive. However, the age of the magnetisation might not be the same as the age of other geologic events in the history of the rock. In plutonic rocks, the magnetisation is acquired after the crystallisation of the rock and before the setting of the K–Ar clock in biotite. In chemically altered rocks, the magnetisation is generally acquired at the time of chemical alteration.

When a reversal occurs, the time required for the change is known from detailed paleomagnetic studies of transition zones to be about 5000 years (Fig. 4.1). Because strata within transition zones cannot generally be correlated globally, the ultimate resolving power of magnetostratigraphy is roughly equal to the duration of transitions. Strata adjacent to a transition zone could thus be correlated globally with a precision of 0.5 per cent for one million year old rocks and 0.005 per cent for 100 million year old rocks were it not for the problems introduced by other factors.

4.1.2 Excursions

Even when the field is not reversing, it undergoes swings in direction with typical amplitudes of 15 degrees and periods of 10^2 to 10^4 years. This *geomagnetic secular variation* is generally too small to be mistaken for the 180 degree changes in field direction which characterise polarity reversals. Occasionally, however, the field appears to undergo an *excursion* characterised by a large change in direction that may approach 180 degrees. Since excursions are thought to have durations of about 1000 years, they offer the potential of providing very sharp stratigraphic markers. However, excursions have proven disappointingly difficult to trace globally for several reasons: excursions are so short that they are missing in many stratigraphic sections; some excursions are probably local rather than global geomagnetic phenomena; and anomalous paleomagnetic directions are sometimes due not to excursions but rather to deformation of the rocks studied (Verosub & Banerjee 1977, Banerjee, Lund & Levi 1979).

With the objective of including in the present report well-documented reversals, but not excursions, the following criteria were used to distinguish between the two phenomena. The first is whether the field changed direction to within several tens of degrees of a complete 180 degree reversal. The second is whether the field remained locked into the reversed direction for a measurable length of time (reversal) or simply moved through the reversed direction (excursion). The third is whether the reversal was recorded at different sites around the world. The duration of the shortest possible polarity interval consistent with these criteria would be somewhat greater than 0.01 Ma, which is twice the length of a polarity transition interval. Whether the Lascamp, Blake, Biwa I (Jamaica), Biwa II (Levantine, Chegan) and Emperor (Ureki) events, which occurred at about 0.05, 0.1, 0.18, 0.28 and 0.47 Ma (Creer, Readman & Jacobs 1980, Rampino 1981, Champion, Dalrymple & Kuntz 1981) are reversals by the above criteria is uncertain. The most likely true subchron of the set is the Emperor (Wilson & Hey 1981, Champion *et al.* 1981).

4.1.3 Polarity intervals, chrons and subchrons

The time interval that elapses between two successive reversals in the polarity of the geomagnetic dynamo is generally referred to as a *polarity interval* (Cox 1968), the latter term being used as a description of a physical phenomenon and not as a chronostratigraphic unit. This usage accords with the statement in the *International stratigraphic guide* that 'interval' may refer to either time or space intervals and therefore should be used as a general term and not as a formal stratigraphic unit (Hedberg 1976, p.15). The lengths of polarity intervals vary from about 0.01 Ma to several tens of millions of years. In the course of paleomagnetic research, long intervals are almost always recognised before short intervals and each new discovery of a short polarity interval changes the local polarity structure.

This may be seen in Fig. 4.1, where prior to the discovery of the short polarity interval labelled τ_3 only one reversed interval would have been recognised spanning the intervals τ_2, τ_3 and τ_4. Therefore in naming or numbering polarity intervals for stratigraphic purposes, a hierarchical set of names is needed that does not change drastically with the discovery of additional short polarity intervals. Fig. 4.1 demonstrates that a scheme of simply numbering polarity intervals in the sequence of their occurrence does not provide such a system.

Figure 4.1. Polarity chrons, polarity subchrons, transition zones, and excursions.

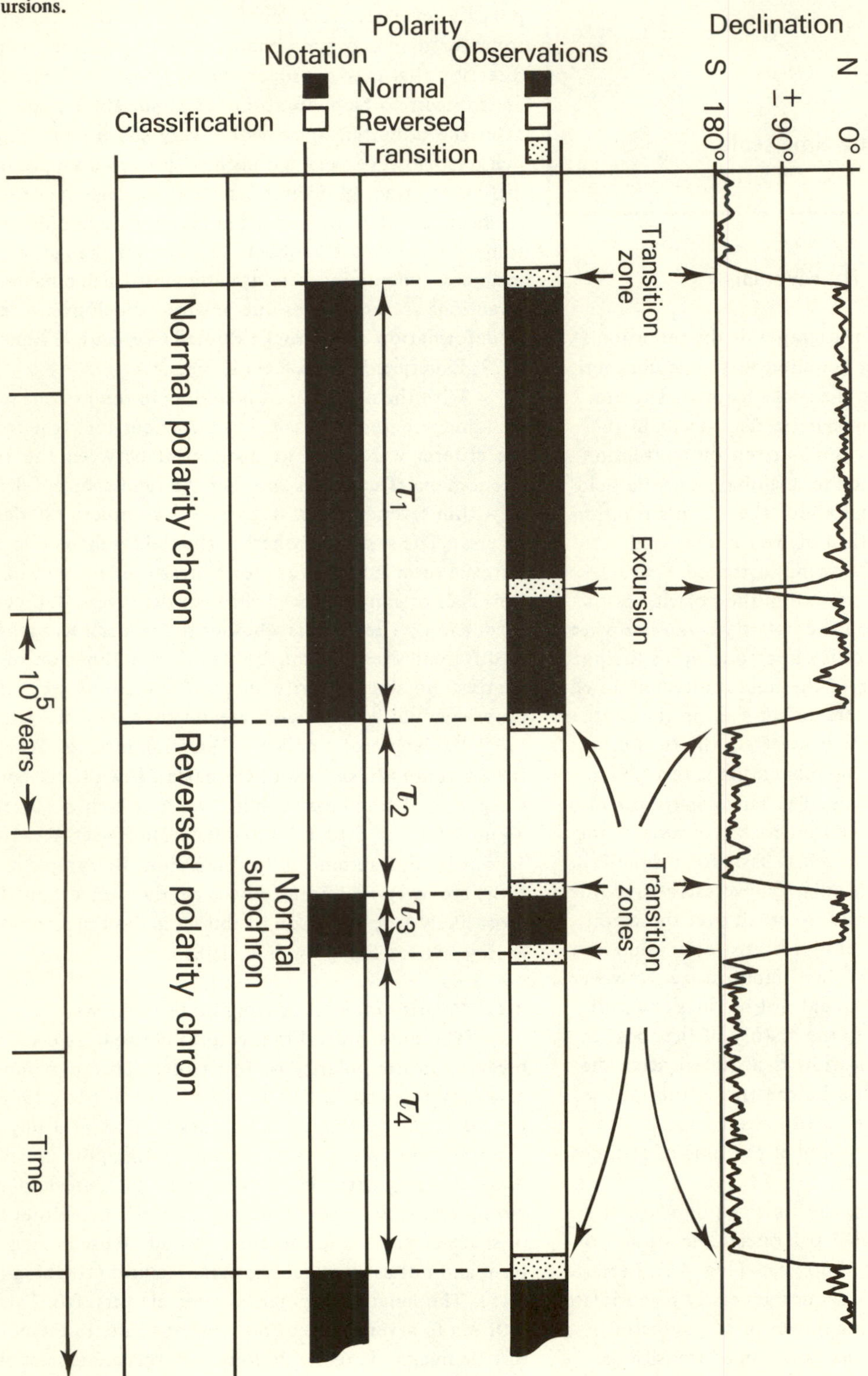

According to the chapter on magnetostratigraphy of the *International stratigraphic guide* prepared by the IUGS International Subcommission on Stratigraphic Classification (ISSC) and the IUGS/IAGA Subcommission on the Magnetic Polarity Time Scale (Anon. 1979), the terms recommended for describing subdivisions of time based on geomagnetic polarity are *polarity subchrons*, *polarity chrons* and *polarity superchrons*. The corresponding chronostratigraphic terms for describing all rocks, magnetic or not, that formed during these time intervals are *polarity subchronozone*, *polarity chronozone* and *polarity superchronozone*. Magnetic lithostratigraphic intervals based on the measured magnetic properties of rocks are *polarity subzone*, *zone* and *superzone*. In earlier reports, the Subcommission recommended the following durations for different levels in the hierarchy (McElhinny 1978).

Term	*Approximate duration*
Polarity subzone	10^4 – 10^5 years
Polarity zone	10^5 – 10^6 years
Polarity superzone	10^6 – 10^7 years
Polarity hyperzone	10^7 – 10^8 years

In accord with these recommendations, we will use *polarity chron* as the geochronologic term to describe the main subdivisions of time recognised on the basis of polarity. For example, the term ‘Matuyama reversed polarity chron’ replaces the earlier term ‘. . . epoch’ (Anon. 1979). The term ‘subchron’ will be used in place of the word ‘event’ to describe very short (< 0.1 Ma) polarity intervals occurring within a chron. For example, the Jaramillo normal subchron occurs within the Matuyama reversed chron (Fig. 4.2).

4.2 Radiometrically dated time scale: 0–5 Ma

For the interval from the present back to 5 Ma, potassium–argon ages and magnetic polarities have been measured for 354 strata of extrusive rocks from many parts of the world. The construction of a self-consistent reversal time scale using these data has firmly established the global character of reversals. In addition, this research has established the existence of very short polarity intervals (subchrons) and the great variation in the lengths of polarity intervals. The current best estimates of the ages of chron boundaries determined from radiometric dating (Mankinen and Dalrymple 1979) are shown in Fig. 4.2. These ages were determined using chronograms similar to those described in Chapter 3 (Cox & Dalrymple 1967, Mankinen & Dalrymple 1979).

4.3 Marine magnetic anomalies: 5–83 Ma

4.3.1 Introduction

Marine magnetic anomalies have provided the richest single source of information about magnetic reversals from the Oxfordian Age to the present, which is the age range of ocean floor that preserves a record of geomagnetic reversals. The main reason for the high fidelity of the marine magnetic record is the remarkable continuity of the geologic processes by which new crust is formed along mid-ocean ridges. The normal and reversed polarity intervals recorded on the ocean floor produce *marine magnetic anomalies* in the form of peaks and troughs that appear on magnetic profiles. Some noise is generally present on these profiles in the form of spurious small anomalies produced by seamounts and other geologic irregularities. Moreover gaps and duplications are commonly present on the profiles because of the jumping of ridges to new positions. In addition, oceanic plates change velocity from time to time, so that marine anomalies must be calibrated using radiometric dating and biostratigraphy. However, despite problems of noise, gaps and non-linearity, the history of geomagnetic polarity comes through more clearly in marine magnetic profiles than in any other type of geologic record.

4.3.2 Resolving power

Although no single magnetic profile provides a perfect record of reversals it has been possible, by comparing profiles from different parts of the world, to identify those anomalies that appear consistently on most high-quality profiles, and in this way to determine which anomalies are due to geologic noise and which anomalies correspond to the history of geomagnetic reversals. All the polarity chrons ($\tau \geqslant 0.1$ Ma) in the marine anomaly record from the Oxfordian Age onwards have probably now been identified, as have many (but not all) of the subchrons ($\tau \leqslant 0.1$ Ma).

The minimum polarity interval length that can be resolved on individual profiles depends upon several factors. Interpretation of a profile begins with the use of a geophysical model, the most common model being a crustal layer made up of a sequence of prisms that are alternately reversely and normally magnetised. The various mathematical procedures for finding the optimum set of prisms to fit a given magnetic profile all yield a sequence of widths w_i for the prisms. These widths, which in stratigraphic terms comprise a set of magnetic polarity zones, are related to the durations τ_i of the corresponding polarity chrons and subchrons by

$$\tau_i = w_i/v$$

where v is the half velocity of ocean-floor spreading at the time the oceanic crust formed. The length of the shortest detectable polarity interval depends upon the rate of spreading, the length of the shortest prism that can be detected on a magnetic profile and the length scale of irregularities in the geologic processes by which new ocean floor forms. At water depths of 3 km the last two lengths are typically about 1 km. For very rapid spreading at a half velocity of 50 km/Ma, this corresponds to a minimum detectable polarity length of 0.02 Ma. This resolution appears to have been achieved for those parts of the reversal time scale for which high-quality profiles are available over ocean floor where spreading was rapid. In a few ideal situations, polarity intervals as short as 0.01 Ma appear to have been captured. Elsewhere the cut-off of observable polarity intervals appears to range from 0.02 to about 0.10 Ma, depending on the rate of ocean-floor spreading and the regularity of the spreading process.

4.3.3 Names and numbers of polarity chrons

Two systems for identifying polarity chrons are widely used. The first is the set of names (Bruhnes, Matuyama, Gauss, Gilbert) used for the radiometrically dated part of the reversal time scale (Fig. 4.2). These have been the standard chron names used for global correlation in Pliocene and Pleistocene stratigraphy for more than a decade and their continued use as informal units is recommended by the IUGS/IAGA Subcommission (Anon. 1979).

The second system is a numerical scheme that began informally when marine geophysicists gave numbers to thirty-two of the most prominent positive anomaly peaks appearing on magnetic profiles over ocean basins. The numbers begin with 1 at mid-oceanic ridges, where new oceanic crust is being generated (Pitman, Herron & Heirtzler 1968). These numbers were then associated with the normally magnetised prisms used to model the positive anomalies (Le Pichon & Heirtzler 1968). In effect, the numbers that had been assigned to anomaly peaks had now become the informal names of magnetic polarity zones. The next step was to divide the widths of the polarity zones

Figure 4.2. Radiometrically-based polarity time scale. Alternative numerical designations based on the numbered sequence of marine anomalies are given in parentheses.

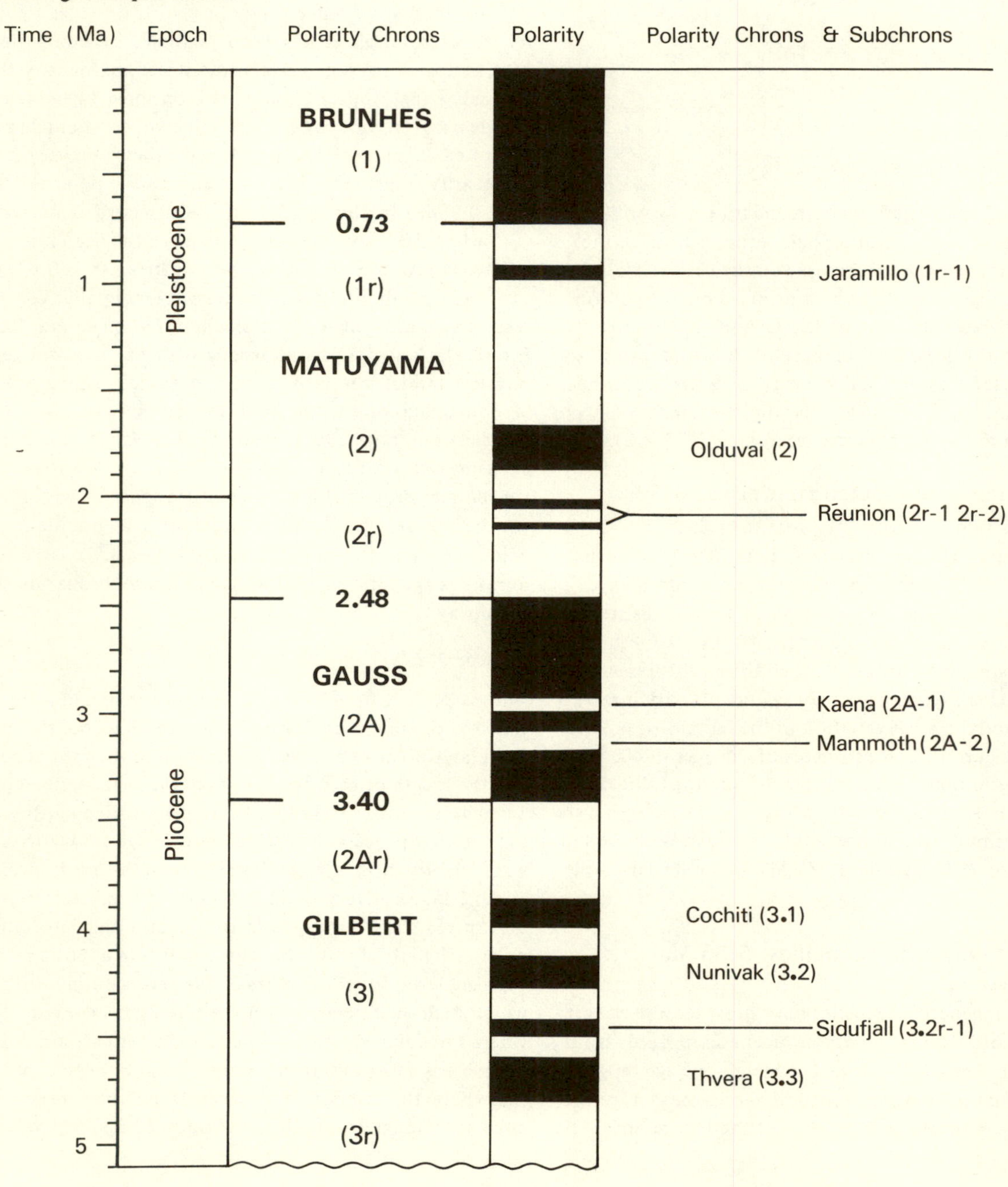

(rock units) by the radiometrically calibrated velocity of spreading to obtain the durations of magnetic polarity chrons (time units) with which the anomaly numbers were then also associated (Heirtzler *et al.* 1968).

The initial set of thirty-two numbered anomalies provided labels for only a fraction of the known chrons, two-thirds of the time covered by the numbering system lying outside any numbered chron. With the growing use of the marine anomaly sequence as a standard for global stratigraphic correlation, several workers have found it useful to label additional chrons by adding letters, decimals and prime marks to the original set of numbers (LaBrecque, Kent & Cande 1977, Ness *et al.* 1980). In this report we have used a composite of these numbering systems, extending them slightly in order to provide numbers for all chrons in a manner that is consistent with prior usage and amenable to change in the course of future discoveries of subchrons. Our approach in extending earlier numbering systems is similar to that used in a library when new books are added. Previously unlabelled reversed chrons are given the number of the next youngest normal chron with the letter 'r' appended (Fig. 4.3). Subdivisions of presently numbered chrons such as 5A are described by additional decimal numbers, e.g. 5A.1, 5A.1r, 5A.2. Unlabelled chrons are described by adding letters to the next youngest numbered chron, e.g. 5A and 5B follow 5 and 5AA and 5AB follows 5A. Finally, subchrons ($\tau \leqslant 0.1$ Ma) are labelled with the number of the chron in which they occur followed by '-1', '-2' etc., the numbers increasing in order of increasing age for presently known subchrons and in order of discovery for subchrons yet to be discovered.

4.3.4 Calibration

Two methods of calibration have been used to check and correct the original marine anomaly reversal time scale of Heirtzler *et al.* (1968) which was based on the assumption that ocean-floor spreading has occurred at a constant rate. The first method is drilling the ocean floor beneath a well defined magnetic anomaly and determining the potassium-argon age of the basalt layer or the biostratigraphic age of the sediment immediately overlying the basalt layer. Since the hiatus between the time of basalt extrusion and the time of first sedimentation on the sea floor is generally no more than a few million years, the sediment age provides a reasonably accurate estimate of the age of the polarity zone unless, of course, the basalt encountered is a sill.

The second approach, which is less direct but potentially more precise, is based on the only property that distinguishes polarity zones from each other, namely their lengths. These lengths vary widely and randomly from one zone to the next. Because of this variation, a sequence of 4 to 6 polarity intervals comprises a magnetic fingerprint or signature which can be correlated with corresponding signatures elsewhere. The widths of the polarity zones in the oceanic crust are vastly different from the thicknesses of the zones in sedimentary sections. However, if rates of ocean-floor spreading and sediment deposition do not vary greatly over intervals of the order of 10 Ma, the ratios of the lengths of polarity zones will be about the same and the magnetic fingerprints will be recognisably similar. Young polarity zones near spreading centres have been dated by matching them directly with the polarity chrons determined by radiometric dating (Vine 1966, Mankinen & Dalrymple

Figure 4.3. Numerical schemes for numbering chrons and subchrons derived from numbered marine magnetic anomalies.

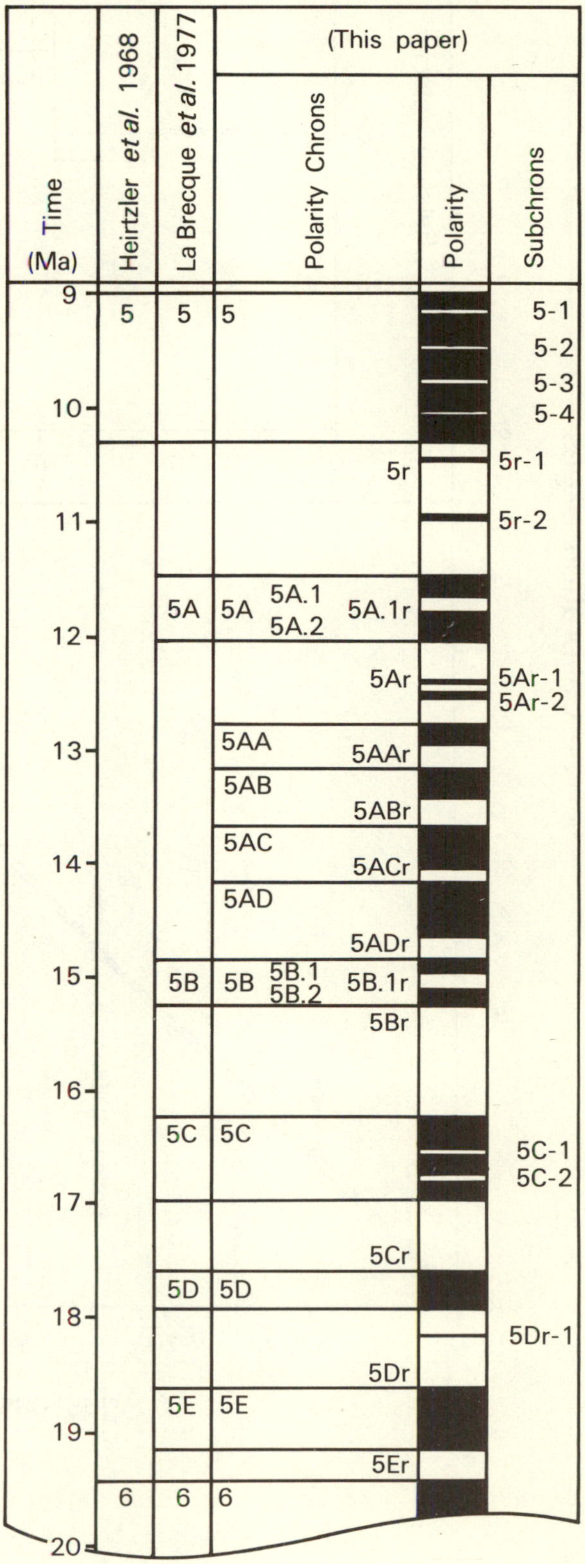

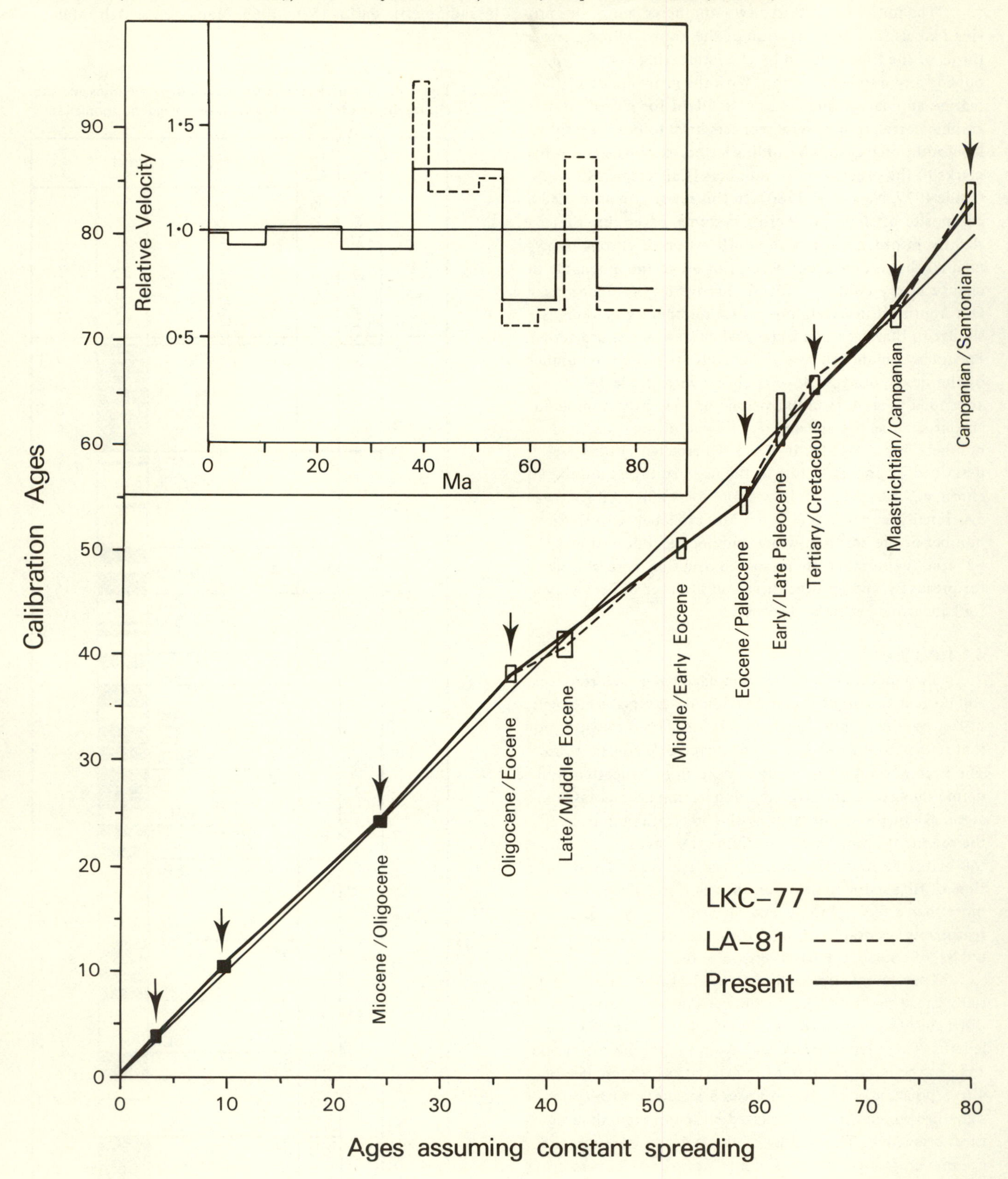

Figure 4.4. Calibration ages versus ages assuming constant velocity of sea-floor spreading. Horizontal axis is age of calibration points on reversal time scale LKC–77 (LaBrecque *et al.* 1977). Vertical axis is age of corresponding calibration points on the reversal time scale LA–81 (Lowrie & Alvarez 1981), connected by dashed lines, and the present time scale, shown as heavy solid lines between the calibration points (arrows) used in the present report. Rectangles are error boxes. Inset shows changes in the apparent velocity of sea-floor spreading corresponding to the LA–81 time scale (dashed line) and present time scale (solid line).

1979). Older polarity zones are currently being dated by correlating them with sedimentary or volcanic sections with well-controlled biostratigraphic ages (Butler *et al.* 1977, Lowrie & Alvarez 1981) and radiometric ages (McDougall *et al.* 1976).

The polarity time scale used in the present report is a modification of the time scale of Lowrie & Alvarez (1981) which uses eleven biostratigraphically controlled calibration points to stretch linearly the reversal time scale of LaBrecque *et al.* (1977). The latter was deduced from marine magnetic anomalies assuming a constant rate of ocean-floor spreading during the Cenozoic and latest Cretaceous. We have modified the time scale of Lowrie & Alvarez (1981) in several minor ways. The first was to drop their calibration points within the Eocene and Paleocene Epochs because these appear to require unrealistically rapid changes in the global rates of ocean-floor spreading, as may be seen in Fig. 4.4 where the calibration ages of Lowrie & Alvarez (1981) are plotted against the corresponding ages of LaBrecque *et al.* (1977). Since the latter correspond to a model of ocean-floor spreading at a nearly constant rate, deviations of the calibration points in Fig. 4.4 from a straight line through the origin imply changes in the velocity of ocean-floor spreading. Our interpretation is that the apparent changes in the velocity of spreading (inset in Fig. 4.4) are artifacts resulting from uncertainties in the radiometric ages of intra-epoch boundaries within the Eocene and Paleocene, which are not well constrained by the available radiometric data (fig. 5 of Hardenbol & Berggren (1978)). We have smoothed the apparent fluctuations in spreading rate by dropping the calibration points within the Eocene and Paleocene while retaining the points at the beginning and end of these epochs, accepting as the dates for these boundaries the values of Hardenbol & Berggren (1978) as corrected for new decay constants by Ness *et al.* (1980). The ages of intra-epoch boundaries were then found by interpolation (Table 4.1) assuming a constant rate of ocean-floor spreading during the Eocene and Paleocene. In passing we note that an age of 57–8 Ma for the Eocene–Paleocene boundary would greatly reduce the apparent speed-up during the Eocene and apparent slow-down during the Paleocene of plate velocities. Radiometric data (Hardenbol & Berggren 1978) indicate, however, that the age of the boundary is probably in the range of 54–6 Ma, as shown in Fig. 4.4.

The same analysis (Fig. 4.4) shows an apparent change in ocean-floor spreading velocity during Late Cretaceous and Early Tertiary time implicit in the reversal time scale of Lowrie & Alvarez (1981). In the present time scale we have assumed that this is due in part to small errors in the radiometric ages used for several calibration points rather than to true variations in spreading velocity. The importance of small differences in calibration ages may be seen by comparing the relative lengths of chrons 32 and 33 in the four time scales shown in Fig. 4.5. In the 'constant velocity' time scales of LaBrecque *et al.* (1977) and Ness *et al.* (1980) the anomaly 33/32 ratio is 3.6 whereas in the time scale of Lowrie & Alvarez (1981) the ratio of durations is 7.1, which would imply nearly a doubling of plate velocity. Accepting Lowrie & Alvarez's stratigraphic correlation of marine polarity zones with biostratigraphically dated magnetostratigraphy, we have reduced the variations in apparent spreading velocity by adjusting the ages of several late Cretaceous boundaries within their ranges of uncertainty. Using dates calculated with new decay constants (Steiger & Jäger 1977), the age of the Cretaceous–Tertiary boundary was taken to be 65.0 Ma, in agreement with Armstrong (1978) and Lanphere & Jones (1978). Adjustments of about 1 Ma were also made in the age of the younger and older boundaries of the Campanian Age, both adjustments being within the range of uncertainty of the radiometric ages of the boundaries. The Late Cretaceous and Cenozoic ages used for the reversal time scale and elsewhere in the report are given in Table 4.1.

Table 4.1. *Cenozoic and Late Cretaceous boundary ages used in this work*

(1) Epoch	(2) Age (Ma) (Ness *et al.* 1980)	(3) Age (Ma) (This report)
Pleistocene		
	2.0	2.0
Pliocene		
	5.1	5.1
Late Miocene		
	11.3	11.3
Middle Miocene		
	14.4	14.4
Early Miocene		
	24.6	24.6
Late Oligocene		
	32.8	32.8
Early Oligocene		
	38.0	38.0
Late Eocene		
	41.0	42.0
Middle Eocene		
	50.3	50.5
Early Eocene		
	54.9	54.9
Late Paleocene		
	61.5	60.2
Early Paleocene		
	66.7	65.0
Maastrichtian		
	72.3	73.0
Campanian		
	84.1	83.0
Santonian		

Column 2: Ages given by Ness *et al.* (1980), who corrected for new K–Ar decay constants the ages given by Berggren & Van Couvering (1974), Hardenbol & Berggren (1978), and Obradovich & Cobban (1975).

Column 3: Modifications which eliminate rapid and probably spurious changes in the apparent rate of sea-floor spreading.

Our time scale from the present to 3.4 Ma is that of Mankinen & Dalrymple (1979). Two recently documented normal subchrons not listed by them or us extend from 0.47 to 0.48 Ma (Wilson & Hey 1981, Champion *et al.* 1981) and from 2.24 to 2.26 Ma (Rea & Blakely 1975, Wilson & Hey 1981). From 3.4 to 10.3 Ma our time scale is that of Ness *et al.* (1980). From 10.3 to 83 Ma our time scale was generated by linearly stretching the time scale of Ness *et al.* (1980) between the eight calibration points T_c' listed in Table 4.2. To find the new age T of a polarity interval boundary with an old age T' on the time scale of Ness *et al.* (1980), let $T_c(y)$ and $T_c(o)$ be the new calibration ages that bracket T and let $T_c'(y)$ and $T_c'(o)$ be the corresponding calibration points on the time scale of Ness *et al.* (1980) that bracket T'. Then,

$$T = T_c(y) + [T' - T_c'(y)]\ [T_c(o) - T_c(y)]/[T_c'(o) - T_c'(y)].$$

The ages of the calibration points T_c' on the time scale of Ness *et al.* (1980) were found by first using the stratigraphic correlation of Lowrie & Alvarez (1981) to identify the polarity interval containing each calibration point. If A and B are the ages of the younger and older boundaries of this interval and if C is the age on the Lowrie & Alvarez (1981) time scale of the calibration point T_c, then the correspond-

Figure 4.5. Comparison of reversal time scales showing large variations in the ratio of the lengths of anomaly chrons 33/32. Time scales are LKC–77 (LaBrecque *et al.* 1977), NLC–80 (Ness *et al.* 1980), LA–81 (Lowrie & Alvarez 1981), and the time scale used in the present work.

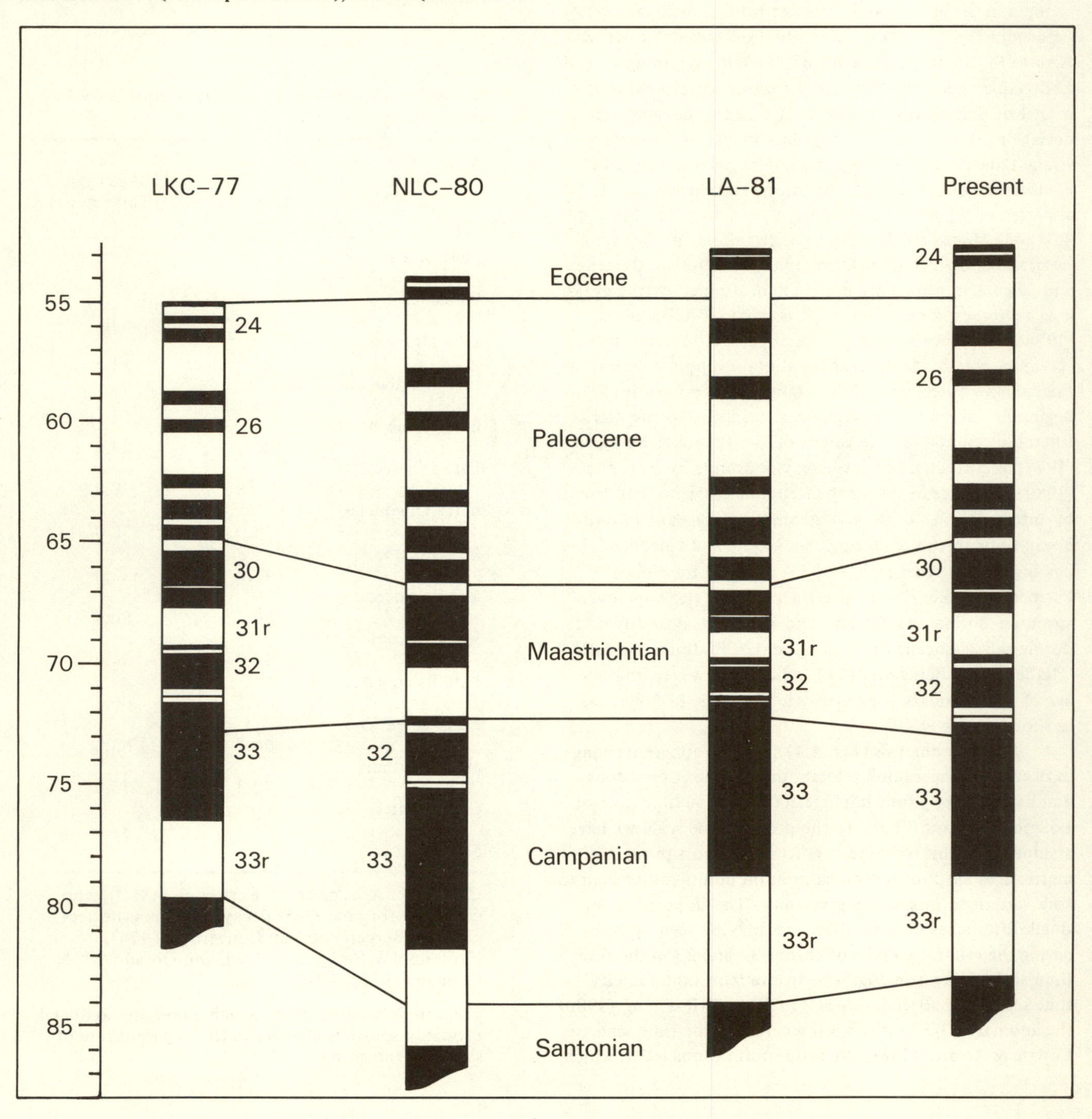

ing calibration age T_c' on the time scale of Ness *et al.* (1980) is given by

$$T_c' = A' + (C - A)\,(B' - A')/(B - A)$$

where A' and B' are the points on the time scale of Ness *et al.* (1980) corresponding to the interval boundaries A and B. The first set of values of T_c' listed in column 6 of Table 4.2 were found in this way and used to generate the figures in this chapter and for the wallchart. An alternative approach which is more consistent with our assumption that rates of ocean-floor spreading have undergone less drastic variations than those implied in the model of Lowrie & Alvarez (1981) is to use the time scale of LaBrecque *et al.* (1977) as the basis for the intra-interval interpolations described by the previous equation. Therefore the interpolation was repeated, using values of A and B from LaBrecque *et al.* (1977) rather than from Lowrie & Alvarez (1981), yielding the values of T_c' shown in column 6 of Table 4.2 and the corresponding values for the ages of polarity boundaries given in Table 4.3. Although differences in the two sets of ages are smaller than uncertainties in the radiometric ages of the calibration points, the latter set of values are logically more consistent with a model of nearly constant ocean-floor spreading. In order to represent the relative lengths of polarity intervals accurately, the ages of the beginning and end of all polarity intervals are listed in Table 4.3 with a precision of 0.01 Ma. The uncertainty in the chronometric ages of the boundaries is about two orders of magnitude greater.

4.3.5 Correlation with biostratigraphy

The field of biostratigraphic–magnetostratigraphic correlation is so large that space permits mention of only a few recent studies. For the Neogene, Berggren & Van Couvering (1978) have reviewed the correlation of magnetostratigraphy with planktonic foraminiferal zonation, calcareous nannoplankton zones, and continental vertebrate faunas. Hardenbol & Berggren (1978) have reviewed similar correlations for the Paleogene. Correlations between the numbered marine anomaly sequence and polarity zones recorded in sediments, which until recently were referred to as named or numbered epochs, are given in these two reviews.

For the Cretaceous and Early Cenozoic the marine anomaly scale has been correlated with marine biostratigraphy in the Umbrian Apennines in Italy (Alvarez & Lowrie 1978, Lowrie, Channell & Alvarez 1980, Lowrie & Alvarez 1981). In the early stages of this research, polarity zones were correlated regionally using the local terms such as 'Gubbio A' and 'Gubbio B'. In recent research, as these polarity zones have become firmly correlated with the marine anomaly sequence, the polarity zones in the Apennines have been correlated both locally and globally using chron numbers (Lowrie *et al.* 1980, Lowrie & Alvarez 1981). The Late Cretaceous and Early Cenozoic marine anomaly sequence has also been correlated with land mammal ages in the south-western United States by Butler *et al.* (1977).

The growing use of chron numbers derived from marine magnetic anomalies rather than names derived from magnetic stratotypes, reflects the limited usefulness of the stratotype concept in magnetostratigraphy. There are two reasons why this has turned out to be the case. The first is that although the ratio of the lengths of polarity zones is the single property upon which identification and correlation of polarity chrons are based, because of variations in rates of deposition, these ratios vary considerably between different sedimentary sections, even amongst those that are very good magnetic recorders (e.g. fig. 1 of Lowrie & Alvarez 1981). Adopting any one sedimentary section as a magnetic stratotype would generally yield as an international standard a set of ratios less accurate than the ratios obtained from a composite based mainly on marine anomaly profiles. The second reason is that because of hiatuses in deposition, some subchrons are commonly absent from the sedimentary record. Adopting any one section as a stratotype would generally result in an international standard lacking a polarity fine structure that is potentially very important for correlation. For these reasons the most generally accepted time scale has evolved as a composite based on the global consistency of many different magnetozones, most of which are found in oceanic crust. The importance of the marine anomaly record may be noted from the observation that it has not proved possible to use stratotypes to deduce a detailed reversal time scale prior to Oxfordian time, the age of the beginning of the marine anomaly record.

4.4 Marine magnetic anomalies: 83–160 Ma

4.4.1 Calibration

Ocean floor that formed from the Aptian to the Santonian is known as the 'Cretaceous quiet zone' because of the absence of globally traceable magnetic anomalies over oceanic crust of this age. The generally accepted explanation is that during this time the polarity of the Earth's field was normal except possibly for a few short reversed polarity intervals, the durations of which were probably less than 0.03 Ma and the ages of which are somewhat uncertain (for review, see Lowrie *et al.* 1980).

The Oxfordian to Barremian interval was a time of rapid reversals. The basic data used for constructing our time scale for this interval are the lengths of polarity intervals given by Cande, Larson & LaBrecque (1978) and Larson & Hilde (1975). Our two calibration points are near the beginning and at the end of the sequence, as described in Table 4.2. We place the end of the sequence at the initial Aptian boundary on the basis of the magnetostratigraphy of Lowrie *et al.* (1980) in the Apennines, although it should be noted that on the basis of deep-sea drilling results an age as young as mid-Aptian cannot be ruled out (Larson, Golovchenko & Pitman 1981). Since the present and previously published reversal time scales were found by linearly stretching the anomaly polarity zones between two calibration points, differences between them are mainly due to changes in the dates assigned to chronostratigraphic age boundaries.

Table 4.2. *Sources of data and calibration points used for the reversal time scale from 165 Ma to the Present*

(1) Time scale ref.	(2) Calibration point	(3) Chron no.	(4)	(5) Age T_c used here	(6) Age T_c' in refs. 3, 9, 10	(7) Age ref.	(8) Corr. ref.
3	Present	1(y)		0.0	0.0		
3	Radiometrically dated polarity time scale	2A(o)	+	3.4	3.4	1	1
3	Magnetostratigraphy of Icelandic lavas with K-Ar dates	5(o)	+	10.3	10.3	2	2
3	Miocene/Oligocene boundary in sediments in Umbrian Apennines, Italy	6Cr	*	24.6	24.0 (24.02)	3, 7	4
3	Oligocene/Eocene boundary in sediments in Umbrian Apennines, Italy	13r	*	38.0	35.86 (35.66)	3, 5	4
3	Eocene/Paleocene boundary in sediments in Umbrian Apennines, Italy	24r	*	54.9	56.65 (57.15)	3, 5	4
3	Paleogene/Cretaceous boundary in sediments in Umbrian Apennines, Italy	29r	*	65.0	66.79 (66.70)	11, 12	4
3	Maastrichtian/Campanian boundary in Umbrian Apennines, Italy	33	*	73.0	75.94 (76.51)	++	4
**	Campanian/Santonian boundary in sediments in Umbrian Apennines, Italy	34	*	83.0	85.92 (85.93)	++	4
9	Aptian/Barremian boundary in sediments in Apennines and southern Alps and DSDP Site 417	M0(o)	+	119.0	109.01	++	6, 8
10	Oxfordian nannoplankton at DSDP Site 105 overlie basalt of polarity zone M25n. Chron M24A(o) placed at middle of Oxfordian (***)	M24A(o)	+	160.0	151.79	++	8, 13
	Time scale of Ref. 10 was stretched by same factor for M25–M29 as time scale in Ref. 9 was for M0–M25	M29(o)	+	165.4	157.43	++	13

Notes

The ages of the boundaries of polarity chrons given in the references listed in column 7 were linearly stretched between the ages listed in column 5 of the stratigraphic tie-points described in column 2.

+ : The boundary of a chron or polarity interval coincides with the age of a calibration point.
* : Age of the calibration falls within a polarity interval.
++ : Age of calibration assigned in this report.
** : Possible short reversed subchrons within normal chron 34 are given in reference 6.
*** : M24A is the middle of the three reversed polarity intervals which comprise chron 24.

Notes to Table 4.2 (cont.)

Conventions
1(y) : The 'y' indicates that this is the younger boundary of chron 1.
2A(o) : The 'o' indicates that this is the older boundary of chron 2A.
6Cr : The 'r' indicates that this is the reversed chron just older than normal chron 6C.

References
1. Mankinen & Dalrymple (1979).
2. McDougall *et al.* (1976).
3. Ness, Levi & Couch (1980).
4. Lowrie & Alvarez (1981).
5. Hardenbol & Berggren (1978).
6. Lowrie, Channell & Alvarez (1980).
7. Berggren & Van Couvering (1978).
8. Van Hinte (1978b).
9. Larson & Hilde (1975).
10. Cande, Larson & LaBrecque (1978).
11. Armstrong (1978).
12. Lanphere & Jones (1978).
13. Larson, Golovchenko & Pitman (1981).

Table 4.3. *Polarity intervals of the KTQ-M Superchron*

Normal			Reversed		
Chrons	Subchrons	Interval boundaries (Ma)	Chrons	Subchrons	Interval boundaries (Ma)
1		0.00- 0.73	1r		0.73- 0.92
	1r-1	0.92- 0.97	1r		0.97- 1.67
2		1.67- 1.87	2r		1.87- 2.01
	2r-1	2.01- 2.04	2r		2.04- 2.12
	2r-2	2.12- 2.14	2r		2.14- 2.48
2A		2.48- 2.92		2A-1	2.92- 3.01
2A		3.01- 3.05		2A-2	3.05- 3.15
2A		3.15- 3.40	2Ar		3.40- 3.86
3.1		3.86- 3.98	3.1r		3.98- 4.12
3.2		4.12- 4.26	3.2r		4.26- 4.41
	3.2r-1	4.41- 4.49	3.2r		4.49- 4.59
3.3		4.59- 4.79	3r		4.79- 5.41
3A		5.41- 5.70		3A-1	5.70- 5.78
3A		5.78- 6.07	3Ar		6.07- 6.42
3B		6.42- 6.55	3Br		6.55- 6.77
4		6.77- 6.86		4-1	6.86- 6.94
4		6.94- 7.34		4-2	7.34- 7.39
4		7.39- 7.44	4r		7.44- 7.81
4A		7.81- 8.18	4Ar		8.18- 8.40
	4Ar-1	8.40- 8.49	4Ar		8.49- 8.80
	4Ar-2	8.80- 8.87	4Ar		8.87- 8.98
5		8.98- 9.13		5-1	9.13- 9.17
5		9.17- 9.47		5-2	9.47- 9.48
5		9.48- 9.75		5-3	9.75- 9.78
5		9.78-10.03		5-4	10.03-10.05
5		10.05-10.30	5r		10.30-10.43
	5r-1	10.43-10.48	5r		10.48-10.91
	5r-2	10.91-10.99	5r		10.99-11.47
5A.1		11.47-11.63	5A.1r		11.63-11.77
5A.2		11.77-12.03	5Ar		12.03-12.36
	5Ar-1	12.36-12.41	5Ar		12.41-12.49
	5Ar-2	12.49-12.54	5Ar		12.54-12.76
5AA		12.76-12.94	5AAr		12.94-13.15
5AB		13.15-13.41	5ABr		13.41-13.65
5AC		13.65-14.04	5ACr		14.04-14.16
5AD		14.16-14.63	5ADr		14.63-14.82

Table 4.3 (*cont.*)

Normal			Reversed		
Chrons	Subchrons	Interval boundaries (Ma)	Chrons	Subchrons	Interval boundaries (Ma)
5B.1		14.82-14.93	5B.1r		14.93-15.09
5B.2		15.09-15.23	5Br		15.23-16.20
5C		16.20-16.50		5C-1	16.50-16.54
5C		16.54-16.72		5C-2	16.72-16.79
5C		16.79-16.98	5Cr		16.98-17.58
5D		17.58-17.91	5Dr		17.91-18.13
	5Dr-1	18.13-18.15	5Dr		18.15-18.59
5E		18.59-19.12	5Er		19.12-19.41
6		19.41-20.50	6r		20.50-20.95
6A.1		20.95-21.22	6A.1r		21.22-21.45
6A.2		21.45-21.78	6Ar		21.78-21.97
6AA		21.97-22.14	6AAr		22.14-22.34
	6AAr-1	22.34-22.43	6AAr		22.43-22.65
6B		22.65-23.06	6Br		23.06-23.37
6C.1		23.37-23.54	6C.1r		23.54-23.76
6C.2		23.76-23.90	6C.2r		23.90-24.15
6C.3		24.15-24.32	6Cr		24.32-25.75
7		25.75-25.88		7-1	25.88-25.94
7		25.94-26.27	7r		26.27-26.74
7A		26.74-26.95	7Ar		26.95-27.27
8		27.27-27.36		8-1	27.36-27.44
8		27.44-28.27	8r		28.27-28.73
9		28.73-29.39		9-1	29.39-29.45
9		29.45-29.91	9r		29.91-30.48
10		30.48-30.84		10-1	30.84-30.90
10		30.90-31.17	10r		31.17-32.19
11		32.19-32.58		11-1	32.58-32.65
11		32.65-33.11	11r		33.11-33.57
12		33.57-34.06	12r		34.06-36.73
13		36.73-36.95		13-1	36.95-37.02
13		37.02-37.40	13r		37.40-38.64
15		38.64-38.80		15-1	38.80-38.83
15		38.83-38.98	15r		38.98-39.30
15A		39.30-39.48	15Ar		39.48-39.60
16		39.60-39.82		16-1	39.82-39.86
16		39.86-40.17	16r		40.17-40.39
17		40.39-41.07		17-1	41.07-41.13
17		41.13-41.29		17-2	41.29-41.34
17		41.34-41.60	17r		41.60-41.74
18		41.74-42.08		18-1	42.08-42.14
18		42.14-42.47		18-2	42.47-42.51
18		42.51-42.84	18r		42.84-43.52
19		43.52-43.87	19r		43.87-44.31
20		44.31-45.49	20r		45.49-47.46
21		47.46-48.69	21r		48.69-49.91
22		49.91-50.43	22r		50.43-51.39
23		51.39-51.57		23-1	51.57-51.60
23		51.60-52.02	23r		52.02-52.22
	23r-1	52.22-52.26	23r		52.26-52.36
24.1		52.36-52.55	24.1r		52.55-52.77
24.2		52.77-53.13	24r		53.13-55.60
25		55.60-56.33	25r		56.33-57.52
26		57.52-58.19	26r		58.19-61.00
27		61.00-61.62	27r		61.62-62.55
28		62.55-63.57	28r		63.57-64.03
29		64.03-64.86	29r		64.86-65.39
30		65.39-66.88	30r		66.88-66.97
31		66.97-67.74	31r		67.74-69.48
32.1		69.48-69.72	32.1r		69.72-69.96
32.2		69.96-71.40	32r		71.40-71.76
	32r-1	71.76-71.81	32r		71.81-72.06
33		72.06-78.53	33r		78.53-82.93

Table 4.4. *Polarity intervals of the JK-M Superchron*

Normal			Reversed		
Chrons	Subchrons	Interval boundaries (Ma)	Chrons	Subchrons	Interval boundaries (Ma)
			M 0		118.21-119.00
M 1n		119.00-122.46	M 1		122.46-122.96
M 2		122.96-123.83	M 3		123.83-126.42
M 4		126.42-127.64	M 5		127.64-128.31
M 6n		128.31-128.49	M 6		128.49-128.63
M 7n		128.63-128.83	M 7		128.83-129.33
M 8n		129.33-129.73	M 8		129.73-130.03
M 9n		130.03-130.38	M 9		130.38-130.96
M10n		130.96-131.39	M10		131.39-131.80
M10Nn		131.80-132.24		M10Nn-1	132.24-132.29
M10Nn		132.29-132.71		M10Nn-2	132.71-132.73
M10Nn		132.73-133.11	M10N		133.11-133.43
M11n		133.43-134.42	M11		134.42-134.98
	M11-1	134.98-135.03	M11		135.03-135.49
M11An		135.49-136.39	M11A		136.39-136.52
M12n		136.52-136.89	M12.1		136.89-137.79
M12.2n		137.79-137.91	M12.2		137.91-138.15
M12An		138.15-138.55	M12A		138.55-138.69
M13n		138.69-138.99	M13		138.99-139.51
M14n		139.51-139.84	M14		139.84-140.85
M15n		140.85-141.64	M15		141.64-142.28
M16n		142.28-144.08	M16		144.08-144.81
M17n		144.81-145.27	M17		145.27-146.94
M18n		146.94-147.57	M18		147.57-148.04
M19n		148.04-148.18		M19n-1	148.18-148.27
M19n		148.27-149.41	M19		149.41-149.93
M20n		149.93-150.27		M20n-1	150.27-150.34
M20n		150.34-151.07	M20		151.07-152.03
M21n		152.03-153.24	M21		153.24-153.77
M22n		153.77-155.51		M22n-1	155.51-155.57
M22n		155.57-155.63		M22n-2	155.63-155.69
M22n		155.69-155.77	M22		155.77-156.71
M22An		156.71-156.86	M22A		156.86-157.06
M23n		157.06-157.47	M23		157.47-157.79
	M23-1	157.79-157.81	M23		157.81-158.51
M24n		158.51-158.89	M24		158.89-159.30
	M24-1	159.30-159.33	M24		159.33-159.56
M24An		159.56-159.70	M24A		159.70-160.00
M24Bn		160.00-160.40	M24B		160.40-160.58
M25n		160.58-160.90	M25		160.90-161.19
M25An		161.19-161.36		M25An-1	161.36-161.45
M25An		161.45-161.56		M25An-2	161.56-161.65
M25An		161.65-161.80	M25A		161.80-161.92
M26n		161.92-162.03		M26n-1	162.03-162.12
M26n		162.12-162.20		M26n-2	162.20-162.29
M26n		162.29-162.38		M26n-3	162.38-162.43
M26n		162.43-162.65	M26		162.65-162.82
M27n		162.82-163.05	M27		163.05-163.22
M28n		163.22-163.55	M28		163.55-163.78
M29n		163.78-164.82	M29		164.82-165.41

4.4.2 Numbers of polarity chrons

The polarity chrons of this sequence are generally described by the designations 'M0' through 'M29' assigned to marine anomalies in order of increasing age by Larson & Pitman (1972), Larson & Hilde (1975) and Cande *et al.* (1978). In contrast with the practice of numbering normal chrons in the Cretaceous-Tertiary-Quaternary reversal sequence, reversed chrons were usually numbered in the Jurassic-Cretaceous reversal sequence. In Table 4.4 and Fig. 4.7 we have extended this numbering system to provide a designation for all chrons in the sequence. Previously unnumbered normal chrons are given the number of the next older reversed chron with 'n' appended, so that M16n occurs after and is stratigraphically above M16. As in the Cretaceous-Tertiary-Quaternary reversal sequence, normal follows reversed in a doublet with the same number.

4.5 Polarity bias superchrons

4.5.1 Phenomenon of polarity bias

If the reversal time scale is viewed through a sliding window 25 Ma wide, the character of the polarity pattern seen in the window undergoes marked changes as the window is moved from the present back to the beginning of geologic time. A typical change occurred at 83 Ma at the end of the Santonian Age (Fig. 4.6). For several tens of millions of years prior to that time the field remained in the normal state with at most a few brief, scattered intervals of reversed polarity. Then, at the end of the Santonian Age, the field began to reverse rapidly and symmetrically, spending approximately equal amounts of time in the normal and reversed polarity states. This pattern has continued to the present.

Paleomagnetic research has shown that throughout geologic time the field has been characterised by long intervals of time during which the *polarity bias* has remained constant. During times of *normal polarity bias* the field remains normal all or almost all of the time. *Reversed polarity bias* describes the opposite state. During times of *mixed polarity* the field alternates symmetrically between the normal and reversed polarity states.

The duration of intervals of constant polarity bias ranges from 30 Ma to about 100 Ma, which is more than an order of magnitude larger than the duration of chrons and subchrons during the Cenozoic. This difference suggests that the physical origin of polarity bias may be different from that of individual reversals. Quite possibly, individual reversals are the result of perturbations in the motion of fluid in the Earth's core, whereas changes in polarity bias reflect longer-term changes in the boundary conditions at the core-mantle interface (Irving & Pullaiah 1976, Cox 1981). Whatever their origin, polarity bias intervals comprise a distinct geomagnetic phenomenon that is useful for global stratigraphic correlation.

Figure 4.6. Reversal time scale from Callovian time to the Present – reduced scale.

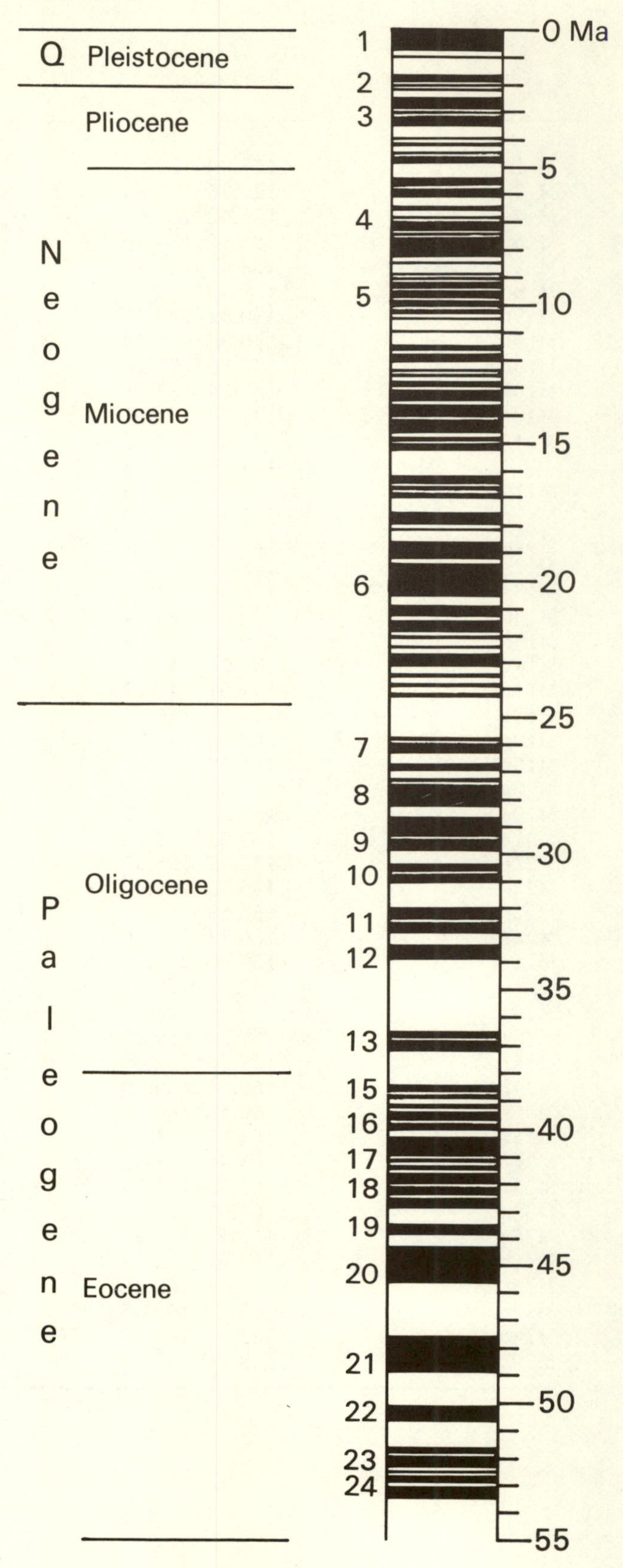

Fig. 4.6. (cont.)

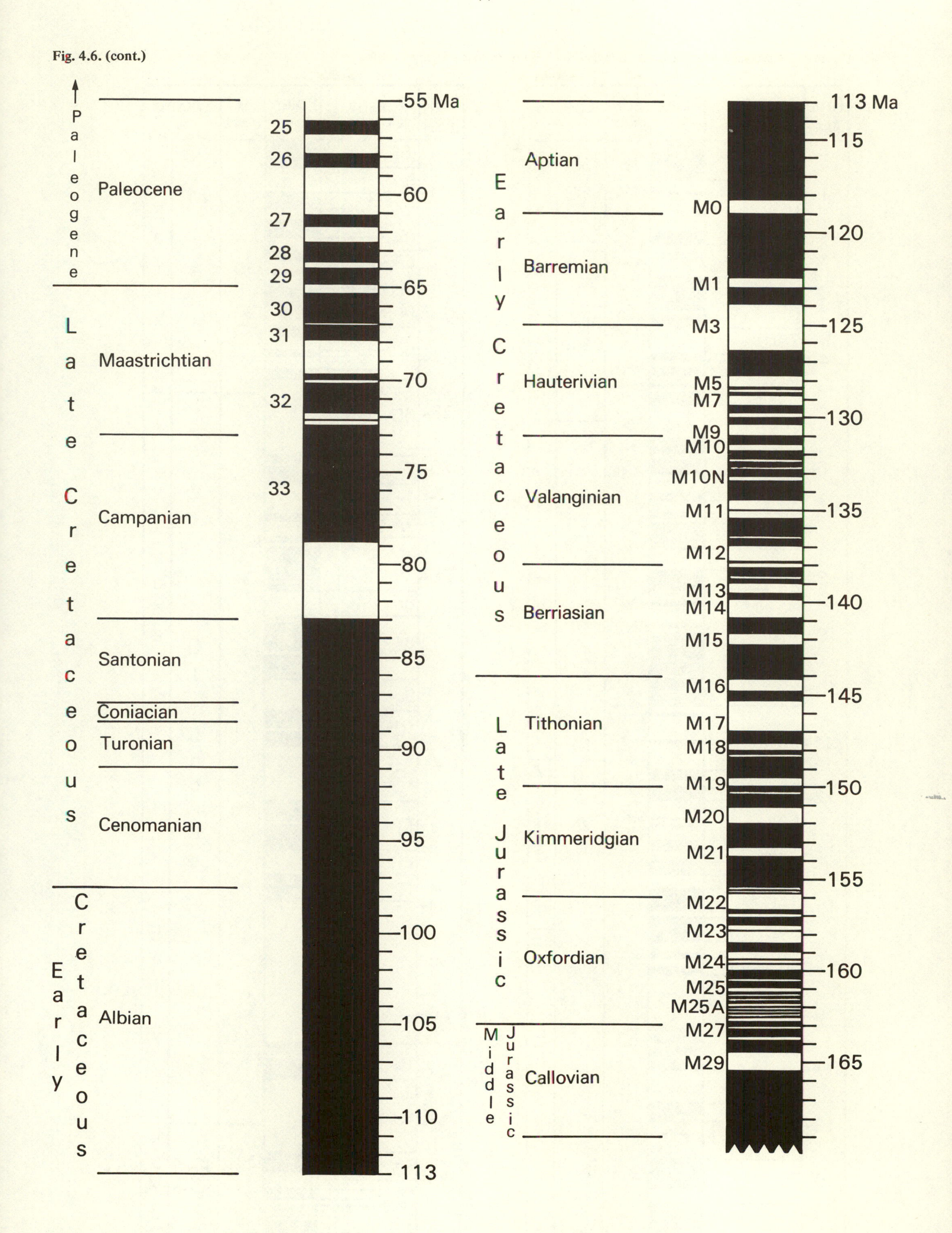
Paleogene
Paleocene
Late Cretaceous
Maastrichtian
Campanian
Santonian
Coniacian
Turonian
Cenomanian
Early Cretaceous
Albian
25
26
27
28
29
30
31
32
33
55 Ma
60
65
70
75
80
85
90
95
100
105
110
113
Early Cretaceous
Aptian
Barremian
Hauterivian
Valanginian
Berriasian
Late Jurassic
Tithonian
Kimmeridgian
Oxfordian
Middle Jurassic
Callovian
M0
M1
M3
M5
M7
M9
M10
M10N
M11
M12
M13
M14
M15
M16
M17
M18
M19
M20
M21
M22
M23
M24
M25
M25A
M27
M29
113 Ma
115
120
125
130
135
140
145
150
155
160
165

Figure 4.7. Reversal time scale from Callovian time to the Present – enlarged scale.

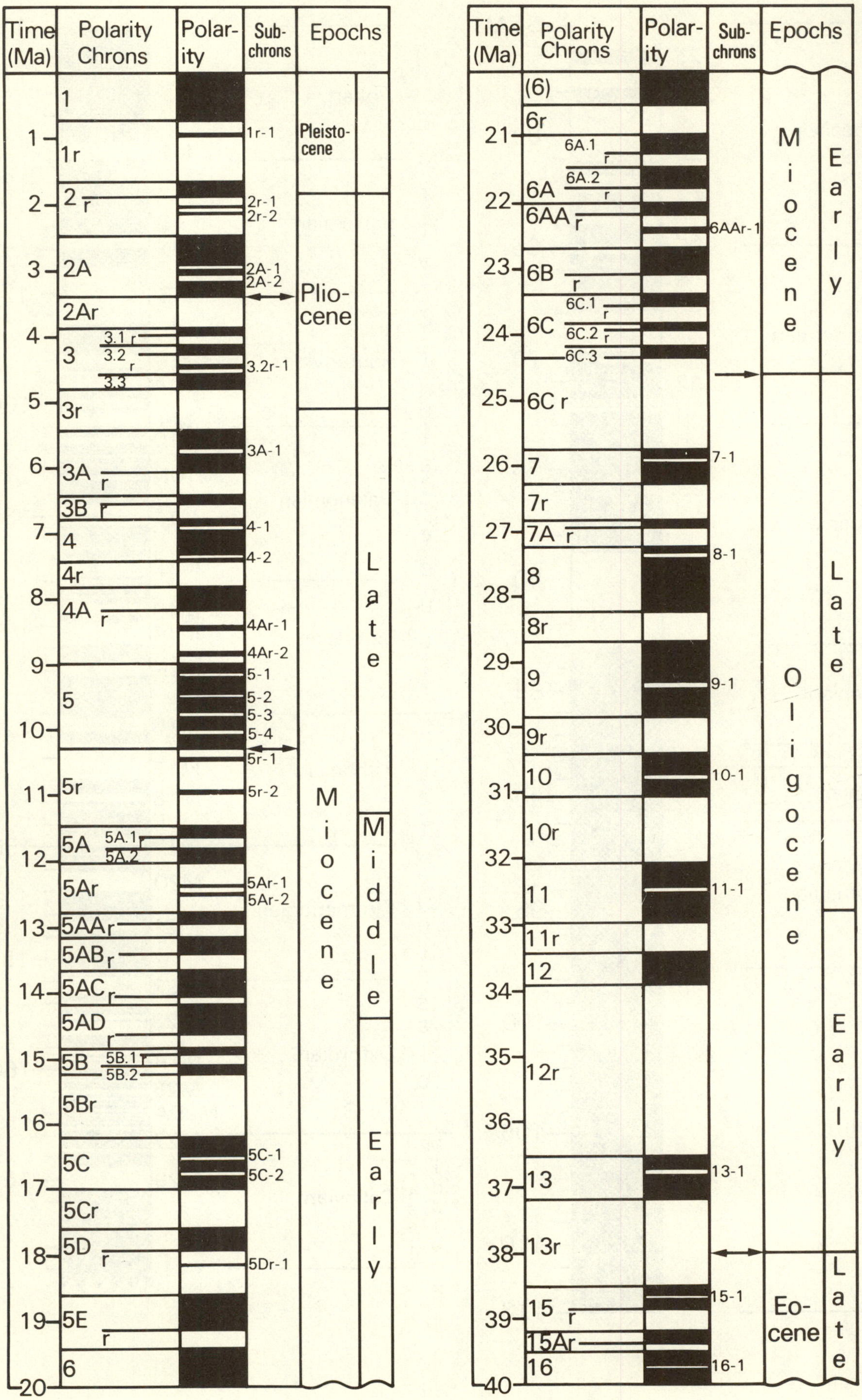

Fig. 4.7. (cont.)

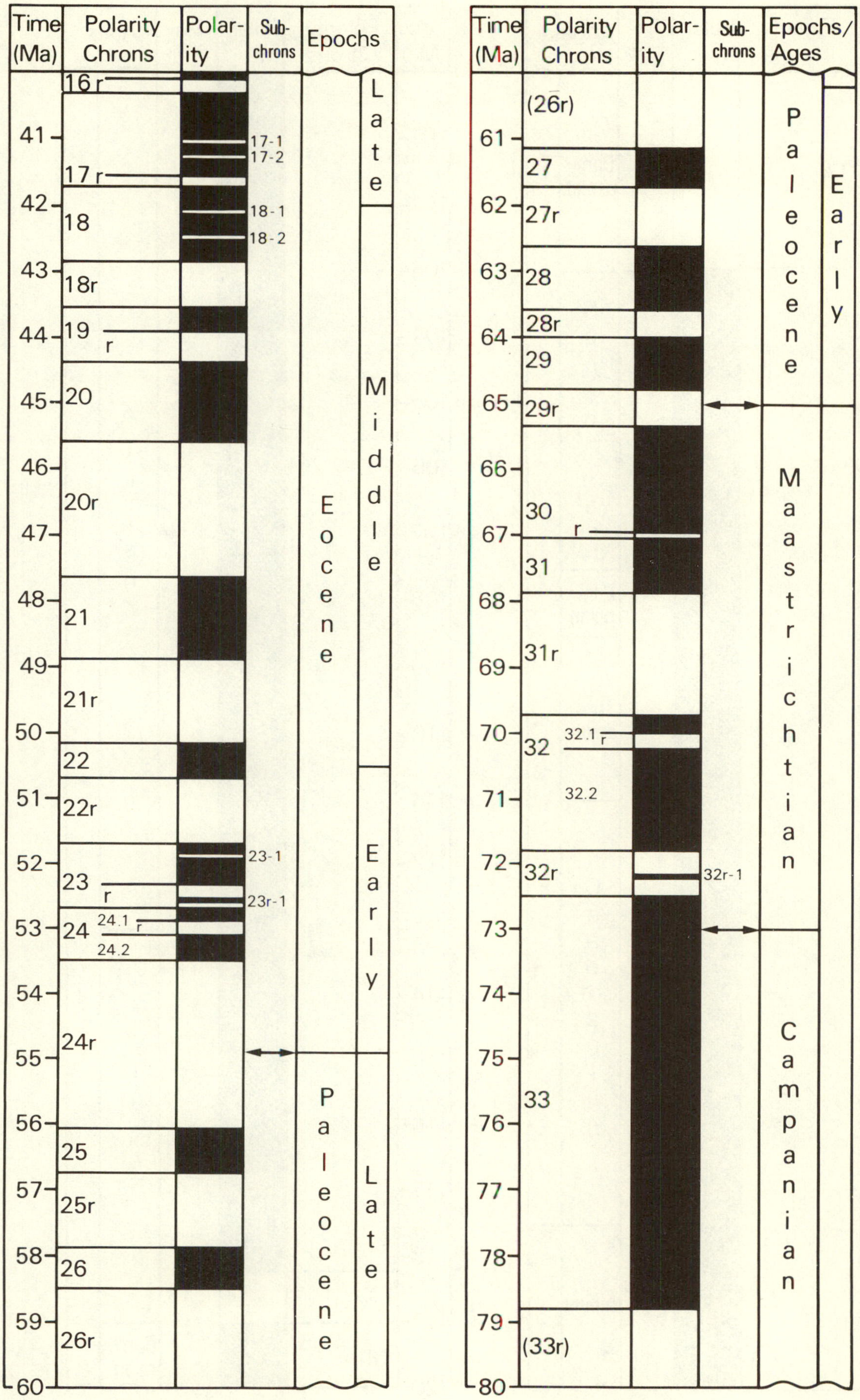

Time (Ma) | Polarity Chrons | Polarity | Subchrons | Epochs

41 – 60 Ma

16r, 17 (17-1, 17-2), 17r, 18 (18-1, 18-2), 18r, 19, 19r, 20, 20r, 21, 21r, 22, 22r, 23 (23-1), 23r (23r-1), 24.1, 24.1r, 24.2, 24r, 25, 25r, 26, 26r

Eocene: Late, Middle, Early

Paleocene: Late

Time (Ma) | Polarity Chrons | Polarity | Subchrons | Epochs/Ages

61 – 80 Ma

(26r), 27, 27r, 28, 28r, 29, 29r, 30, r, 31, 31r, 32.1, r, 32, 32.2, 32r (32r-1), 33, (33r)

Paleocene: Early

Maastrichtian

Campanian

Fig. 4.7. (cont.)

Time (Ma)	Polarity Chrons	Polarity	Subchrons	Ages
81	33r			Campanian
82				
83			↔	
84				Santonian
85				
86				
87				
88				Coniacian
89				
90				Turonian
91				
92				Cenomanian
93				
94				
95				
96				
97				
98				Albian
99				
100				

Time (Ma)	Polarity Chrons	Polarity	Subchrons	Ages
101				
102				
103				
104				
105				Albian
106				
107				
108				
109				
110				
111				
112				
113				
114				
115				Aptian
116				
117				
118				
	MO			
119				
	(M1n)			Barremian
120				

Fig. 4.7. (cont.)

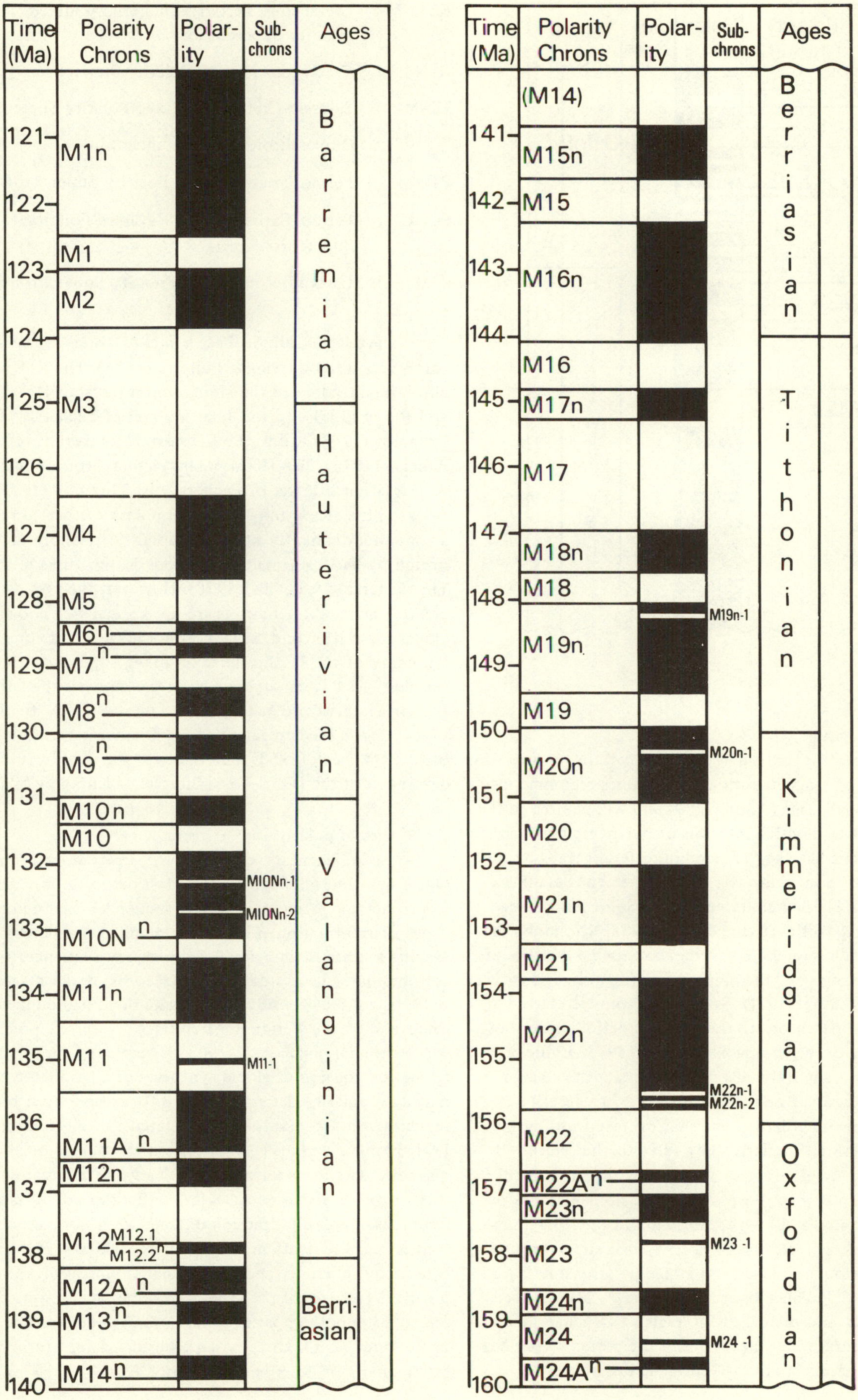

Time (Ma) | Polarity Chrons | Polar-ity | Sub-chrons | Ages

121 122 123 124 125 126 127 128 129 130 131 132 133 134 135 136 137 138 139 140

M1n
M1
M2
M3
M4
M5
M6n
M7n
M8n
M9n
M10n
M10
M10N n
M11n
M11
M11A n
M12n
M12 M12.1 M12.2n
M12A n
M13n
M14n

M10Nn-1
M10Nn-2
M11-1

Barremian
Hauterivian
Valanginian
Berriasian

Time (Ma) | Polarity Chrons | Polar-ity | Sub-chrons | Ages

141 142 143 144 145 146 147 148 149 150 151 152 153 154 155 156 157 158 159 160

(M14)
M15n
M15
M16n
M16
M17n
M17
M18n
M18
M19n
M19
M20n
M20
M21n
M21
M22n
M22
M22An
M23n
M23
M24n
M24
M24An

M19n-1
M20n-1
M22n-1
M22n-2
M23-1
M24-1

Berriasian
Tithonian
Kimmeridgian
Oxfordian

Fig. 4.7. (cont.)

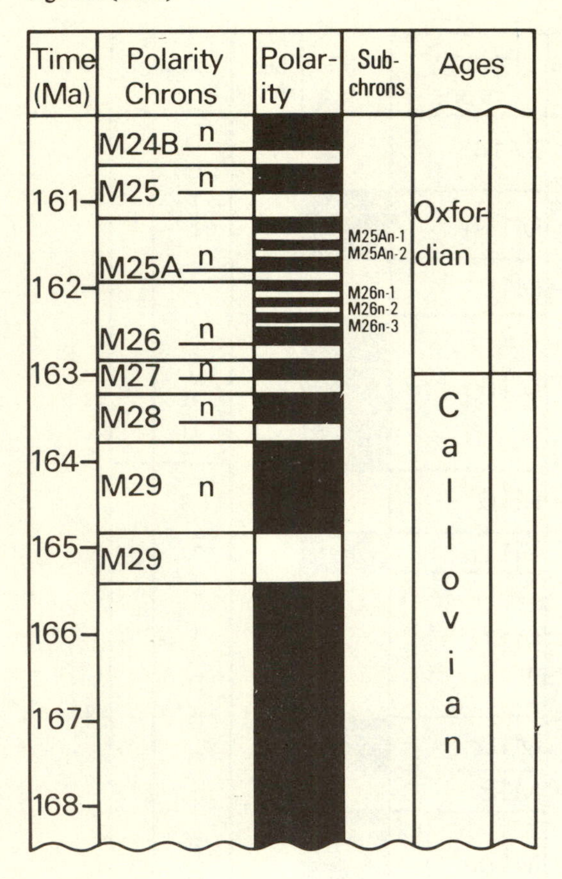

4.5.2 Nomenclature

The name 'superchron' is the next level above 'chron' in the hierarchy of magnetostratigraphic names recommended for international usage (Anon. 1979) and will be used in the present report to describe intervals of polarity bias. Polarity bias superchrons have been named in different ways by different workers. Some have left them unnamed (Sasajima & Shimada 1966, Helsley & Steiner 1969). Some have named them for type localities (Irving & Parry 1963, Khramov 1967, Pechersky & Khramov 1973). Some have used a mixture of type locality names and the names of distinguished scientists (McElhinny & Burek 1971). Some have simply referred polarity bias superchrons to the geologic period or periods in which they occur (Irving & Couillard 1973, Irving & Pullaiah 1976). The latter use the expression 'Cretaceous normal quiet interval' and the symbol 'KN' to describe the interval of normal polarity bias that occurred during the Cretaceous Period. Similarly they refer to the 'Permo-Carboniferous quiet reversed interval' with the symbol 'PCR'. In the present report we use the nomenclature of Irving & Pullaiah (1976) with several minor modifications and extensions. Three types of polarity bias superchrons are recognised: normal, reversed and mixed polarity superchrons (versus Irving & Pullaiah's normal quiet intervals, reversed quiet intervals, and disturbed intervals). Particular superchrons are identified by the names of the period or periods in which they occur, as in the following examples.

Present

KTQ-M	Cretaceous-Tertiary-Quaternary Mixed Polarity Superchron
K-N	Cretaceous Normal Polarity Superchron
JK-M	Jurassic-Cretaceous Mixed Polarity Superchron
	(Superchron sequence uncertain)
PTr-M	Permo-Triassic Mixed Polarity Superchron
PC-R	Permo-Carboniferous Reversed Polarity Superchron
C-M	Carboniferous Mixed Polarity Superchron

The relationship of these names to others in the literature is reviewed by Irving & Pullaiah (1976). The K-N superchron is equivalent to the Mercanton interval of McElhinny and Burek (1971), to the Jalal interval of Pechersky & Khramov (1973) and to the KN normal interval of Irving & Pullaiah (1976). The JK-M polarity superchron is equivalent to the Hissar interval of Pechersky & Khramov (1973). Between the JK-M and the PTr-M polarity superchrons was a time predominantly of mixed polarity, possibly interrupted by short normal superchrons in the Jurassic and Triassic (Irving & Pullaiah 1976). However, the ages of these possible normal superchrons are not known well enough for global correlation and, in fact, even the age of the upper boundary of the Permo-Triassic mixed superchron is uncertain. The PTr-M superchron is the equivalent of the Illawarra interval of Pechersky & Khramov (1973), the PC-R superchron is the equivalent of the Kiaman reversed interval of Irving & Parry (1963), and the C-M superchron is the equivalent of the Debal Tseva interval of Khramov (1967).

4.5.3 Ages of polarity superchrons

Our fragmentary knowledge of superchrons prior to Oxfordian time (the age of the oldest ocean floor preserving a decipherable magnetic anomaly sequence) is based on two types of information. The first is detailed magnetostratigraphic studies of individual sedimentary sequences with continuous or nearly continuous records of the magnetic field. It was such a study in Australia that resulted in the discovery of the Permo-Carboniferous reversed polarity superchron (Irving & Parry 1963). The second type of information consists of global syntheses of all available paleomagnetic polarity data including information from radiometrically and paleontologically dated rocks (McElhinny 1971, Irving & Pullaiah 1976). These global sets of paleomagnetic data are analysed statistically by calculating for rocks within a specified age window the fraction of samples (and presumably the fraction of time) with normal polarity. Values near 1 indicate normal polarity, values near 1/2 indicate mixed polarity, and values near 0 indicate reversed polarity. The ages of pre-Oxfordian superchrons were determined using both approaches, as described below. The ages of the three post-Oxfordian superchrons were determined by the methods described previously.

Figure 4.8. Polarity bias superchrons.

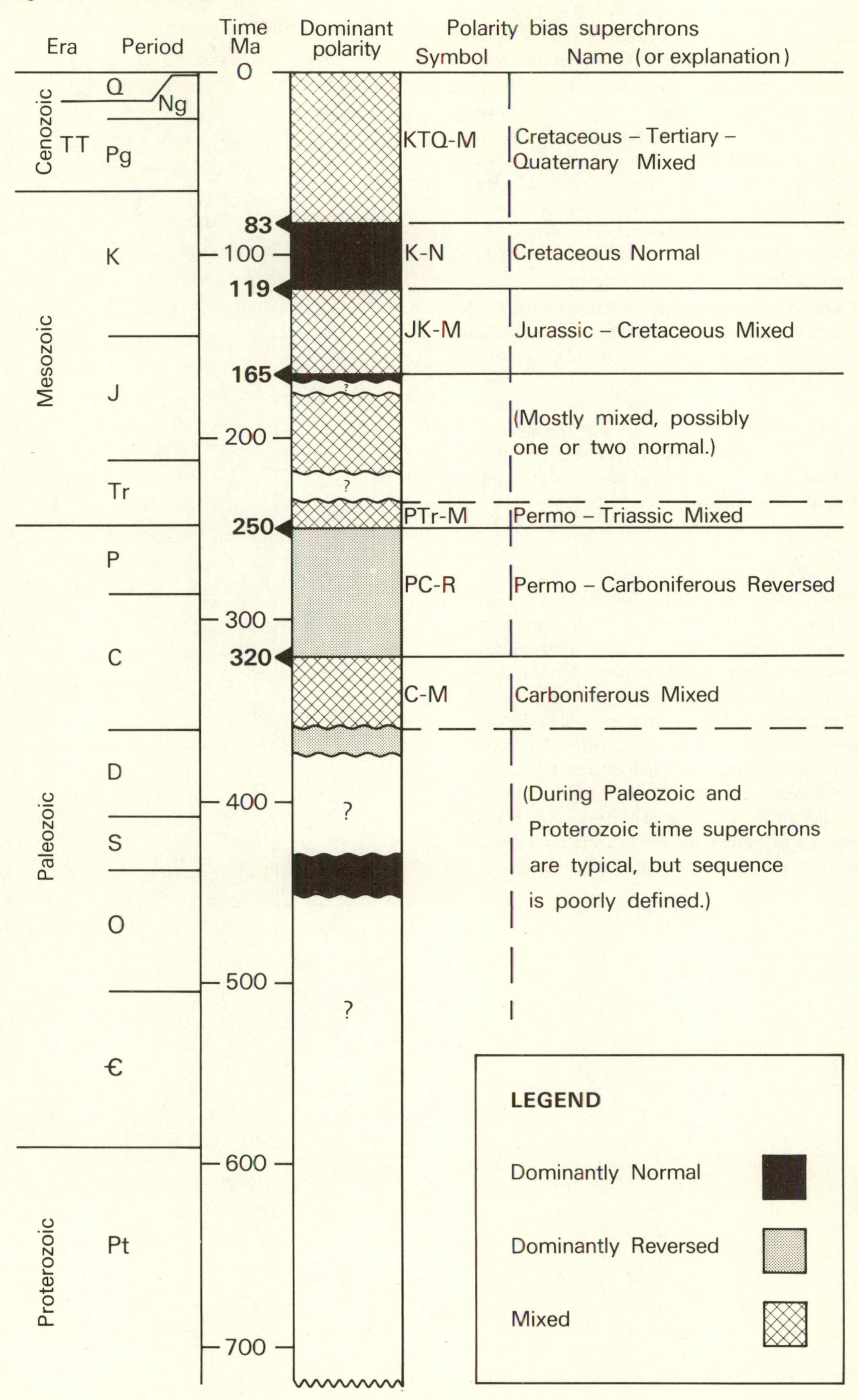

The initial and terminal boundaries of the Permo-Carboniferous reversed superchron are well dated by stratigraphic studies in Australia, North America and the USSR. The initial boundary is either within the Namurian Epoch or between the Namurian and Westphalian Epochs and the terminal boundary is either late or mid Tatarian (Irving & Pullaiah 1976). In the present analysis we have placed the initial boundary as Namurian at 320 Ma and the terminal boundary at the midpoint of the Tatarian Age at 250 Ma (Fig. 4.8). The difference between these values and the dates of 313 and 227 Ma used by Irving & Pullaiah (1976) reflects revisions in the geologic time scale and not different interpretations of the paleomagnetic data.

The Permo-Carboniferous reversed superchron PC-R is preceded by the Carboniferous mixed polarity superchron C-M, the beginning of which is not well dated. The early Paleozoic paleomagnetic record, although fragmentary, points to the presence of superchrons similar to those in the younger part of the geologic record. The Devonian field appears to have been predominantly reversed and the Silurian and Ordovician fields to have been predominantly normal (Irving & Pullaiah 1976).

The phenomenon of polarity bias also appears to have occurred during Proterozoic time but as yet the polarity structure is poorly defined. Paleomagnetic data from Laurentia (Irving & McGlynn 1976) suggest that during Proterozoic time the field was 'normal' 77 per cent of the time. However, knowing whether a given Proterozoic paleomagnetic direction corresponds to the polarity we now call normal depends upon tracing continuously back from the present to Proterozoic time the path along which the spin axis has moved as viewed from North America. There are several gaps in the record across which different pole paths can be drawn, corresponding to different polarity histories. Therefore, although the available data demonstrates that polarity bias occurred during the Proterozoic Eon, it is still uncertain whether the bias was normal or reversed.

5

Linear time scale plots

5.1 Introduction

The numerous and changing subdivisions of geologic time and geologic history are difficult to conceive without some visual aid, and the demand for visual aids is demonstrated by the readiness with which suitable charts are acquired and displayed. In recent years the most popular wallchart in many countries has been the colourful version edited by Van Eysinga and published by Elsevier (3rd edn 1975). Like BP, many oil companies have developed their own charts but these have rarely been available to the scientific community in general. Many organisations and authors around the world have produced a variety of time and stratigraphic charts, but these have frequently been constrained by page size or by local requirements.

A wallchart displaying the essential data was obviously an essential companion to this book and this chapter explains that wallchart (also by Harland *et al.* and published by CUP in 1982). It is based on the two charts, from BP and CASP, revised and supplemented by the linear geologic time scale from 4600 Ma to Present on the right-hand side, accompanied by our 'events' columns. The wallchart is printed with colours approximating to an international standard (see below, Section 5.3). Whether or not the product turns out to be a significant advance, the evidence and reasoning that have contributed to it have been recorded in this book. So that the information on the wallchart is available in this volume we have essentially repeated the different parts of the wallchart as figures in this chapter.

Previous chapters have discussed the chronostratic divisions of geologic time, the chronometric dating of their boundaries (and their uncertainties) and paleomagnetic reversal events. Ancient sea levels and the use of colour for time intervals are discussed briefly below. The overall dimensions of the wallchart (100 cm x 60 cm) were chosen to allow an adequate level of detail within a sheet of good proportion and manageable size. The divisions of time into four equal-sized segments, Precambrian, Paleozoic, Mesozoic and Cenozoic, which resulted in the four different linear time scales (highlighted in colour), was done to permit detailed subdivision within each segment and the addition of a fair level of detail in magnetostratigraphy and sea-level curves. These four segments of the wallchart are also reproduced in this chapter as Figs. 5.2–5.8.

On the right-hand side of the wallchart we have added another column showing some facets of Earth history at a single linear scale, along with a few Lunar events. The two first-order Phanerozoic sea-level cycles (Vail, Mitchum & Thompson 1977) are shown here, as are Precambrian events, and this column is reproduced as Fig. 5.9.

The lack of a Precambrian magnetostratigraphy and sea-level curve and the transfer of Precambrian events to the Earth history column on the right allowed space for the insertion of a Quaternary panel plotted with some simplification from M. J. Hambrey's Chart 2.17 in this book.

5.2 Eustatic changes

The subject of global sea-level changes through geologic time has long aroused interest. As a result of recent advances in the interpretation of reflection seismic records the subject has become topical, and we have therefore decided to attempt a summary of the present state of knowledge on sea-level variations through Phanerozoic time.What follows is not likely to be construed as presenting a consensus, least of all by active workers in the field, and it is likely to change over the next few years. Even so, there is a stimulus value in the exercise.

Changes in sea level have been invoked to explain geologic phenomena since the days of Hutton in the eighteenth century (Fairbridge 1961). The timing and magnitude of these changes have been estimated from the following.

(1) The areas of continents flooded at intervals in the past (Egyed 1956, Wise 1974, Hallam 1977, Cogley 1981).

(2) The volume changes of spreading ocean ridges with time (Hays & Pitman 1973, Flemming & Roberts 1973). Errors in this approach arise due to the use of only one hypsometric curve throughout time (Bond 1978, Harrison *et al.* 1981); this also at least partly explains the difference between results obtained by this method and (1) above.

(3) Seismic stratigraphy (Vail *et al.* 1977, Vail & Todd 1981). It is claimed that this approach, developed by Vail and his colleagues in the Exxon Corporation using a wealth of proprietary data, provides a powerful stratigraphic tool for hydrocarbon exploration. Certain successions of stratigraphic features are recognised in reflection seismic records, related to the changes of sea level which caused them, and thus dated by correlation with the global cycles of sea level established by previous work. In this way sedimentary basin sequences can be interpreted and dated prior to exploratory drilling.

Vail and his co-workers (papers in Payton 1977) built up their knowledge of changing sea levels through a series of steps. They first interpreted seismic sections from all the continents (except Antarctica) and their adjacent seas as chronostratigraphic charts using available stratigraphic controls. From the chronostratigraphic charts they built up charts, for each region studied, of the cycles of relative changes of sea level. Finally, they produced an integrated chart of global cycles which they described as 'simply a modal average of the correlative regional cycles'.

Second-order cycles, or supercycles, of 10–80 Ma durations and third-order cycles of 1–10 Ma durations were recognised on the regional and global cycle charts, and these in turn defined two first-order cycles. The latter are shown on the right-hand, linear, portion of our wallchart; a composite curve of the second and third order cycles is presented in the main chart.

On their global cycle chart Vail *et al.* (1977) showed relative changes of sea level on a scale delineated by the levels of maximum highstand at the end of Cretaceous time and minimum lowstand in the Oligocene Epoch. They also calibrated this chart using data from Hays & Pitman (1973) and Pitman (1978) and presented a chart showing actual eustatic changes of sea level. Prime features of these two charts were very gradual rises in sea level and relatively rapid falls, which appear to be instantaneous at the scale in which they are presented.

Criticisms of the work of Vail *et al.* (1977) focussed on these apparently very rapid falls of sea level. Some workers (e.g. Brown & Fisher 1979, Pitman 1978) disputed the significance of the features on the seismic records interpreted as indicating rapid falls; others were unable to identify mechanisms to account for the rate of falls (e.g. Pitman 1978, Donovan & Jones 1979). Vail & Todd (1981) noted these points and have admitted that their original chart of 'relative changes in sea level' is better described as a chart of 'relative changes of coastal onlap'. They have redrawn the interval from latest Triassic to Neocomian time in their chart as a curve of 'estimated global sea level' showing fewer 'instantaneous' falls in sea level and with falls in general being less abrupt. Fig. 5.1 also compares curves of relative coastal onlap and estimated eustatic sea-level changes for the Mesozoic and Cenozoic intervals. On the main chart we used these more refined 'estimated sea-level curves' from Fig. 5.1; the Paleozoic curve is a primary coastal onlap curve from Vail *et al.* (1977).

The first-order cycles of Vail *et al.* (1977) are in broad agreement with other eustatic sea-level curves (e.g. Egyed 1956, Hallam 1977). The second- and third-order cycles also seem to be in general agreement with the few comparable data available (e.g. Hallam 1978, Hancock & Kauffman 1979), but not all (e.g. Cooper 1977). This suggests that there is a movement towards a consensus on eustatic changes in outline if not in detail, although at least one worker considers this a fruitless exercise as he believes that only regional, not global changes can be inferred (Mörner 1981).

The debate on the origin of patterns of onlap sedimentation, whether due to global eustatic sea-level changes or to tectonic subsidence and flexure at passive margins, continues. Recently, Watts (1982) has concluded that many of the variations in coastal onlap used by Vail *et al.* to define second-order changes in sea level probably have a tectonic rather than a eustatic control. For additional and new information on the Exxon position on eustatic cycles and seismic stratigraphic interpretation the reader is referred to a recent paper by Beard, Sangree & Smith (1982).

5.3 International colour scheme

From the outset, we decided to print the chart in colour for both clarity and interest and to restate the desirability of a standard international colour scale for time stratigraphy.

The choice of colours was less easy. We are familiar with several coloured charts, e.g. the Shell Oil Company Legend, a Total Oil Company chart, and Van Eysinga's wallchart which has become perhaps the best known of late, but these all differ in their use of colour.

The Commission for the Geological Map of the World (CGMW) was conceived at the Second International Geological Congress, Bologna, 1881, and the choice of colours for (largely) time-stratigraphic purposes is discussed at length in the Congress Proceedings. Apart from colour plates in the Proceedings volume, an early example of the agreed standard colour scheme is presented by the *Geological map of Europe 1:500 000* prepared in accordance with Congress's resolutions. It seemed appropriate in 1981 to consider adopting the 1881 scheme, but this idea was abandoned at an early stage when we found the original descriptions of the colours to be imprecise while, after the passage of nearly 100 years, the colours in the Proceedings volume and on two separate examples of the European map no longer matched and no 'standard' colours could therefore be determined.

The work of the CGMW has continued, mainly since the Eleventh International Geological Congress in Stockholm, 1910, with interruptions in the war years 1914–18 and 1939–45, and is pursued in Paris where it has been connected with the International Union of Geological Sciences, and with Unesco which publishes its maps. The Unesco *Geological world atlas* (Choubert & Faure-Muret 1976) represents the culmination of decades of trials and the first sheets were produced in 1974. Vojacek (1979) outlined the history of this project and provided useful background technical information about the printing of the *Atlas*. He also noted (1979) that 'with this project Unesco also hopes to set international standards for the presentation of geological symbols and for the colour designs of geological maps'.

This seemed a laudable aim, and sufficient justification for adopting the Unesco colours, but there proved to be a serious commercial obstacle to this course of action. Unesco

Figure 5.1. (a)(i) Mesozoic cycles of relative changes of coastal onlap (Vail *et al.* 1977, as redesignated with revisions (dashed) by Vail & Todd 1981). (ii)–(iv). Estimated global eustatic changes in sea-level during the Mesozoic (Vail *et al.* 1977, Vail & Todd 1981, Hallam 1981).
(b)(i) Cenozoic cycles of relative changes of coastal onlap (Vail *et al.* 1977, as redesignated by Vail & Todd 1981). (ii) Estimated global eustatic changes of sea-level during the Cenozoic (Vail *et al.* 1977). The Norian–present curve of estimated eustatic changes of sea-level on the Geological Time Scale Chart is a composite of (b)(ii), (a)(ii) and (a)(iii).
Figure compiled by A.J. Fleet.

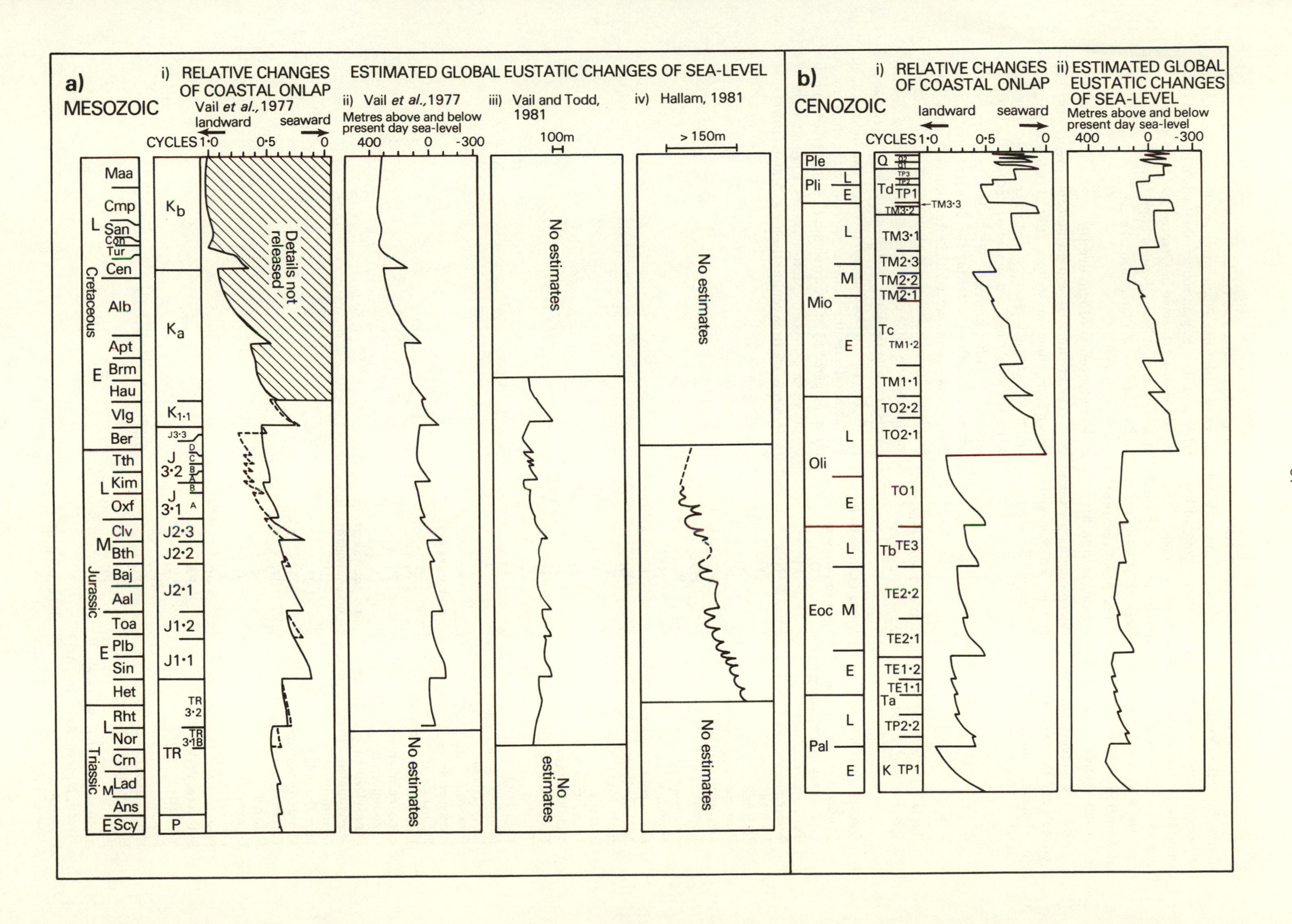

a)
MESOZOIC
i) RELATIVE CHANGES OF COASTAL ONLAP
Vail et al., 1977
landward
seaward
CYCLES
1·0
0·5
0
Details not released
ESTIMATED GLOBAL EUSTATIC CHANGES OF SEA-LEVEL
ii) Vail et al., 1977
Metres above and below present day sea-level
400
0
-300
No estimates
iii) Vail and Todd, 1981
100m
No estimates
No estimates
iv) Hallam, 1981
> 150m
No estimates
No estimates
Cretaceous
Jurassic
Triassic
Maa
Cmp
San
Con
Tur
Cen
Alb
Apt
Brm
Hau
Vlg
Ber
Tth
Kim
Oxf
Clv
Bth
Baj
Aal
Toa
Plb
Sin
Het
Rht
Nor
Crn
Lad
Ans
Scy
L
E
M
Kb
Ka
K1·1
J3·3
J 3·2
J 3·1
J2·3
J2·2
J2·1
J1·2
J1·1
TR 3·2
TR 3·1
TR
P
b)
CENOZOIC
i) RELATIVE CHANGES OF COASTAL ONLAP
landward
seaward
CYCLES
1·0
0·5
0
ii) ESTIMATED GLOBAL EUSTATIC CHANGES OF SEA-LEVEL
Metres above and below present day sea-level
400
0
-300
Ple
Pli
Mio
Oli
Eoc
Pal
Q
Td
TP1
TM3·3
TM3·2
TM3·1
TM2·3
TM2·2
TM2·1
Tc
TM1·2
TM1·1
TO2·2
TO2·1
TO1
Tb
TE3
TE2·2
TE2·1
TE1·2
TE1·1
Ta
TP2·2
K TP1

used no less than thirty-five colours (including black) in printing their atlas (Vojacek 1979). This was impractical for the low-cost commercial printing operation envisaged by Cambridge University Press, and represents a standard unlikely to be emulated by many organisations, governmental or private. We are indebted to H. J. Vojacek of Mercury–Walch in Hobart, Tasmania for providing proof sheets of the Unesco scheme (personal communication with BP).

We have matched the Unesco legend colours as closely as possible and the equivalents between our wallchart and their colours are shown in Table 5.1. Some modifications have been necessary as the Unesco legend has no subdivisions for Cenozoic, Phanerozoic and Priscoan and uses a scheme of Precambrian subdivisions which has been superseded. Also as explained in Chapter 2 we have used Mississippian and Pennsylvanian in preference to Dinantian and Silesian for the international scale.

Table 5.1. *Wallchart colour equivalents with Unesco international colour scheme for geologic maps*

Our colour block	Unesco colour code
Cenozoic	no equivalent
Quaternary and Pleistogene	Q
Tertiary	TT
Neogene	N
Pliocene	N_2
Miocene	N_1
Paleogene	PG
Oligocene	PG3
Eocene	PG2
Paleocene	PG1
Mesozoic	Mz
Cretaceous	K
Late Cretaceous	K_2
Early Cretaceous	K_1
Jurassic	J
Late Jurassic	J_3
Middle Jurassic	J_2
Early Jurassic	J_1
Triassic	T
Late Triassic	T_3
Middle Triassic	T_2
Early Triassic	T_1
Paleozoic	Pz
Phanerozoic	no equivalent
Permian	P
Late Permian	P_2
Early Permian	P_1
Carboniferous	C
Pennsylvanian	no equivalent
Mississippian	no equivalent
Devonian	D
Late Devonian	D_3
Middle Devonian	D_2
Early Devonian	D_1
Silurian	S
Ordovician	O
Cambrian	Є
Proterozoic	PD
Sinian (& Pt_3)	PA
Vendian	PA_1
Sturtian	PA_2
Riphean (& R_3, R_2, R_1)	PC
Huronian (& Pt_1)	PD_1
Pt_2	PB
Archean (Randian, Swazian, Isuan, Hadean, Ar_3, Ar_2, Ar_1)	A
Priscoan	no equivalent

5.4 Selected geologic events

The Precambrian and Phanerozoic events have been drawn from numerous sources. Stille's (1924) four orogenies and their constituent tectonic phases are included because some are widely referred to and he defined their stratigraphic age rather precisely. We do not, however, recommend their use, preferring the Standard Stratigraphic Scale for precision. Indeed it is often not clear what is the tectonic value of some of these names and so we place an asterisk before Stille's names to distinguish them. Glacial events are from Hambrey & Harland (1981) and Harland (in press). Faunal and floral appearances are also indicated, but for further details the reader is referred to the chart 'Evolution' prepared by M. Fewtrell Smith and published in 1981 by the Open University, Milton Keynes, England. Ocean-floor events are from Smith, Hurley & Briden (1981).

Figures 5.2–5.9 (see following pages). Time scales and events (from wallchart).

Legend

Tie points	●
Individual boundary calculations	○
Tillites & glacial episodes	▲
Last appearance	LA
First appearance	FA

Scales

Scale	Era	
0 5 10 15	Cenozoic	1 cm = 3 Ma
0 10 20 30 40	Mesozoic	1 cm = 8 Ma
0 20 40 60 80	Paleozoic	1 cm = 16 Ma
0 200 400 600 800 1000	Precambrian	1 cm = 200 Ma

Figure 5.2. Cenozoic time scale.

Eon	Era	Sub-era	Period	Epoch		Age	Picks (Ma)
Phanerozoic (Ph)	Cenozoic (Cz)	Quat. (Q)	Pleistogene (Ptg)	Pleistocene (Ple)			2·0
		Tertiary (TT)	Neogene (Ng)	Pliocene (Pli)	L	Piacenzian (Pia)	3·3
					E	Zanclian (Zan)	5·1
				Miocene (Mio)	L	Messinian (Mes)	6·5
						Tortonian (Tor)	11·3
					M	Serravallian (Srv)	
						Late Langhian (Lan)	14·4
					E	Early Langhian (Lan)	
						Burdigalian (Bur)	19
						Aquitanian (Aqt)	24·6
			Paleogene (Pg)	Oligocene (Oli)	L	Chattian (Cht)	32·8
					E	Rupelian (Rup)	38·0
				Eocene (Eoc)	L	Priabonian (Prb)	42·0
					M	Bartonian (Brt)	44
						Lutetian (Lut)	50·5
					E	Ypresian (Ypr)	54·9
				Paleocene (Pal)	L	Thanetian (Tha)	60·2
					E	Danian (Dan)	65·0
	Mesozoic (Mz)		Cretaceous (K)	Late	Senonian	Maastrichtian (Maa)	73
						Campanian (Cmp)	83
						Santonian (San)	87·5
						Coniacian (Con)	88·5
						Turonian (Tur)	91
						Cenomanian (Cen)	97·5

Age in Ma scale: 5, 10, 15, 20, 25, 30, 35, 40, 45, 50, 55, 60, 65, 100

Error in ages (Ma): 5, 10, 15

Figure 5.3. Cenozoic events.

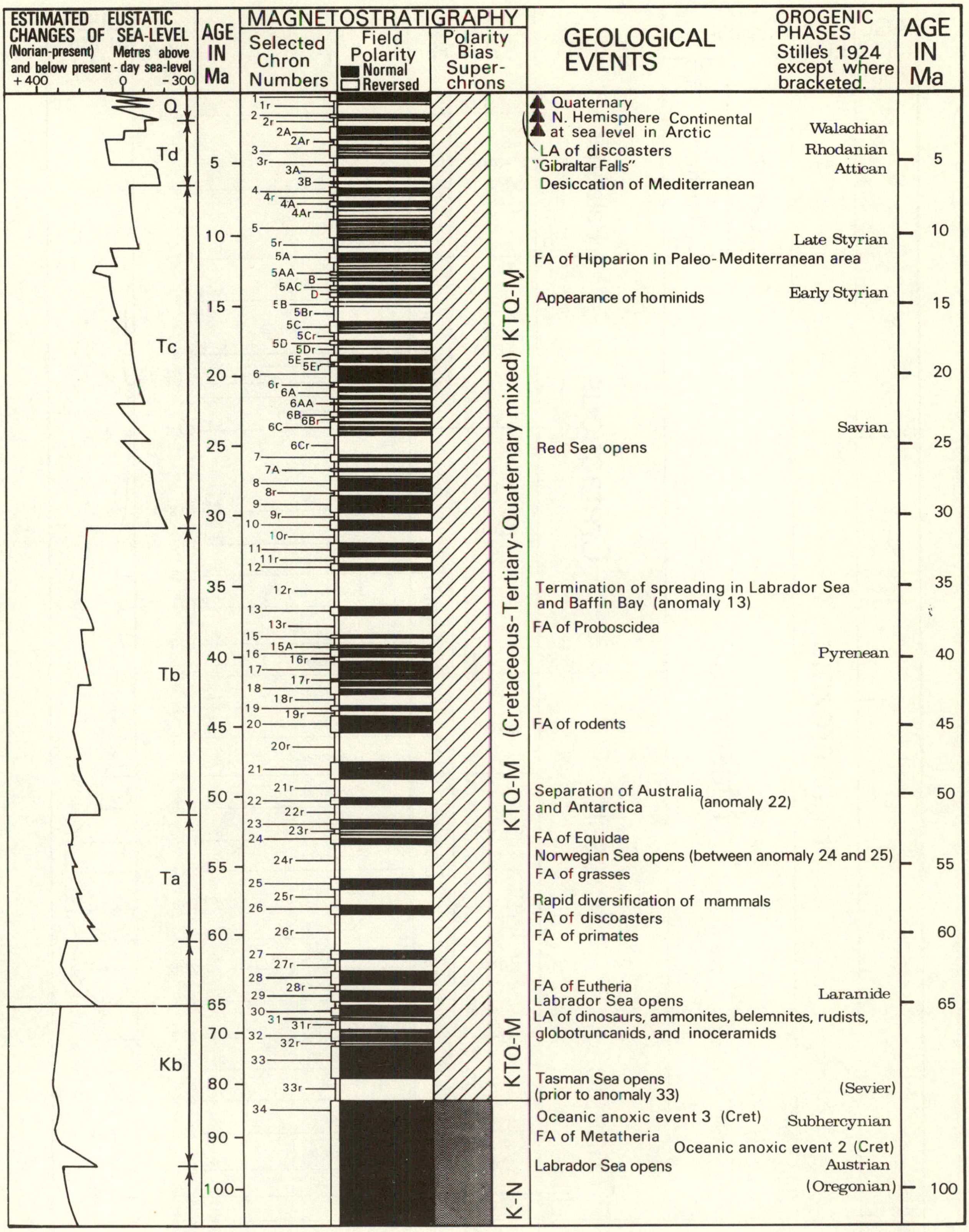

ESTIMATED EUSTATIC CHANGES OF SEA-LEVEL (Norian-present) Metres above and below present-day sea-level +400 0 -300	AGE IN Ma	MAGNETOSTRATIGRAPHY: Selected Chron Numbers	Field Polarity (Normal / Reversed)	Polarity Bias Superchrons	GEOLOGICAL EVENTS	OROGENIC PHASES Stille's 1924 except where bracketed.	AGE IN Ma
Q		1, 1r, 2, 2r, 2A, 2Ar		KTQ-M	Quaternary; N. Hemisphere Continental at sea level in Arctic	Walachian	
Td	5	3, 3r, 3A, 3B, 4, 4r, 4A, 4Ar			LA of discoasters; "Gibraltar Falls"; Desiccation of Mediterranean	Rhodanian; Attican	5
	10	5, 5r, 5A, 5AA, 5AC, B, D			FA of Hipparion in Paleo-Mediterranean area	Late Styrian	10
Tc	15	5B, 5Br, 5C, 5Cr, 5D, 5Dr, 5E, 5Er			Appearance of hominids	Early Styrian	15
	20	6, 6r, 6A, 6AA, 6B, 6Br, 6C		(Cretaceous-Tertiary-Quaternary mixed)			20
	25	6Cr, 7, 7A, 8, 8r			Red Sea opens	Savian	25
	30	9, 9r, 10, 10r, 11, 11r, 12					30
Tb	35	12r, 13, 13r			Termination of spreading in Labrador Sea and Baffin Bay (anomaly 13); FA of Proboscidea		35
	40	15, 15A, 16, 16r, 17, 17r, 18, 18r, 19, 19r				Pyrenean	40
	45	20, 20r, 21			FA of rodents		45
	50	21r, 22, 22r, 23, 23r, 24		KTQ-M	Separation of Australia and Antarctica (anomaly 22); FA of Equidae		50
Ta	55	24r, 25, 25r, 26			Norwegian Sea opens (between anomaly 24 and 25); FA of grasses; Rapid diversification of mammals; FA of discoasters		55
	60	26r, 27, 27r, 28, 28r, 29			FA of primates; FA of Eutheria; Labrador Sea opens	Laramide	60
Kb	65	30, 31, 31r, 32		KTQ-M	LA of dinosaurs, ammonites, belemnites, rudists, globotruncanids, and inoceramids		65
	70	32r, 33					
	80	33r			Tasman Sea opens (prior to anomaly 33)	(Sevier)	
	90	34		K-N	Oceanic anoxic event 3 (Cret); FA of Metatheria; Oceanic anoxic event 2 (Cret); Labrador Sea opens	Subhercynian; Austrian	
	100					(Oregonian)	100

Figure 5.4. Mesozoic time scale.

AGE IN Ma	EON	ERA	SUB-ERA	PERIOD	EPOCH		AGE		PICKS (Ma)
55	Phanerozoic (Ph)	Cenozoic (Cz)	Tertiary (TT)	Paleogene (Pg)	Eoc				54·9
60					Paleocene (Pal)	L	Thanetian	Tha	60·2
65						E	Danian	Dan	65·0
		Mesozoic (Mz)		Cretaceous (K)	Late	Senonian	Maastrichtian	Maa	73
							Campanian	Cmp	83
							Santonian	San	87·5
							Coniacian	Con	88·5
							Turonian	Tur	91
100							Cenomanian	Cen	97·5
					Early		Albian	Alb	113
							Aptian	Apt	119
							Barremian	Brm	125
						Neocomian	Hauterivian	Hau	131
							Valanginian	Vlg	138
							Berriasian	Ber	144
150				Jurassic (J)	Late	Malm	Tithonian	Tth	150
							Kimmeridgian	Kim	156
							Oxfordian	Oxf	163
					Middle	Dogger	Callovian	Clv	169
							Bathonian	Bth	175
							Bajocian	Baj	181
							Aalenian	Aal	188
					Early	Lias	Toarcian	Toa	194
200							Pliensbachian	Plb	200
							Sinemurian	Sin	206
							Hettangian	Het	213
				Triassic (Tr)	Late		Rhaetian	Rht	219
							Norian	Nor	225
							Carnian	Crn	231
					Middle		Ladinian	Lad	238
							Anisian	Ans	243
248					Scythian	Scy			248
250		Paleozoic (Pz)		Permian (P)	Late		Tatarian	Tat	253
							Kazanian	Kaz	
							Ufimian	Ufi	258
					Early		Kungurian	Kun	263
							Artinskian	Art	268
							Sakmarian	Sak	
							Asselian	Ass	286

ERROR IN AGES (Ma): 5 10 15

Figure 5.5. Mesozoic events.

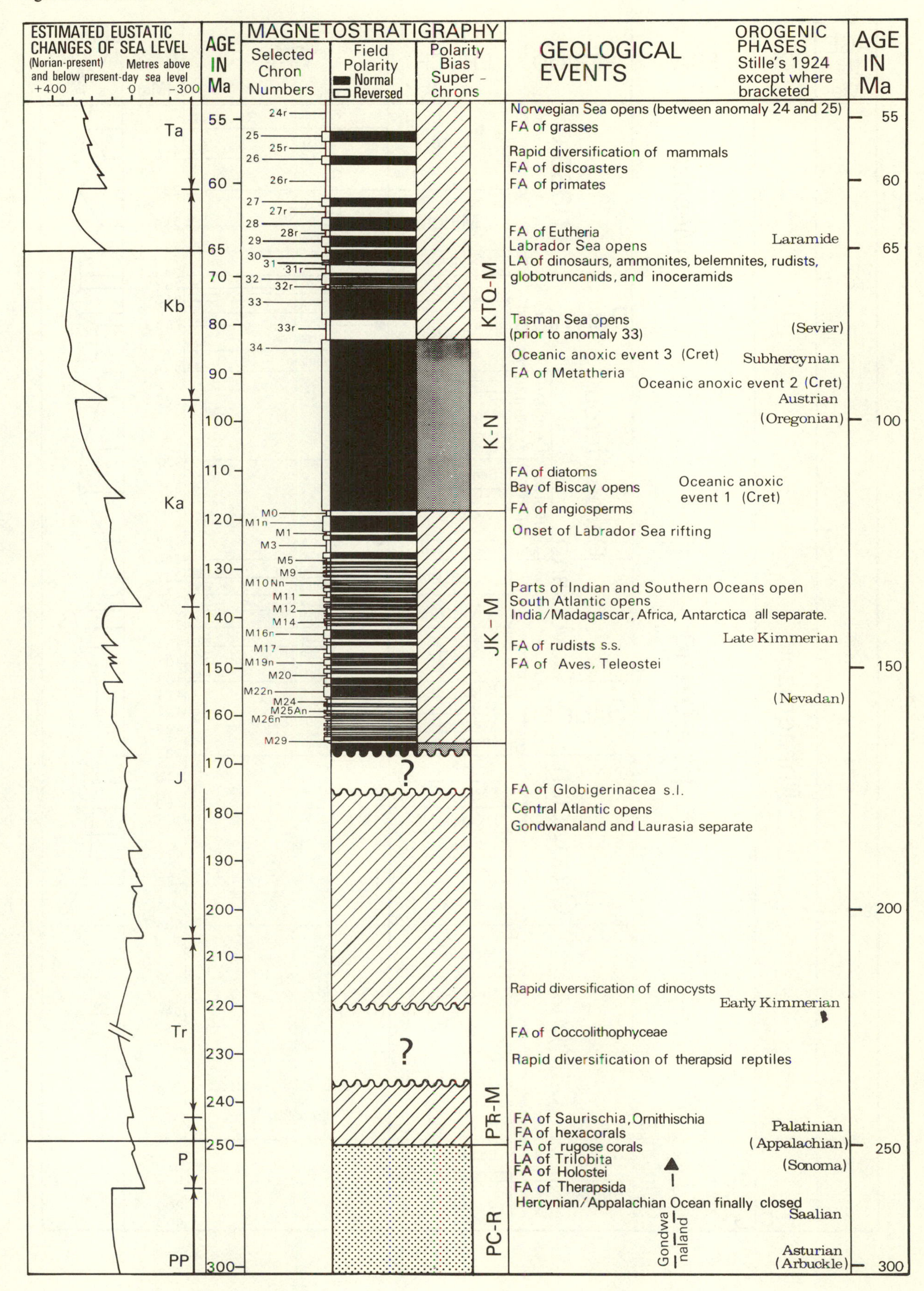

Figure 5.6. Paleozoic time scale.

AGE IN Ma	EON	ERA	PERIOD	EPOCH	AGE	PICKS (Ma)
	Phanerozoic (Ph)	Mesozoic (Mz)	Triassic (Tr)	Late	Rhaetian (Rht)	219
					Norian (Nor)	225
					Carnian (Crn)	231
				Middle	Ladinian (Lad) ●	238
					Anisian (Ans)	243
248				Scythian (Scy)		248
250		Paleozoic (Pz)	Permian (P)	Late	Tatarian (Tat)	253
					Kazanian (Kaz)	
					Ufimian (Ufi)	258
				Early	Kungurian (Kun)	263
					Artinskian (Art) ●	268
					Sakmarian (Sak)	
					Asselian (Ass) ●	286
			Carboniferous (C) – Pennsylvanian (Pen)	Gzelian (Gze)		
300				Kasimovian (Kas)		296
				Moscovian (Mos)		
				Bashkirian (Bsh)		315
			Carboniferous (C) – Mississippian (Mis)	Serpukhovian (Spk)		320
				Visean (Vis)		333
350				Tournaisian (Tou) ●		352
			Devonian (D)	Late	Famennian (Fam)	360
					Frasnian (Frs)	367
				Middle	Givetian (Giv)	374
					Eifelian (Eif)	380
				Early	Emsian (Ems)	387
400					Siegenian (Sig)	394
					Gedinnian (Ged)	401
			Silurian (S)	Pridoli (Prd)		408
				Ludlow (Lud)		414
				Wenlock (Wen) ●		421
				Llandovery (Lly)		428
			Ordovician (O)	Ashgill (Ash)		438
450				Caradoc (Crd)		448
				Llandeilo (Llo)		458
				Llanvirn (Llv)		468
				Arenig (Arg)	● Poor	478
500				Tremadoc (Tre)		488
			Cambrian (Є)	Merioneth (Mer)		505
				St David's (St D)	● Poor	525
550				Caerfai (Crf)		540
					— — —?— — — —	570
590					Tommotian (Tom)	590
600	P-Є	Pt3	Sinian (Z) – Vendian (V)	Ediacaran (Edi)		630
				Varangian (Var)		670
			Sturtian (U)			800
		?				900

ERROR IN AGES Ma: 5 10 15

Figure 5.7. Paleozoic events.

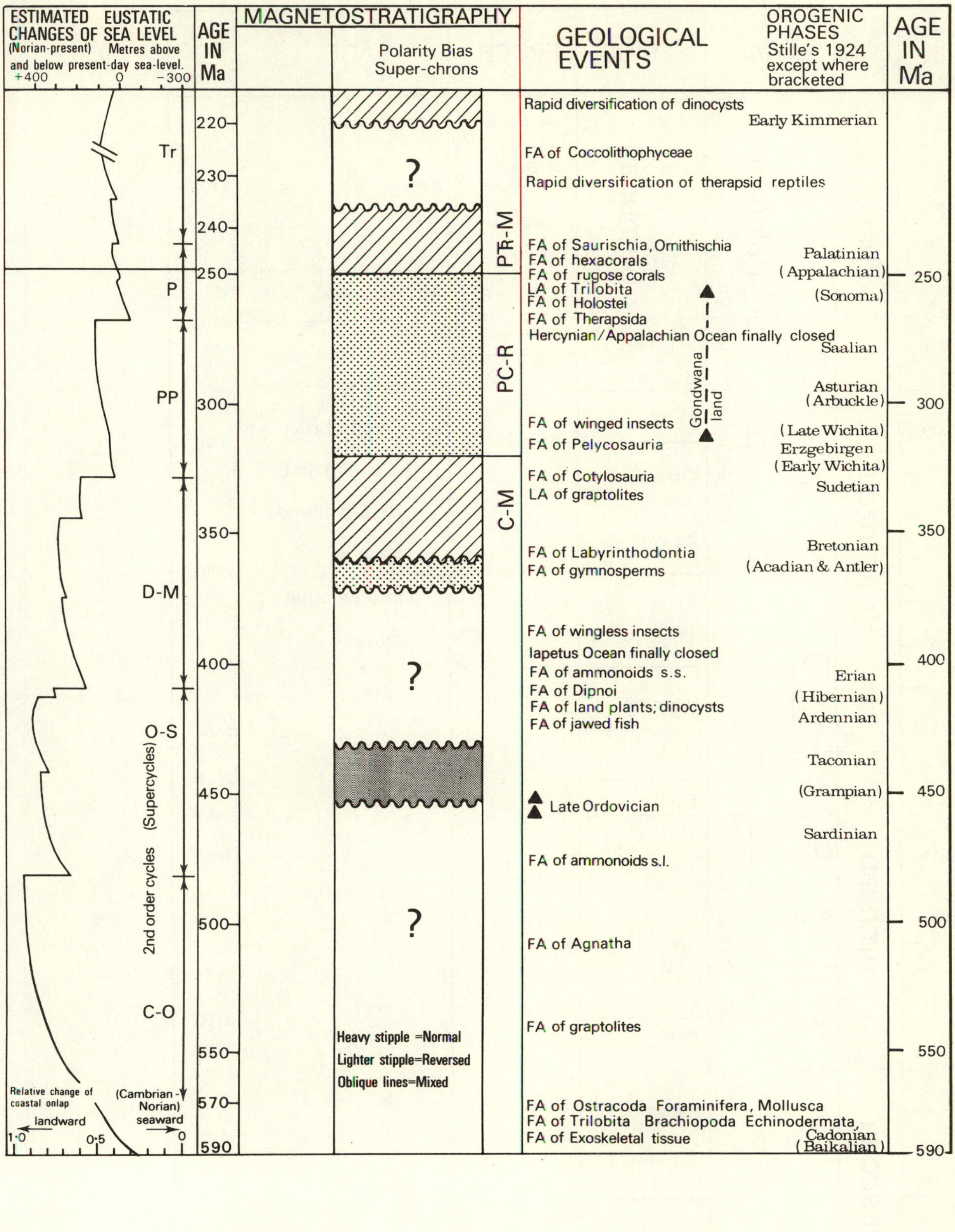

ESTIMATED EUSTATIC CHANGES OF SEA LEVEL (Norian-present) Metres above and below present-day sea-level. +400 0 −300

AGE IN Ma

MAGNETOSTRATIGRAPHY

Polarity Bias Super-chrons

GEOLOGICAL EVENTS

OROGENIC PHASES Stille's 1924 except where bracketed

AGE IN Ma

Tr

P

PP

D-M

O-S

C-O

2nd order cycles (Supercycles)

(Cambrian - Norian)

Relative change of coastal onlap

landward seaward

1·0 0·5 0

220 230 240 250 300 350 400 450 500 550 570 590

?

PTr-M

PC-R

C-M

?

?

Heavy stipple =Normal

Lighter stipple=Reversed

Oblique lines=Mixed

Rapid diversification of dinocysts

Early Kimmerian

FA of Coccolithophyceae

Rapid diversification of therapsid reptiles

FA of Saurischia, Ornithischia

FA of hexacorals

FA of rugose corals

LA of Trilobita

FA of Holostei

FA of Therapsida

Hercynian/Appalachian Ocean finally closed

Gondwana land

Palatinian

(Appalachian)

(Sonoma)

Saalian

Asturian

(Arbuckle)

FA of winged insects

FA of Pelycosauria

(Late Wichita)

Erzgebirgen

(Early Wichita)

FA of Cotylosauria

LA of graptolites

Sudetian

FA of Labyrinthodontia

FA of gymnosperms

Bretonian

(Acadian & Antler)

FA of wingless insects

Iapetus Ocean finally closed

FA of ammonoids s.s.

FA of Dipnoi

FA of land plants; dinocysts

FA of jawed fish

Erian

(Hibernian)

Ardennian

Taconian

(Grampian)

Late Ordovician

Sardinian

FA of ammonoids s.l.

FA of Agnatha

FA of graptolites

FA of Ostracoda Foraminifera, Mollusca

FA of Trilobita Brachiopoda Echinodermata,

FA of Exoskeletal tissue

Cadonian

(Baikalian)

250 300 350 400 450 500 550 590

Figure 5.8. Precambrian time scale.

AGE IN Ma	EON	ERA		PERIOD	EPOCH	AGE	PICKS (Ma)
	Phanerozoic Ph	Paleozoic Pz		Cambrian Є	Merioneth Mer		525
					St Davids St D	● Poor	540
550					Caerfai Crf	— — —?— — — —	570
						Tommotian Tom	590
590	Proterozoic	Pt3	Sinian Z	Vendian V	Ediacaran Edi	Precambrian Sub-Era and Period names have no international status.	630
600					Varangian Var		670
				Sturtian U			800
		—?—	Riphean R				900
1000		Pt2					1050
				Yurmatin Y			1350
				Burzyan B			1650
		—?—					
2000		Pt1					2100
				Huronian H			2400
	Pt	—?—					2500
	Archean	Ar3					2630
			Swazian Sw	Randian Ran			2800
3000		—?—					3000
		Ar2					
		—?—					3500
		Ar1		Isuan I			3750
	Ar						3900
4000	Priscoan			Hadean Hde			4000
5000	Pr						

Figure 5.9. Global time scale and events.

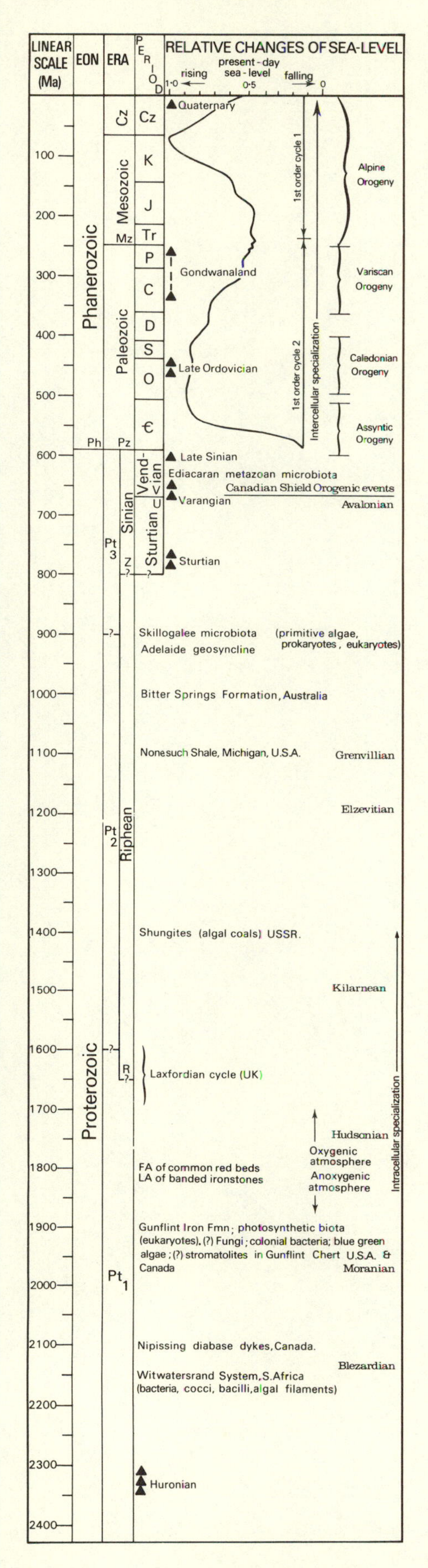

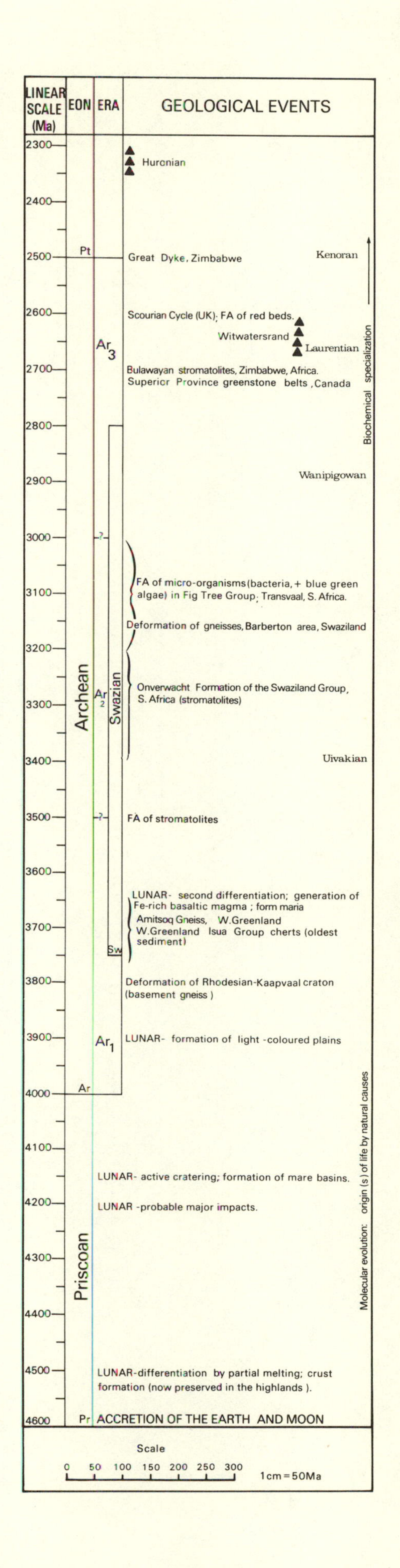

APPENDIX 1

Conversion of ages using new constants

Introduction

In general a radiometric age is obtained from the following equation:

$$t = (1/\lambda) \log_e[(D/P)+1]$$

where t = age; λ = decay constant; D = concentration of daughter atoms in the mineral due to radioactive decay; P = concentration of parent atoms in the mineral. In cases where some daughter atoms are already present, as in Rb–Sr dating, it is assumed that these have been subtracted from the concentration of the daughter atoms first to give the concentration due solely to radioactive decay.

Refinements in mass spectrometers and in the measurement of radioactive decay have led to improvements in the precision with which isotopic abundances and decay constants have been determined. For Rb–Sr and U–Pb ages the relationship between two ages calculated by different decay constants is the same as the ratio of the decay constants. For K–Ar ages, the relationship is more complex because ^{40}K decays in two separate ways, each with its own decay constant. We therefore provide separate tables for standardisation of K–Ar ages.

At the IUGS meeting in Sydney in 1976, the Subcommission on Geochronology recommended that the isotopic abundance ratios and decay constants should be standardised to permit comparisons of results from different laboratories (Steiger & Jäger 1977, 1978). Isotopic abundances are known with much greater precision than most decay constants. Consequently, the standards adopted may need future modification if experimental determinations require them to change.

Virtually all isotopic ages published before 1978 use non-standard constants. This appendix outlines how a non-standard age may be standardised. The dating methods used for the ages in the data file (Table 3.1) are coded according to the following conventions:

1 = Old Western K–Ar and Ar–Ar (4.72,0.584,1.19)

2 = Old USSR K–Ar and Ar–Ar (4.720, 0.557,1.19)

3 = New K–Ar and Ar–Ar (4.962,0.581,1.167)

4 = Old Rb–Sr (1.39)

5 = Old Rb–Sr (1.47)

6 = New Rb–Sr (1.42)

7 = Old U–Pb (0.154,0.971)

8 = New or corrected U–Pb (0.155 125 5,0.984 85)

9 = Old fission-track

10 = New fission-track

All fission-track ages used in this time scale use the standard constants.

Rb–Sr ages

The code numbers in Table 3.1 are 4 (old decay constant of 1.39×10^{-11} a^{-1}); 5 (old decay constant of 1.47×10^{-11} a^{-1}) and 6 (standard decay constant of 1.42×10^{-11} a^{-1}).

The corrections are simple. The standardised age is given by:

$$\text{standardised age} = \text{published age} \times (\text{old decay constant}/1.42 \times 10^{-11})$$

Example. Published age is 190 Ma using 1.39 as decay constant. Standardised age must be 190 x (1.39/1.42) = 186.0 Ma (186 Ma). Published age is 780 Ma using 1.47 decay constant. Standardised age must be 780 x (1.47/1.42) = 807.5 Ma (807 Ma).

Note. Only a few Rb–Sr ages use decay constants differing from these values.

U–Pb ages

U–Pb ages are assumed to be the mean of a ^{238}U–^{206}Pb and a ^{235}U–^{207}Pb age, both of which are given equal weight.

These are coded 7 (ages using old constants) and 8 (standard). The standardised age is given by:

$$\text{standardised age} = \text{published age} \times 0.99$$

K–Ar ages

The table below gives standardised ages for ages that use the constants most common in Western laboratories (code number 1) or Soviet laboratories (code number 2). To find a standardised Western age, look down the age tables until the published age is found; the second figure in the column gives the standardised age. To find a standardised Soviet age, look down the age tables until the published age is found; the third figure in the column gives the standardised age.

Example. Published Western age = 31 Ma; standardised Western age = 31.76 Ma, (31.8 Ma). Published Soviet age = 690 Ma; standardised Soviet age = 671.6 Ma (672 Ma).

Note. Ages published prior to about 1960 may use constants other than those used to standardise the age. Such ages need to be recalculated individually and are not discussed here. Some Western laboratories used a constant of 0.585 instead of 0.584 for the beta-decay of ^{40}K: the standardised ages are not significantly different from those provided by the table.

Old age	*New Western*	*New Soviet*	*Old age*	*New Western*	*New Soviet*	*Old age*	*New Western*	*New Soviet*	*Old age*	*New Western*	*New Soviet*
0.1	0.10	0.10	5.1	5.23	4.98				51	52.23	49.84
0.2	0.20	0.19	5.2	5.33	5.08				52	53.25	50.81
0.3	0.30	0.29	5.3	5.43	5.18				53	54.28	51.79
0.4	0.41	0.39	5.4	5.53	5.28				54	55.30	52.77
0.5	0.51	0.49	5.5	5.63	5.37				55	56.33	53.75
0.6	0.61	0.58	5.6	5.74	5.47				56	57.35	54.72
0.7	0.72	0.68	5.7	5.84	5.57				57	58.37	55.70
0.8	0.82	0.78	5.8	5.94	5.67				58	59.39	56.68
0.9	0.92	0.88	5.9	6.05	5.76				59	60.42	57.65
1.0	1.02	0.98	6.0	6.15	5.86				60	61.44	58.63
1.1	1.13	1.07	6.1	6.25	5.96	11	11.27	10.75	61	62.46	59.61
1.2	1.23	1.17	6.2	6.35	6.06	12	12.30	11.73	62	63.48	60.58
1.3	1.33	1.27	6.3	6.46	6.16	13	13.32	12.71	63	64.51	61.56
1.4	1.43	1.37	6.4	6.56	6.26	14	14.35	13.68	64	65.53	62.54
1.5	1.54	1.47	6.5	6.66	6.35	15	15.37	14.66	65	66.55	63.51
1.6	1.64	1.56	6.6	6.76	6.45	16	16.40	15.64	66	67.58	64.49
1.7	1.74	1.66	6.7	6.87	6.55	17	17.42	16.61	67	68.60	65.47
1.8	1.84	1.76	6.8	6.97	6.65	18	18.44	17.59	68	69.62	66.44
1.9	1.95	1.86	6.9	7.07	6.74	19	19.47	18.57	69	70.64	67.42
2.0	2.05	1.95	7.0	7.17	6.84	20	20.49	19.55	70	71.67	68.39
2.1	2.15	2.05	7.1	7.28	6.94	21	21.52	20.52	71	72.69	69.37
2.2	2.25	2.15	7.2	7.38	7.04	22	22.54	21.50	72	73.71	70.35
2.3	2.36	2.25	7.3	7.48	7.13	23	23.57	22.48	73	74.73	71.33
2.4	2.46	2.34	7.4	7.58	7.23	24	24.59	23.46	74	75.76	72.30
2.5	2.56	2.44	7.5	7.69	7.33	25	25.62	24.43	75	76.78	73.28
2.6	2.66	2.54	7.6	7.79	7.43	26	26.64	25.41	76	77.80	74.26
2.7	2.76	2.64	7.7	7.89	7.53	27	27.66	26.39	77	78.82	75.23
2.8	2.87	2.74	7.8	7.99	7.62	28	28.69	27.37	78	79.85	76.21
2.9	2.97	2.83	7.9	8.10	7.72	29	29.71	28.34	79	80.87	77.19
3.0	3.07	2.93	8.0	8.20	7.82	30	30.73	29.32	80	81.89	78.16
3.1	3.17	3.03	8.1	8.30	7.92	31	31.76	30.30	81	82.91	79.14
3.2	3.28	3.13	8.2	8.40	8.02	32	32.78	31.27	82	83.93	80.11
3.3	3.38	3.22	8.3	8.51	8.11	33	33.81	32.25	83	84.95	81.09
3.4	3.48	3.32	8.4	8.61	8.21	34	34.83	33.23	84	85.98	82.07
3.5	3.59	3.42	8.5	8.71	8.31	35	35.86	34.21	85	87.00	83.04
3.6	3.69	3.52	8.6	8.81	8.41	36	36.88	35.18	86	88.02	84.02
3.7	3.79	3.62	8.7	8.92	8.50	37	37.90	36.16	87	89.04	85.00
3.8	3.89	3.71	8.8	9.02	8.60	38	38.93	37.14	88	90.07	85.97
3.9	4.00	3.81	8.9	9.12	8.70	39	39.95	38.11	89	91.09	86.95
4.0	4.10	3.91	9.0	9.22	8.80	40	40.97	39.09	90	92.11	87.93
4.1	4.20	4.01	9.1	9.33	8.89	41	42.00	40.07	91	93.13	88.90
4.2	4.30	4.10	9.2	9.43	8.99	42	43.02	41.05	92	94.15	89.88
4.3	4.41	4.20	9.3	9.53	9.09	43	44.04	42.02	93	95.17	90.85
4.4	4.51	4.30	9.4	9.63	9.19	44	45.07	43.00	94	96.20	91.83
4.5	4.61	4.40	9.5	9.73	9.28	45	46.09	43.98	95	97.22	92.81
4.6	4.71	4.50	9.6	9.84	9.38	46	47.12	44.95	96	98.24	93.78
4.7	4.82	4.59	9.7	9.94	9.48	47	48.14	45.93	97	99.26	94.76
4.8	4.92	4.69	9.8	10.04	9.58	48	49.16	46.91	98	100.28	95.74
4.9	5.02	4.79	9.9	10.14	9.68	49	50.18	47.88	99	101.30	96.71
5.0	5.12	4.89	10.0	10.25	9.77	50	51.21	48.86	100	102.33	97.69

Old age	*New Western*	*New Soviet*	*Old age*	*New Western*	*New Soviet*	*Old age*	*New Western*	*New Soviet*	*Old age*	*New Western*	*New Soviet*
			141	144.2	137.7	191	195.1	186.5	241	246.0	235.2
			142	145.2	138.7	192	196.1	187.4	242	247.0	236.2
			143	146.2	139.7	193	197.2	188.4	243	248.0	237.2
			144	147.2	140.6	194	198.2	189.4	244	249.0	238.1
			145	148.3	141.6	195	199.2	190.4	245	250.1	239.1
			146	149.3	142.6	196	200.2	191.3	246	251.1	240.1
			147	150.3	143.6	197	201.2	192.3	247	252.1	241.1
			148	151.3	144.5	198	202.3	193.3	248	253.1	242.0
			149	152.3	145.5	199	203.3	194.3	249	254.1	243.0
			150	153.4	146.5	200	204.3	195.2	250	255.1	244.0
101	103.3	98.7	151	154.4	147.5	201	205.3	196.2	251	256.2	245.0
102	104.4	99.6	152	155.4	148.4	202	206.3	197.2	252	257.2	245.9
103	105.4	100.6	153	156.4	149.4	203	207.3	198.2	253	258.2	246.9
104	106.4	101.6	154	157.4	150.4	204	208.4	199.1	254	259.2	247.9
105	107.4	102.6	155	158.4	151.4	205	209.4	200.1	255	260.2	248.9
106	108.5	103.5	156	159.5	152.3	206	210.4	201.1	256	261.2	249.8
107	109.5	104.5	157	160.5	153.3	207	211.4	202.1	257	262.3	250.8
108	110.5	105.5	158	161.5	154.3	208	212.4	203.0	258	263.3	251.8
109	111.5	106.5	159	162.5	155.3	209	213.4	204.0	259	264.3	252.7
110	112.5	107.4	160	163.5	156.2	210	214.5	205.0	260	265.3	253.7
111	113.6	108.4	161	164.6	157.2	211	215.5	206.0	261	266.3	254.7
112	114.6	109.4	162	165.6	158.2	212	216.5	206.9	262	267.3	255.7
113	115.6	110.4	163	166.6	159.2	213	217.5	207.9	263	268.3	256.6
114	116.6	111.4	164	167.6	160.1	214	218.5	208.9	264	269.4	257.6
115	117.6	112.3	165	168.6	161.1	215	219.6	209.9	265	270.4	258.6
116	118.7	113.3	166	169.7	162.1	216	220.6	210.8	266	271.4	259.6
117	119.7	114.3	167	170.7	163.1	217	221.6	211.8	267	272.4	260.5
118	120.7	115.3	168	171.7	164.0	218	222.6	212.8	268	273.4	261.5
119	121.7	116.2	169	172.7	165.0	219	223.6	213.8	269	274.4	262.5
120	122.7	117.2	170	173.7	166.0	220	224.6	214.7	270	275.5	263.5
121	123.8	118.2	171	174.8	167.0	221	225.7	215.7	271	276.5	264.4
122	124.8	119.2	172	175.8	167.9	222	226.7	216.7	272	277.5	265.4
123	125.8	120.1	173	176.8	168.9	223	227.7	217.7	273	278.5	266.4
124	126.8	121.1	174	177.8	169.9	224	228.7	218.6	274	279.5	267.4
125	127.8	122.1	175	178.8	170.9	225	229.7	219.6	275	280.5	268.3
126	128.9	123.1	176	179.8	171.8	226	230.7	220.6	276	281.5	269.3
127	129.9	124.0	177	180.9	172.8	227	231.8	221.6	277	282.6	270.3
128	130.9	125.0	178	181.9	173.8	228	232.8	222.5	278	283.6	271.3
129	131.9	126.0	179	182.9	174.8	229	233.8	223.5	279	284.6	272.2
130	133.0	127.0	180	183.9	175.7	230	234.8	224.5	280	285.6	273.2
131	134.0	127.9	181	184.9	176.7	231	235.8	225.5	281	286.6	274.2
132	135.0	128.9	182	186.0	177.7	232	236.8	226.4	282	287.6	275.2
133	136.0	129.9	183	187.0	178.7	233	237.9	227.4	283	288.7	276.1
134	137.0	130.9	184	188.0	179.6	234	238.9	228.4	284	289.7	277.1
135	138.1	131.8	185	189.0	180.6	235	239.9	229.4	285	290.7	278.1
136	139.1	132.8	186	190.0	181.6	236	240.9	230.3	286	291.7	279.0
137	140.1	133.8	187	191.1	182.6	237	241.9	231.3	287	292.7	280.0
138	141.1	134.8	188	192.1	183.5	238	242.9	232.3	288	293.7	281.0
139	142.1	135.8	189	193.1	184.5	239	244.0	233.3	289	294.7	282.0
140	143.2	136.7	190	194.1	185.5	240	245.0	234.2	290	295.8	282.9

Old age	*New Western*	*New Soviet*	*Old age*	*New Western*	*New Soviet*	*Old age*	*New Western*	*New Soviet*	*Old age*	*New Western*	*New Soviet*
291	296.8	283.9	341	347.5	332.6	391	398.1	381.2	441	448.6	429.9
292	297.8	284.9	342	348.5	333.6	392	399.1	382.2	442	449.7	430.8
293	298.8	285.9	343	349.5	334.5	393	400.1	383.2	443	450.7	431.8
294	299.8	286.8	344	350.5	335.5	394	401.1	384.2	444	451.7	432.8
295	300.8	287.8	345	351.5	336.5	395	402.1	385.1	445	452.7	433.8
296	301.8	288.8	346	352.5	337.5	396	403.2	386.1	446	453.7	434.7
297	302.9	289.8	347	353.6	338.4	397	404.2	387.1	447	454.7	435.7
298	303.9	290.7	348	354.6	339.4	398	405.2	388.1	448	455.7	436.7
299	304.9	291.7	349	355.6	340.4	399	406.2	389.0	449	456.7	437.6
300	305.9	292.7	350	356.6	341.4	400	407.2	390.0	450	457.7	438.6
301	306.9	293.7	351	357.6	342.3	401	408.2	391.0	451	458.7	439.6
302	307.9	294.6	352	358.6	343.3	402	409.2	391.9	452	459.8	440.6
303	309.0	295.6	353	359.6	344.3	403	410.2	392.9	453	460.8	441.5
304	310.0	296.6	354	360.6	345.2	404	411.2	393.9	454	461.8	442.5
305	311.0	297.6	355	361.7	346.2	405	412.3	394.9	455	462.8	443.5
306	312.0	298.5	356	362.7	347.2	406	413.3	395.8	456	463.8	444.5
307	313.0	299.5	357	363.7	348.2	407	414.3	396.8	457	464.8	445.4
308	314.0	300.5	358	364.7	349.1	408	415.3	397.8	458	465.8	446.4
309	315.0	301.4	359	365.7	350.1	409	416.3	398.8	459	466.8	447.4
310	316.1	302.4	360	366.7	351.1	410	417.3	399.7	460	467.8	448.3
311	317.1	303.4	361	367.7	352.1	411	418.3	400.7	461	468.8	449.3
312	318.1	304.4	362	368.7	353.0	412	419.3	401.7	462	469.9	450.3
313	319.1	305.3	363	369.8	354.0	413	420.4	402.6	463	470.9	451.3
314	320.1	306.3	364	370.8	355.0	414	421.4	403.6	464	471.9	452.2
315	321.1	307.3	365	371.8	356.0	415	422.4	404.6	465	472.9	453.2
316	322.1	308.3	366	372.8	356.9	416	423.4	405.6	466	473.9	454.2
317	323.1	309.2	367	373.8	357.9	417	424.4	406.5	467	474.9	455.1
318	324.2	310.2	368	374.8	358.9	418	425.4	407.5	468	475.9	456.1
319	325.2	311.2	369	375.8	359.8	419	426.4	408.5	469	476.9	457.1
320	326.2	312.2	370	376.8	360.8	420	427.4	409.5	470	477.9	458.1
321	327.2	313.1	371	377.9	361.8	421	428.4	410.4	471	478.9	459.0
322	328.2	314.1	372	378.9	362.8	422	429.4	411.4	472	479.9	460.0
323	329.2	315.1	373	379.9	363.7	423	430.5	412.4	473	481.0	461.0
324	330.2	316.1	374	380.9	364.7	424	431.5	413.3	474	482.0	461.9
325	331.3	317.0	375	381.9	365.7	425	432.5	414.3	475	483.0	462.9
326	332.3	318.0	376	382.9	366.7	426	433.5	415.3	476	484.0	463.9
327	333.3	319.0	377	383.9	367.6	427	434.5	416.3	477	485.0	464.9
328	334.3	319.9	378	384.9	368.6	428	435.5	417.2	478	486.0	465.8
329	335.3	320.9	379	386.0	369.6	429	436.5	418.2	479	487.0	466.8
330	336.3	321.9	380	387.0	370.5	430	437.5	419.2	480	488.0	467.8
331	337.3	322.9	381	388.0	371.5	431	438.5	420.1	481	489.0	468.7
332	338.4	323.8	382	389.0	372.5	432	439.6	421.1	482	490.0	469.7
333	339.4	324.8	383	390.0	373.5	433	440.6	422.1	483	491.0	470.7
334	340.4	325.8	384	391.0	374.4	434	441.6	423.1	484	492.1	471.7
335	341.4	326.8	385	392.0	375.4	435	442.6	424.0	485	493.1	472.6
336	342.4	327.7	386	393.0	376.4	436	443.6	425.0	486	494.1	473.6
337	343.4	328.7	387	394.1	377.4	437	444.6	426.0	487	495.1	474.6
338	344.4	329.7	388	395.1	378.3	438	445.6	427.0	488	496.1	475.6
339	345.4	330.7	389	396.1	379.3	439	446.6	427.9	489	497.1	476.5
340	346.5	331.6	390	397.1	380.3	440	447.6	428.9	490	498.1	477.5

Old age	*New Western*	*New Soviet*	*Old age*	*New Western*	*New Soviet*	*Old age*	*New Western*	*New Soviet*	*Old age*	*New Western*	*New Soviet*
491	499.1	478.5	541	549.5	527.0	591	599.8	575.6	641	650.1	624.1
492	500.1	479.4	542	550.5	528.0	592	600.9	576.5	642	651.1	625.1
493	501.1	480.4	543	551.5	529.0	593	601.9	577.5	643	652.1	626.0
494	502.1	481.4	544	552.5	529.9	594	602.9	578.5	644	653.1	627.0
495	503.2	482.4	545	553.5	530.9	595	603.9	579.5	645	654.1	628.0
496	504.2	483.3	546	554.6	531.9	596	604.9	580.4	646	655.1	628.9
497	505.2	484.3	547	555.6	532.9	597	605.9	581.4	647	656.1	629.9
498	506.2	485.3	548	556.6	533.8	598	606.9	582.4	648	657.1	630.9
499	507.2	486.2	549	557.6	534.8	599	607.9	583.3	649	658.1	631.9
500	508.2	487.2	550	558.6	535.8	600	608.9	584.3	650	659.1	632.8
501	509.2	488.2	551	559.6	536.7	601	609.9	585.3	651	660.2	633.8
502	510.2	489.2	552	560.6	537.7	602	610.9	586.3	652	661.2	634.8
503	511.2	490.1	553	561.6	538.7	603	611.9	587.2	653	662.2	635.7
504	512.2	491.1	554	562.6	539.7	604	612.9	588.2	654	663.2	636.7
505	513.2	492.1	555	563.6	540.6	605	613.9	589.2	655	664.2	637.7
506	514.2	493.0	556	564.6	541.6	606	614.9	590.1	656	665.2	638.6
507	515.3	494.0	557	565.6	542.6	607	615.9	591.1	657	666.2	639.6
508	516.3	495.0	558	566.6	543.5	608	616.9	592.1	658	667.2	640.6
509	517.3	496.0	559	567.6	544.5	609	617.9	593.0	659	668.2	641.6
510	518.3	496.9	560	568.7	545.5	610	619.0	594.0	660	669.2	642.5
511	519.3	497.9	561	569.7	546.5	611	620.0	595.0	661	670.2	643.5
512	520.3	498.9	562	570.7	547.4	612	621.0	596.0	662	671.2	644.5
513	521.3	499.8	563	571.7	548.4	613	622.0	596.9	663	672.2	645.4
514	522.3	500.8	564	572.7	549.4	614	623.0	597.9	664	673.2	646.4
515	523.3	501.8	565	573.7	550.3	615	624.0	598.9	665	674.2	647.4
516	524.3	502.8	566	574.7	551.3	616	625.0	599.8	666	675.2	648.3
517	525.3	503.7	567	575.7	552.3	617	626.0	600.8	667	676.2	649.3
518	526.3	504.7	568	576.7	553.2	618	627.0	601.8	668	677.2	650.3
519	527.4	505.7	569	577.7	554.2	619	628.0	602.7	669	678.2	651.3
520	528.4	506.6	570	578.7	555.2	620	629.0	603.7	670	679.2	652.2
521	529.4	507.6	571	579.7	556.2	621	630.0	604.7	671	680.2	653.2
522	530.4	508.6	572	580.7	557.1	622	631.0	605.7	672	681.2	654.2
523	531.4	509.6	573	581.7	558.1	623	632.0	606.6	673	682.2	655.1
524	532.4	510.5	574	582.7	559.1	624	633.0	607.6	674	683.2	656.1
525	533.4	511.5	575	583.7	560.0	625	634.0	608.6	675	684.2	657.1
526	534.4	512.5	576	584.8	561.0	626	635.0	609.5	676	685.3	658.0
527	535.4	513.4	577	585.8	562.0	627	636.0	610.5	677	686.3	659.0
528	536.4	514.4	578	586.8	563.0	628	637.0	611.5	678	687.3	660.0
529	537.4	515.4	579	587.8	563.9	629	638.1	612.5	679	688.3	660.9
530	538.4	516.4	580	588.8	564.9	630	639.1	613.4	680	689.3	661.9
531	539.4	517.3	581	589.8	565.9	631	640.1	614.4	681	690.3	662.9
532	540.4	518.3	582	590.8	566.8	632	641.1	615.4	682	691.3	663.9
533	541.5	519.3	583	591.8	567.8	633	642.1	616.3	683	692.3	664.8
534	542.5	520.2	584	592.8	568.8	634	643.1	617.3	684	693.3	665.8
535	543.5	521.2	585	593.8	569.8	635	644.1	618.3	685	694.3	666.8
536	544.5	522.2	586	594.8	570.7	636	645.1	619.2	686	695.3	667.7
537	545.5	523.1	587	595.8	571.7	637	646.1	620.2	687	696.3	668.7
538	546.5	524.1	588	596.8	572.7	638	647.1	621.2	688	697.3	669.7
539	547.5	525.1	589	597.8	573.6	639	648.1	622.2	689	698.3	670.6
540	548.5	526.1	590	598.8	574.6	640	649.1	623.1	690	699.3	671.6

Old age	*New Western*	*New Soviet*	*Old age*	*New Western*	*New Soviet*	*Old age*	*New Western*	*New Soviet*	*Old age*	*New Western*	*New Soviet*
691	700.3	672.6	741	750.4	721.1	791	800.5	769.5	841	850.5	817.9
692	701.3	673.6	742	751.4	722.0	792	801.5	770.5	842	851.5	818.9
693	702.3	674.5	743	752.4	723.0	793	802.5	771.4	843	852.5	819.9
694	703.3	675.5	744	753.4	724.0	794	803.5	772.4	844	853.5	820.8
695	704.3	676.5	745	754.4	724.9	795	804.5	773.4	845	854.5	821.8
696	705.3	677.4	746	755.4	725.9	796	805.5	774.3	846	855.5	822.8
697	706.3	678.4	747	756.4	726.9	797	806.5	775.3	847	856.5	823.7
698	707.3	679.4	748	757.4	727.8	798	807.5	776.3	848	857.5	824.7
699	708.3	680.3	749	758.4	728.8	799	808.5	777.3	849	858.5	825.7
700	709.3	681.3	750	759.5	729.8	800	809.5	778.2	850	859.5	826.6
701	710.3	682.3	751	760.5	730.7	801	810.5	779.2	851	860.5	827.6
702	711.3	683.3	752	761.5	731.7	802	811.5	780.2	852	861.5	828.6
703	712.3	684.2	753	762.5	732.7	803	812.5	781.1	853	862.5	829.5
704	713.3	685.2	754	763.5	733.7	804	813.5	782.1	854	863.5	830.5
705	714.3	686.2	755	764.5	734.6	805	814.5	783.1	855	864.5	831.5
706	715.3	687.1	756	765.5	735.6	806	815.5	784.0	856	865.5	832.4
707	716.4	688.1	757	766.5	736.6	807	816.5	785.0	857	866.5	833.4
708	717.4	689.1	758	767.5	737.5	808	817.5	786.0	858	867.5	834.4
709	718.4	690.0	759	768.5	738.5	809	818.5	786.9	859	868.5	835.4
710	719.4	691.0	760	769.5	739.5	810	819.5	787.9	860	869.5	836.3
711	720.4	692.0	761	770.5	740.4	811	820.5	788.9	861	870.5	837.3
712	721.4	692.9	762	771.5	741.4	812	821.5	789.8	862	871.5	838.3
713	722.4	693.9	763	772.5	742.4	813	822.5	790.8	863	872.5	839.2
714	723.4	694.9	764	773.5	743.3	814	823.5	791.8	864	873.5	840.2
715	724.4	695.9	765	774.5	744.3	815	824.5	792.7	865	874.5	841.2
716	725.4	696.8	766	775.5	745.3	816	825.5	793.7	866	875.5	842.1
717	726.4	697.8	767	776.5	746.3	817	826.5	794.7	867	876.5	843.1
718	727.4	698.8	768	777.5	747.2	818	827.5	795.7	868	877.5	844.1
719	728.4	699.7	769	778.5	748.2	819	828.5	796.6	869	878.5	845.0
720	729.4	700.7	770	779.5	749.2	820	829.5	797.6	870	879.5	846.0
721	730.4	701.7	771	780.5	750.1	821	830.5	798.6	871	880.5	847.0
722	731.4	702.6	772	781.5	751.1	822	831.5	799.5	872	881.5	847.9
723	732.4	703.6	773	782.5	752.1	823	832.5	800.5	873	882.5	848.9
724	733.4	704.6	774	783.5	753.0	824	833.5	801.5	874	883.5	849.9
725	734.4	705.5	775	784.5	754.0	825	834.5	802.4	875	884.5	850.8
726	735.4	706.5	776	785.5	755.0	826	835.5	803.4	876	885.5	851.8
727	736.4	707.5	777	786.5	755.9	827	836.5	804.4	877	886.5	852.8
728	737.4	708.5	778	787.5	756.9	828	837.5	805.3	878	887.5	853.7
729	738.4	709.4	779	788.5	757.9	829	838.5	806.3	879	888.5	854.7
730	739.4	710.4	780	789.5	758.8	830	839.5	807.3	880	889.5	855.7
731	740.4	711.4	781	790.5	759.8	831	840.5	808.2	881	890.5	856.6
732	741.4	712.3	782	791.5	760.8	832	841.5	809.2	882	891.5	857.6
733	742.4	713.3	783	792.5	761.8	833	842.5	810.2	883	892.5	858.6
734	743.4	714.3	784	793.5	762.7	834	843.5	811.1	884	893.5	859.6
735	744.4	715.2	785	794.5	763.7	835	844.5	812.1	885	894.5	860.5
736	745.4	716.2	786	795.5	764.7	836	845.5	813.1	886	895.5	861.5
737	746.4	717.2	787	796.5	765.6	837	846.5	814.1	887	896.5	862.5
738	747.4	718.1	788	797.5	766.6	838	847.5	815.0	888	897.5	863.4
739	748.4	719.1	789	798.5	767.6	839	848.5	816.0	889	898.5	864.4
740	749.4	720.1	790	799.5	768.5	840	849.5	817.0	890	899.5	865.4

Old age	*New Western*	*New Soviet*	*Old age*	*New Western*	*New Soviet*	*Old age*	*New Western*	*New Soviet*	*Old age*	*New Western*	*New Soviet*
891	900.5	866.3	941	950.4	914.7	991	1000.2	963.1	1155	1163	1121
892	901.5	867.3	942	951.3	915.7	992	1001.2	964.0	1160	1168	1126
893	902.5	868.3	943	952.3	916.6	993	1002.2	965.0	1165	1173	1131
894	903.5	869.2	944	953.3	917.6	994	1003.2	966.0	1170	1178	1136
895	904.5	870.2	945	954.3	918.6	995	1004.2	966.9	1175	1183	1140
896	905.5	871.2	946	955.3	919.5	996	1005.2	967.9	1180	1188	1145
897	906.4	872.1	947	956.3	920.5	997	1006.2	968.9	1185	1193	1150
898	907.4	873.1	948	957.3	921.5	998	1007.2	969.8	1190	1198	1155
899	908.4	874.1	949	958.3	922.4	999	1008.2	970.8	1195	1202	1160
900	909.4	875.0	950	959.3	923.4	1000	1009.2	971.8	1200	1207	1165
901	910.4	876.0	951	960.3	924.4				1205	1212	1169
902	911.4	877.0	952	961.3	925.4				1210	1217	1174
903	912.4	877.9	953	962.3	926.3				1215	1222	1179
904	913.4	878.9	954	963.3	927.3				1220	1227	1184
905	914.4	879.9	955	964.3	928.3				1225	1232	1189
906	915.4	880.8	956	965.3	929.2				1230	1237	1193
907	916.4	881.8	957	966.3	930.2				1235	1242	1198
908	917.4	882.8	958	967.3	931.2				1240	1247	1203
909	918.4	883.7	959	968.3	932.1				1245	1252	1208
910	919.4	884.7	960	969.3	933.1				1250	1257	1213
911	920.4	885.7	961	970.3	934.1	1005	1014	976	1255	1262	1218
912	921.4	886.7	962	971.3	935.0	1010	1019	981	1260	1267	1222
913	922.4	887.6	963	972.3	936.0	1015	1024	986	1265	1272	1227
914	923.4	888.6	964	973.3	937.0	1020	1029	991	1270	1277	1232
915	924.4	889.6	965	974.3	937.9	1025	1034	995	1275	1282	1237
916	925.4	890.5	966	975.3	938.9	1030	1039	1000	1280	1287	1242
917	926.4	891.5	967	976.3	939.9	1035	1043	1005	1285	1292	1247
918	927.4	892.5	968	977.3	940.8	1040	1048	1010	1290	1297	1251
919	928.4	893.4	969	978.3	941.8	1045	1053	1015	1295	1302	1256
920	929.4	894.4	970	979.3	942.8	1050	1058	1020	1300	1307	1261
921	930.4	895.4	971	980.3	943.7	1055	1063	1024	1305	1311	1266
922	931.4	896.3	972	981.3	944.7	1060	1068	1029	1310	1316	1271
923	932.4	897.3	973	982.3	945.7	1065	1073	1034	1315	1321	1275
924	933.4	898.3	974	983.2	946.6	1070	1078	1039	1320	1326	1280
925	934.4	899.2	975	984.2	947.6	1075	1083	1044	1325	1331	1285
926	935.4	900.2	976	985.2	948.6	1080	1088	1049	1330	1336	1290
927	936.4	901.2	977	986.2	949.5	1085	1093	1053	1335	1341	1295
928	937.4	902.1	978	987.2	950.5	1090	1098	1058	1340	1346	1300
929	938.4	903.1	979	988.2	951.5	1095	1103	1063	1345	1351	1304
930	939.4	904.1	980	989.2	952.4	1100	1108	1068	1350	1356	1309
931	940.4	905.0	981	990.2	953.4	1105	1113	1073	1355	1361	1314
932	941.4	906.0	982	991.2	954.4	1110	1118	1078	1360	1366	1319
933	942.4	907.0	983	992.2	955.3	1115	1123	1082	1365	1371	1324
934	943.4	907.9	984	993.2	956.3	1120	1128	1087	1370	1376	1329
935	944.4	908.9	985	994.2	957.3	1125	1133	1092	1375	1381	1333
936	945.4	909.9	986	995.2	958.2	1130	1138	1097	1380	1386	1338
937	946.4	910.8	987	996.2	959.2	1135	1143	1102	1385	1391	1343
938	947.4	911.8	988	997.2	960.2	1140	1148	1107	1390	1396	1348
939	948.4	912.8	989	998.2	961.1	1145	1153	1111	1395	1401	1353
940	949.4	913.7	990	999.2	962.1	1150	1158	1116	1400	1405	1357

Old age	*New Western*	*New Soviet*	*Old age*	*New Western*	*New Soviet*	*Old age*	*New Western*	*New Soviet*	*Old age*	*New Western*	*New Soviet*
1405	1410	1362	1655	1657	1603	1905	1903	1844	2155	2147	2084
1410	1415	1367	1660	1662	1608	1910	1907	1848	2160	2152	2089
1415	1420	1372	1665	1667	1613	1915	1912	1853	2165	2157	2093
1420	1425	1377	1670	1672	1618	1920	1917	1858	2170	2162	2098
1425	1430	1382	1675	1677	1622	1925	1922	1863	2175	2167	2103
1430	1435	1386	1680	1682	1627	1930	1927	1868	2180	2172	2108
1435	1440	1391	1685	1686	1632	1935	1932	1872	2185	2177	2113
1440	1445	1396	1690	1691	1637	1940	1937	1877	2190	2182	2117
1445	1450	1401	1695	1696	1642	1945	1942	1882	2195	2186	2122
1450	1455	1406	1700	1701	1646	1950	1947	1887	2200	2191	2127
1455	1460	1410	1705	1706	1651	1955	1952	1892	2205	2196	2132
1460	1465	1415	1710	1711	1656	1960	1956	1896	2210	2201	2137
1465	1470	1420	1715	1716	1661	1965	1961	1901	2215	2206	2141
1470	1475	1425	1720	1721	1666	1970	1966	1906	2220	2211	2146
1475	1480	1430	1725	1726	1670	1975	1971	1911	2225	2216	2151
1480	1484	1435	1730	1731	1675	1980	1976	1916	2230	2221	2156
1485	1489	1439	1735	1736	1680	1985	1981	1920	2235	2226	2160
1490	1494	1444	1740	1741	1685	1990	1986	1925	2240	2230	2165
1495	1499	1449	1745	1745	1690	1995	1991	1930	2245	2235	2170
1500	1504	1454	1750	1750	1695	2000	1996	1935	2250	2240	2175
1505	1509	1459	1755	1755	1699	2005	2001	1940	2255	2245	2180
1510	1514	1463	1760	1760	1704	2010	2005	1944	2260	2250	2184
1515	1519	1468	1765	1765	1709	2015	2010	1949	2265	2255	2189
1520	1524	1473	1770	1770	1714	2020	2015	1954	2270	2260	2194
1525	1529	1478	1775	1775	1719	2025	2020	1959	2275	2265	2199
1530	1534	1483	1780	1780	1723	2030	2025	1964	2280	2270	2204
1535	1539	1488	1785	1785	1728	2035	2030	1968	2285	2274	2208
1540	1544	1492	1790	1790	1733	2040	2035	1973	2290	2279	2213
1545	1549	1497	1795	1795	1738	2045	2040	1978	2295	2284	2218
1550	1554	1502	1800	1800	1743	2050	2045	1983	2300	2289	2223
1555	1558	1507	1805	1804	1747	2055	2050	1988	2305	2294	2228
1560	1563	1512	1810	1809	1752	2060	2054	1992	2310	2299	2232
1565	1568	1516	1815	1814	1757	2065	2059	1997	2315	2304	2237
1570	1573	1521	1820	1819	1762	2070	2064	2002	2320	2309	2242
1575	1578	1526	1825	1824	1767	2075	2069	2007	2325	2313	2247
1580	1583	1531	1830	1829	1771	2080	2074	2012	2330	2318	2252
1585	1588	1536	1835	1834	1776	2085	2079	2016	2335	2323	2256
1590	1593	1541	1840	1839	1781	2090	2084	2021	2340	2328	2261
1595	1598	1545	1845	1844	1786	2095	2089	2026	2345	2333	2266
1600	1603	1550	1850	1849	1791	2100	2094	2031	2350	2338	2271
1605	1608	1555	1855	1854	1795	2105	2098	2036	2355	2343	2276
1610	1613	1560	1860	1858	1800	2110	2103	2040	2360	2348	2280
1615	1618	1565	1865	1863	1805	2115	2108	2045	2365	2352	2285
1620	1622	1569	1870	1868	1810	2120	2113	2050	2370	2357	2290
1625	1627	1574	1875	1873	1815	2125	2118	2055	2375	2362	2295
1630	1632	1579	1880	1878	1820	2130	2123	2060	2380	2367	2300
1635	1637	1584	1885	1883	1824	2135	2128	2065	2385	2372	2304
1640	1642	1589	1890	1888	1829	2140	2133	2069	2390	2377	2309
1645	1647	1593	1895	1893	1834	2145	2138	2074	2395	2382	2314
1650	1652	1598	1900	1898	1839	2150	2142	2079	2400	2387	2319

Old age	*New Western*	*New Soviet*	*Old age*	*New Western*	*New Soviet*	*Old age*	*New Western*	*New Soviet*	*Old age*	*New Western*	*New Soviet*
2405	2391	2324	2655	2635	2563	2905	2878	2803	3155	3120	3042
2410	2396	2328	2660	2640	2568	2910	2883	2807	3160	3125	3047
2415	2401	2333	2665	2645	2573	2915	2888	2812	3165	3130	3052
2420	2406	2338	2670	2650	2578	2920	2892	2817	3170	3135	3056
2425	2411	2343	2675	2654	2582	2925	2897	2822	3175	3140	3061
2430	2416	2348	2680	2659	2587	2930	2902	2827	3180	3145	3066
2435	2421	2352	2685	2664	2592	2935	2907	2831	3185	3150	3071
2440	2426	2357	2690	2669	2597	2940	2912	2836	3190	3154	3075
2445	2430	2362	2695	2674	2602	2945	2917	2841	3195	3159	3080
2450	2435	2367	2700	2679	2606	2950	2922	2846	3200	3164	3085
2455	2440	2372	2705	2684	2611	2955	2926	2851	3205	3169	3090
2460	2445	2376	2710	2688	2616	2960	2931	2855	3210	3174	3095
2465	2450	2381	2715	2693	2621	2965	2936	2860	3215	3179	3099
2470	2455	2386	2720	2698	2626	2970	2941	2865	3220	3183	3104
2475	2460	2391	2725	2703	2630	2975	2946	2870	3225	3188	3109
2480	2465	2396	2730	2708	2635	2980	2951	2874	3230	3193	3114
2485	2469	2400	2735	2713	2640	2985	2956	2879	3235	3198	3118
2490	2474	2405	2740	2718	2645	2990	2960	2884	3240	3203	3123
2495	2479	2410	2745	2722	2649	2995	2965	2889	3245	3208	3128
2500	2484	2415	2750	2727	2654	3000	2970	2894	3250	3212	3133
2505	2489	2419	2755	2732	2659	3005	2975	2898	3255	3217	3138
2510	2494	2424	2760	2737	2664	3010	2980	2903	3260	3222	3142
2515	2499	2429	2765	2742	2669	3015	2985	2908	3265	3227	3147
2520	2504	2434	2770	2747	2673	3020	2990	2913	3270	3232	3152
2525	2508	2439	2775	2752	2678	3025	2994	2918	3275	3237	3157
2530	2513	2443	2780	2757	2683	3030	2999	2922	3280	3242	3162
2535	2518	2448	2785	2761	2688	3035	3004	2927	3285	3246	3166
2540	2523	2453	2790	2766	2693	3040	3009	2932	3290	3251	3171
2545	2528	2458	2795	2771	2697	3045	3014	2937	3295	3256	3176
2550	2533	2463	2800	2776	2702	3050	3019	2941	3300	3261	3181
2555	2538	2467	2805	2781	2707	3055	3023	2946	3305	3266	3185
2560	2543	2472	2810	2786	2712	3060	3028	2951	3310	3271	3190
2565	2547	2477	2815	2791	2717	3065	3033	2956	3315	3275	3195
2570	2552	2482	2820	2795	2721	3070	3038	2961	3320	3280	3200
2575	2557	2487	2825	2800	2726	3075	3043	2965	3325	3285	3205
2580	2562	2491	2830	2805	2731	3080	3048	2970	3330	3290	3209
2585	2567	2496	2835	2810	2736	3085	3053	2975	3335	3295	3214
2590	2572	2501	2840	2815	2740	3090	3057	2980	3340	3300	3219
2595	2577	2506	2845	2820	2745	3095	3062	2985	3345	3304	3224
2600	2581	2511	2850	2825	2750	3100	3067	2989	3350	3309	3228
2605	2586	2515	2855	2829	2755	3105	3072	2994	3355	3314	3233
2610	2591	2520	2860	2834	2760	3110	3077	2999	3360	3319	3238
2615	2596	2525	2865	2839	2764	3115	3082	3004	3365	3324	3243
2620	2601	2530	2870	2844	2769	3120	3087	3008	3370	3329	3248
2625	2606	2535	2875	2849	2774	3125	3091	3013	3375	3333	3252
2630	2611	2539	2880	2854	2779	3130	3096	3018	3380	3338	3257
2635	2616	2544	2885	2859	2784	3135	3101	3023	3385	3343	3262
2640	2620	2549	2890	2863	2788	3140	3106	3028	3390	3348	3267
2645	2625	2554	2895	2868	2793	3145	3111	3032	3395	3353	3272
2650	2630	2558	2900	2873	2798	3150	3116	3037	3400	3358	3276

Old age	*New Western*	*New Soviet*	*Old age*	*New Western*	*New Soviet*	*Old age*	*New Western*	*New Soviet*	*Old age*	*New Western*	*New Soviet*
3405	3363	3281	3655	3604	3520	3905	3846	3759	4155	4087	3998
3410	3367	3286	3660	3609	3525	3910	3851	3764	4160	4092	4002
3415	3372	3291	3665	3614	3530	3915	3855	3768	4165	4096	4007
3420	3377	3295	3670	3619	3534	3920	3860	3773	4170	4101	4012
3425	3382	3300	3675	3624	3539	3925	3865	3778	4175	4106	4017
3430	3387	3305	3680	3628	3544	3930	3870	3783	4180	4111	4022
3435	3392	3310	3685	3633	3549	3935	3875	3788	4185	4116	4026
3440	3396	3315	3690	3638	3553	3940	3879	3792	4190	4121	4031
3445	3401	3319	3695	3643	3558	3945	3884	3797	4195	4125	4036
3450	3406	3324	3700	3648	3563	3950	3889	3802	4200	4130	4041
3455	3411	3329	3705	3653	3568	3955	3894	3807	4205	4135	4045
3460	3416	3334	3710	3657	3573	3960	3899	3811	4210	4140	4050
3465	3421	3338	3715	3662	3577	3965	3904	3816	4215	4145	4055
3470	3425	3343	3720	3667	3582	3970	3908	3821	4220	4149	4060
3475	3430	3348	3725	3672	3587	3975	3913	3826	4225	4154	4064
3480	3435	3353	3730	3677	3592	3980	3918	3831	4230	4159	4069
3485	3440	3358	3735	3682	3596	3985	3923	3835	4235	4164	4074
3490	3445	3362	3740	3686	3601	3990	3928	3840	4240	4169	4079
3495	3450	3367	3745	3691	3606	3995	3933	3845	4245	4174	4084
3500	3454	3372	3750	3696	3611	4000	3937	3850	4250	4178	4088
3505	3459	3377	3755	3701	3616	4005	3942	3854	4255	4183	4093
3510	3464	3381	3760	3706	3620	4010	3947	3859	4260	4188	4098
3515	3469	3386	3765	3711	3625	4015	3952	3864	4265	4193	4103
3520	3474	3391	3770	3715	3630	4020	3957	3869	4270	4198	4107
3525	3479	3396	3775	3720	3635	4025	3961	3874	4275	4203	4112
3530	3483	3401	3780	3725	3639	4030	3966	3878	4280	4207	4117
3535	3488	3405	3785	3730	3644	4035	3971	3883	4285	4212	4122
3540	3493	3410	3790	3735	3649	4040	3976	3888	4290	4217	4127
3545	3498	3415	3795	3739	3654	4045	3981	3893	4295	4222	4131
3550	3503	3420	3800	3744	3659	4050	3986	3897	4300	4227	4136
3555	3508	3424	3805	3749	3663	4055	3990	3902	4305	4231	4141
3560	3512	3429	3810	3754	3668	4060	3995	3907	4310	4236	4146
3565	3517	3434	3815	3759	3673	4065	4000	3912	4315	4241	4150
3570	3522	3439	3820	3764	3678	4070	4005	3916	4320	4246	4155
3575	3527	3444	3825	3768	3682	4075	4010	3921	4325	4251	4160
3580	3532	3448	3830	3773	3687	4080	4015	3926	4330	4256	4165
3585	3537	3453	3835	3778	3692	4085	4019	3931	4335	4260	4169
3590	3541	3458	3840	3783	3697	4090	4024	3936	4340	4265	4174
3595	3546	3463	3845	3788	3702	4095	4029	3940	4345	4270	4179
3600	3551	3467	3850	3793	3706	4100	4034	3945	4350	4275	4184
3605	3556	3472	3855	3797	3711	4105	4039	3950	4355	4280	4189
3610	3561	3477	3860	3802	3716	4110	4043	3955	4360	4284	4193
3615	3566	3482	3865	3807	3721	4115	4048	3959	4365	4289	4198
3620	3570	3487	3870	3812	3725	4120	4053	3964	4370	4294	4203
3625	3575	3491	3875	3817	3730	4125	4058	3969	4375	4299	4208
3630	3580	3496	3880	3822	3735	4130	4063	3974	4380	4304	4212
3635	3585	3501	3885	3826	3740	4135	4068	3979	4385	4308	4217
3640	3590	3506	3890	3831	3745	4140	4072	3983	4390	4313	4222
3645	3595	3510	3895	3836	3749	4145	4077	3988	4395	4318	4227
3650	3599	3515	3900	3841	3754	4150	4082	3993	4400	4323	4232

Old age	New Western	New Soviet	Old age	New Western	New Soviet	Old age	New Western	New Soviet	Old age	New Western	New Soviet
4405	4328	4236	4555	4472	4379	4705	4617	4523	4855	4761	4666
4410	4333	4241	4560	4477	4384	4710	4621	4527	4860	4766	4670
4415	4337	4246	4565	4482	4389	4715	4626	4532	4865	4771	4675
4420	4342	4251	4570	4487	4394	4720	4631	4537	4870	4775	4680
4425	4347	4255	4575	4491	4399	4725	4636	4542	4875	4780	4685
4430	4352	4260	4580	4496	4403	4730	4641	4546	4880	4785	4690
4435	4357	4265	4585	4501	4408	4735	4645	4551	4885	4790	4694
4440	4361	4270	4590	4506	4413	4740	4650	4556	4890	4795	4699
4445	4366	4274	4595	4511	4418	4745	4655	4561	4895	4799	4704
4450	4371	4279	4600	4516	4422	4750	4660	4566	4900	4804	4709
4455	4376	4284	4605	4520	4427	4755	4665	4570	4905	4809	4713
4460	4381	4289	4610	4525	4432	4760	4670	4575	4910	4814	4718
4465	4386	4294	4615	4530	4437	4765	4674	4580	4915	4819	4723
4470	4390	4298	4620	4535	4441	4770	4679	4585	4920	4823	4728
4475	4395	4303	4625	4540	4446	4775	4684	4589	4925	4828	4732
4480	4400	4308	4630	4544	4451	4780	4689	4594	4930	4833	4737
4485	4405	4313	4635	4549	4456	4785	4694	4599	4935	4838	4742
4490	4410	4317	4640	4554	4461	4790	4698	4604	4940	4843	4747
4495	4414	4322	4645	4559	4465	4795	4703	4608	4945	4847	4752
4500	4419	4327	4650	4564	4470	4800	4708	4613	4950	4852	4756
4505	4424	4332	4655	4568	4475	4805	4713	4618	4955	4857	4761
4510	4429	4337	4660	4573	4480	4810	4718	4623	4960	4862	4766
4515	4434	4341	4665	4578	4484	4815	4722	4628	4965	4867	4771
4520	4439	4346	4670	4583	4489	4820	4727	4632	4970	4872	4775
4525	4443	4351	4675	4588	4494	4825	4732	4637	4975	4876	4780
4530	4448	4356	4680	4593	4499	4830	4737	4642	4980	4881	4785
4535	4453	4360	4685	4597	4504	4835	4742	4647	4985	4886	4790
4540	4458	4365	4690	4602	4508	4840	4746	4651	4990	4891	4794
4545	4463	4370	4695	4607	4513	4845	4751	4656	4995	4896	4799
4550	4467	4375	4700	4612	4518	4850	4756	4661	5000	4900	4804

APPENDIX 2

List of formations

Reprinted with permission from *General stratigraphy* by J. W. Gregory & B. H. Barrett, Methuen's Geological Series (General editor J. W. Gregory), London 1931, pages 240–57

Name	System	Author	Year	Locality or Derivation	Reference	Use
Aalenian	M. Jur.	Mayer-Eymar	1864	Aalen, Wurtemburg	Tabl. synchr.	
Acadian	M. Camb.	Dawson	1855	Acadia, Canada	Acad. Geol.	
Acheulian	L. Pleist.	Mortillet	1878	St. Acheul, France	Congr. géol. Paris, p.179	
Aftonian	Pleist.	Chamberlin and Salisbury	1906	Afton, Iowa	Geol., iii, p.383	1st American interglacial
Albian	M. Cret.	d'Orbigny	1842	Aube, France	Pal. fr., Crét., ii	
Alexandrian	L. Silur.	Savage	1908	Alexander Co., U.S.A.	Am. Jour. Sci (4), 25, pp.433–4	= Llandovery
Algonkian	U. Pampal.	Van Hise	1892	Algonquin, Canada	Bull. U.S.G.S., 86, p.475	
Alleghenian	M. Carb.	Prosser	1901	Allegheny, U.S.A.	Am. Jour. Sci. (4), xi, p.199	pt. Moscovian
Animikean	U. Pampal	J. S. Hunt	1873	Animikie, L. Superior	Tr. Amer. Inst. Min. Eng., i, pp.331–45; ii, pp.58–9	
Anisian	M. Trias.	Waagen and Diener	1895		Akad. Wiss. Wien, civ.	
Anversian	U. Mio.	Cogels	1879	Antwerp, Belgium	Explic. planchettes d'Hoboken, etc., Carte géol. Belg.	
Aptian	L. Cret.	d'Orbigny	1843	Aptia, France	Pal. fr., Crét., ii	
Aquilonian	U. Jur.	Pavlov	1892	Aquilo, France	Argile de Speeton, p.192	
Aquitanian	U. Olig.	Mayer-Eymar	1858	Aquitaine, France	Acta Schw. Nat. Ges. Trogen, p.188	
Arenig	L. Ord.	Sedgwick	1847	Arenig, Wales		
Argovian	M. Jur.	Marcou	1848	Argovie, Switz.	Jura Salinois, p.116	= Corallian
Arnusian	L. Pleist.	Mayer-Eymar	1884	Arno, Italy	Classif. de Terre	= Sicilian
Artinskian	L. Perm.	Karpinsky	1874	Artinsk, Russia	Gorn. Journ., ii	
Ashgillian	U. Ord.	Marr	1905	Ashgill, Lake District	Q. J. G. S., lxi, p.lxxxiv	
Astian	M. Plio.	Rouville	1853	Asti, Italy	Descr. géol. Montpellier, p.155	
Aturian	U. Cret.	de Lapparent and Munier	1893	Aturia, Spain	Geol., 3rd ed., p.1150.	Campanian
Autunian	L. Perm.	de Lapparent	1893	Autun, France	*ibid.*, p.886	= Artinskian
Auversian	U. Eoc.	Dollfus	1880	Auvers, France	Expos. géol.	= Ledian
Bajocian	M. Jur.	d'Orbigny	1847	Bayeux, France	Pal. fr., Jura, i, p.606	Inf. Ool.
Balcombian	Olig.	Hall and Pritchard	1902	Balcombe, Vict.	Proc. R. S. Vict., n.s. xiv, p.78	
Barremian	L. Cret.	Coquand	1861	Barrême, France	Mem. Soc. Emul. Provence, i, p.127	
Bartonian	U. Eoc.	Mayer-Eymar	1857	Barton, England	Verh. Schweiz. Nat. Ges., Trogen, p.178	
Bathonian	M. Jur.	Omalius	1843	Bath, England	Précis Géol., p.470	
Bedoulian	Cret.	Toucas	1888	La Bedoule, France	Bull. Soc. géol. Fr. (3), xvi, p.921	
Bernician	L. Carb.	Woodward	1856	Bernicia, S. Scotland	Man. Moll.	
Berriasian	L. Cret.	Coquand	1876	Berrias, France	Bull. Soc. géol. Fr. (3), iii, p.685	
Bolderian	U. Mio.	Dumont	1849	Bolderberg, Belgium	Bull. Acad. Sci., Belg.	
Bononian	U. Jur.	Blake	1888	Bononia, Boulogne	Cited in de Lapparent, Bull. Soc. Géol. Fr. (3), xxi, p.462	Portlandian

Name	System	Author	Year	Locality or derivation	Reference	Use
Bradfordian	M. Jur.	Desor	1859	Bradford, Wilts.	Etud. Jura Neuchat, p.85	L. Bath.
Burdigalian	L. Mio.	Depéret	1892	Bordeaux, France	Bull. Soc. géol. Fr. (3), xx, p.155; xxi, p.170	
Butleyan	M. Plio.	Harmer	1900	Butley, England	Q. J. G. S., xvi, p.721	= U. Astian
Callovian	M. Jur.	d'Orbigny	1849	Kelloway, England	Pal. fr., Jura, i, p.608	
Campanian	Cret.	Coquand	1857	Campania, N. France	Bull. géol. Fr. (2), xiv, p.887	
Canadian	L. Ord.	Dana	1874	Canada	Am. Jour. Sci. (3), viii, p.214	Arenig
Caradocian	U. Ordov.	Murchison	1839	Caradoc, Wales	Silurian System	
Cartennian	L. Mio.	Pomel	1858	Tenes, Algeria	C. R. Acad. Sci., 1858, p.480	Burdigalian
Casselian	U. Olig.	Dollfus	1910	Cassel, Hesse	Bull. Soc. géol. Fr. (4), x, p.582	Chattian
Casterlian	L. Plio.	Dumont; van den Broeck	1874	Casterlé, Belgium	Ann. Soc. R. Mal. Belg., xvii, iii–viii	= Plaisancian
Cayugan	U. Silur.	Clark and Schuchert	1898	L. Cayuga, N. Y.	Science, n.s., x, p.876	
Cenomanian	L. Cret.	d'Orbigny	1852	Le Mans, France	Cours él., Pal., ii	
Champlainian	M. Ord.	Emmons	1842	L. Champlain, U.S.A.	Geol. N. Y., pp.100–1	
Charmouthian	L. Jur.	Mayer-Eymar	1864	Charmouth, England	Tabl. synchr.	
Chatauquan	U. Dev.	Clark and Schuchert	1898		Science, n.s., x, pp.874–78	
Chattian	U. Olig.	Fuchs	1894	Chatti, tribe in Hesse	K.-ungar. geol. Anstalt. Mitt. x, p.173	
Chillesfordian	L. Pleist.	Prestwich	1849	Chillesford, Suffolk	Q. J. G. S. v, p.345	
Cincinnatian	U. Ord.	Meek and Worthen	1865	Cincinnati, U.S.A.	Pr. Acad. Nat. Sci. Philad., xvii, p.155	
Coblenzian	L. Dev.	Dumont	1848	Coblenz, Germany	Mem. Ter. Ard., 2nd pt., p.183	
Comanchean	L. Cret.	R. T. Hill	1887	Comanche, U.S.A.	Am. Jour. Sci. (3), xxxiv, pp.287–309	
Conemaughan	U. Carb.	Prosser	1901	Conemaugh, U.S.A.	*Ibid.*, (4), xi, p.199	pt. Uralian
Coniacian	U. Cret.	Coquand	1857	Cognac, France	Bull. Soc. géol. Fr. (2), xiv, p.882	
Corallian	M. Jur.	Thurmann	1832	England		
Couvinian	M. Dev.	Dupont	1885	Couvin, Belgium	Carte géol. Belge	
Croixian	U. Camb.	Walcott	1912	St. Croix, Min.	Smithson, Misc. Coll., 57, x, pp.306–7	Pacific coast
Cromerian	L. Pleist.	Harmer	1900	Cromer, England	Q. J. G. S., lvi, 725	= Cromer Forest, Bed.
Cuisian	L. Eoc.	Dollfus	1880	Cuise-la-Motte, Fr.	Soc. géol. Norm., p.589	= London Clay
Danian	U. Cret.	Desor	1850	Denmark		
Delémontian	U. Olig.	Greppin	1867	Bernese Jura	Essai sur Jura, p.128	Aquitanian
Demetian	M. Carb.	S. P. Woodward	1856		Man. Moll., p.409	Coal Meas. and M. Grit.
Deutozoic	U. Paleoz.	Lapworth	1888	*Deuteros* = second	Intro. Textbk. Geol., p.512	
Diestian	L. Plio.	Dumont	1839	Diest, Belgium	Acad. Sci. Belg.	Plaisancian
Dimetian	Pampal.	Hicks	1878	Dimetia, Wales	Rep. Brit. Assoc.	
Dinantian	L. Carb.	Lapparent	1893	Dinant, Belgium	Traité Géol., p.819	
Dittonian	Up. Dev.	W. W. King	1921		Proc. G.S., 1921, p.124	Up. O.R.S.
Divesian	M. Jur.	Renevier	1874	Dives, France	Tabl. Terr.	Oxfordian
Doinyan	Eoc.	Gregory	1896	Doinyo Lersubugo, Kenya Col.	Great Rift Valley, p.235	
Domerian	L. Jur.	Bonarelli	1894	Mt. Domero, Italy	Acad. Turin, xxx	
Donzérian	Cret.	Torcapel	1882	Donzère, France	Urg. du Languedoc, p.4	Barremian
Dordonian	U. Cret.	Coquand	1857	Dordogne, France	Bull. Soc. géol. Fr. (2), xiv, p.882	Danian
Downtonian	U. Silur.	Lapworth	1879	Downton Castle, Eng.	Ann. Mag. N. H. (5), iii, p.455	
Dubisian	U. Jur.	Desor	1859	Doubs, Switzerland	Jura Neuch., p.45	Purbeckian
Durntenian	M. Pleist.	Mayer-Eymar	1881	Durnten, Switzerland	Classif. internat.	Interglacial
Dyas	Permian	Marcou	1859		Bibliot. Univers. Geneve, 1859	
Eifelian	M. Dev.	Dumont	1848	Eifel, Germany	Mem. Ard., p.382	
Elberfeldian	U. Carb.	Mayer-Eymar	1881	Elberfeld, Germany	Classif. internat.	Uralian
Emscherian	U. Cret.	de Lapparent	1893		Géol., 3rd ed., p.1150	
Erian	Dev.	Dawson	1871	L. Erie, N. America	Rep. Geol. Surv. Can., p.10	
Etcheminian	L. Carb.			Indian Tribe		
Etrurian	L. Olig.	Pareto	1865	Etruria, Italy	Bull. Soc. géol. Fr., xxii, p.215	Tongrian
Falunian	Mio.	d'Orbigny	1852	Faluns of France	Cours. élém., ii, p.775	
Famennian	U. Dev.	Gosselet	1880	Fammene, Belgium	Esq. géol. N. France, p.107	
Firmitian	U. Olig.	Dollfus	1880	Ferté-Alais, France	Expos. géol. Havre, p.600	Aquitanian
Flandrian	L. Eoc.	Mayer-Eymar	1881	Flanders	Classif. internat.	Montian
Forestian	U. Pleist.	Jas. Geikie	1895	Forest Beds	Q. J. G. S., p.250	4th and 5th Interglacials
Fossanian	U. Plio.	Sacco	1886	Fossans, Italy	Bull. Soc. géol. Fr. (3), xv, p.27	Astian

Name	System	Author	Year	Locality or derivation	Reference	Use
Franconian	M. Trias.	de Lapparent	1883	Franconia, Germany	Géol. p.793	Muschelkalk
Frasnian	U. Dev.	Gosselet	1880	Frasne, Belgium	Esq. géol. N. France, p.95	
Gargasian	L. Cret.	Kilian	1887	Gargas, France	Ann. géol. univers., iii, p.314	Aptian
Garumnian	U. Cret.	Leymerie	1862	Garonne, France	Bull. Soc. géol. Fr. (2), xix, p.1107	Part of Danian
Gedgravian	M. Plio.	Harmer	1900	Gedgrave	Q. J. G. S. lvi, p.707	Plaisancian
Gedinnian	L. Dev.	Dumont	1848	Gédinne, Belgium	Mem. Terr. Ard., p.167	
Georgian	L. Camb.	Hitchcock	1861	Georgia, N. America	Bull. U.S. Geol. Surv., No. 81	
Givetian	M. Dev.	Gosselet	1880	Givet, France	Esq. géol. N. France, p.88	
Glyptician	M. Jur.	Etallon	1861	Zone of *Glypticus hieroglyphicus*	Mem. Emul. Doubs, vi, p.53	Corallian
Gothlandian	Silur.	de Lapparent	1893	Gothland, Baltic	Géol., p.748	
Gshelian	U. Carb.	Nikitin	1890	Gshel, Russia	Mem. géol. Russ., v, p.156	Uralian
Guadalupian	L. Perm.	Girty	1902	Guadalupe Mts. U.S.A.	Am. Jour. Sci. (4), xiv, p.368	= Artinskian
Hauterivian	L. Cret.	Renevier	1874	Hauterive, Switz.	Tabl. Terr. sed.	Part of Neocomian
Heathcotian	Up. Camb.	Gregory	1902	Heathcote, Victoria	Proc. R.S. Vict., n.s., xv, p.148	
Heersian	L. Eocene	Dumont	1851	Heers, Belgium		Thanetian
Helderbergian	L. Dev.	Clark and Schuchert	1898	Helderberg Mts., N.Y.	Science, n.s., x, pp.874–78	
Helvetian	M. Mio.	Mayer-Eymar	1857	Helvetia, Switzerland	Verh. Schweiz. Nat. Ges., Trogen, Table	
Helvetian	Pleist.	J. Geikie	1895	Helvetia, Switzerland	Jour. Géol., p.248	
Hettangian	L. Jur.	Renevier	1864	Hettange, France	Not. Alp. vaud., i, p.51	
Humbletonian	U. Perm.	Mayer-Eymar	1888	Humbleton, Yorks	Tabl. Terr. séd.	Thuringian
Hunsruckian	L. Dev.	Dumont	1848	Hunsruck, Germany	Mem. Terr. Ard., p.194	Siegenian
Huronian	U. Pampal.	Logan	1850	L. Huron, Canada	Rep. Geol. Surv., Canada	
Icenian	U. Plio.	Harmer	1900	Norwich Crag	Q. J. G. S., lvi, p.721	= Sicilian
Igualadian	M. Eoc.	Vézian	1858	Igualada, Spain	Bull. Soc. géol. Fr. (2), xv, p.438	Lutetian
Illinoian	Pleisto.	Chamberlin and Salisbury	1906	Illinois	Geol., iii, p.383	= Rissian
Iowan	Pleisto.	Chamberlin and Salisbury	1906	Iowa	*Ibid.*, iii, p.383	4th Amer. Glaciation
Jacksonian	U. Eocene	Heilprin	1888	Jackson, Alabama	Congr. géol. inter. Rep. Amer. Com., p.814	
Janjucian	Mio.	Hall and Pritchard	1902	Janjuc, Victoria	Proc. R. Soc. Vict., n.s., xiv, p.78	
Jeurian	M. Olig.	Dollfus	1880	Jeurre, France	Expos. géol. Havre, p.599	Tongrian
Jovarian	M. Olig.	Dollfus	1880	Jonarre, France	Expos. géol. Havre, p.590	Tongrian
Juvavian	U. Trias.	Mojsisovics	1892	Juvavo, Salzburg	Sitzb. Ak. Wiss. Wien, p.777	
Kansan	Pleisto.	Chamberlin and Salisbury	1906	Kansas	Geol., iii, p.383	= Mindelian
Kapitian	Cret. Eoc.	Gregory	1896	Kapiti, Kenya Col.	Great Rift Valley, p.235	
Karnian	U. Trias.	Mojsisovics	1869	Carnic Alps	Verh. géol. Reichs., p.65	
Keewatin	Pampal.	Lawson	1888	Indian word for W. wind	Archaean Geol., p.70	
Keweenawan	Pampal.	Brooks	1876	Keweenaw, America	Am. Jour. Sci. (3), xi, pp.206–11	
Kimeridgian	U. Jur.	Thurmann	1832	Kimeridge, England	*Cf.* d'Orbigny in Pal. fr. Jur., i, p.610	
Ladinian	L. Trias.	Bittner	1892	Ladini, people of	Jahr. Reichs. Wien, xlii, p.387	Bunter: pt
Laedonian	M. Jur.	Marcou	1846	Laedo, Switzerland	Jura Salin., p.70	Bajocian
Laekenian	M. Eoc.	Dumont	1851	Laeken, Belgium		Lutetian
Laikipian	Mio.	Gregory	1896	Laikipia, Kenya Col.	Great Rift Valley, p.235; Rift Valley and E. Afr., p.204	
Lanarkian	M. Carb.	Kidston	1905	Lanark, Scotland	Q. J. G. S., lxi, pp.308–21	Pt. Moscovian
Landenian	L. Eoc.	Dumont	1849	Landen, Belgium	Bull. Acad. Belg., xvi, p.16	= Thanetian
Langhian	M. Mio.	Pareto	1865	Langhe, Italy	Bull. Soc. geol. Fr. (2), xxii, p.229	= Burdigalian
Latdorfian	L. Olig.	Mayer-Eymar	1863	Latdorf, Germany	*Ibid.*, xxi, p.7	Tongrian
Lausannian	M. Mio.	Rollier	1892	Lausanne, Switz.	Eclog. geol. Helv., iii, p.83	Burdigalian
Ledburian	U. Silur.	Renevier	1874	Ledbury, England	Tabl. Terr.	Downtonian
Ledian	U. Eoc.	Mourlon	1880	Lede, Belgium	Soc. malac. Belg., xviii, p.10	Bartonian
Lenhamian	L. Plioc.	Harmer	1900	Lenham, Kent	Q. J. G. S., lvi, p.708	= Diestian
Lennoxian	U. Pampal.	Gregory	1928	Lennox, Scotland	Trans. Geol. Soc. Glasgow, xviii, p.305	
Liburnian	U. Cret.	Stache	1889	Dalmatia	Abh. geol. Reichs.	= Danian
Ligerian	M. Mio.	Rouville	1853	Loire, France	Géol. Montpellier, p.180	Helvetian
Ligurian	L. Olig.	Mayer-Eymar	1857	Liguria, Italy	Verh. Schweiz. Nat. Ges. Trogen, p.182	Tongrian

Name	System	Author	Year	Locality or derivation	Reference	Use
Lingulian	U. Camb.	Renevier	1874	*Lingula*	Tabl. Terr. séd., 1st ed.	
Llandeilian	M. Ord.	Murchison	1839	Llandeilo, Wales	Silurian System	
Llandoverian	L. Sil.	Murchison	1839	Llandovery, Wales	*Ibid.*	
Llanvirnian	L. Ord.	Marr	1905	Llanvirn, S. Wales	Q. J. G. S., lxi, Proc. lxxxi	
Lodevian	M. Perm.	Renevier	1874	Lodève, France	Tabl. Terr. séd. 1st ed.	Punjabian
Loganian	Pampal.	Lawson	1913	Sir William Logan	Congr. géol. inter. Canada	
Londinian	L. Eoc.	Mayer-Eymar	1857	London, England	Verh. Schweiz. Nat. Ges., Trogen, p.175	
Longmyndian	L. Camb.	Mayer-Eymar	1874	Longmynd, England	Class. Méthod.	
Lotharingian	L. Jur.	Haug	1911	Lorraine, France	Traité, p.961	Pt. Charmouthian
Ludian	L. Olig.	de Lapparent	1893	Ludes, France	Géol., 3rd edit., p.1219	Tongrian
Ludlovian	U. Silur.	Murchison	1839	Ludlow, England	Silurian System	
Lusitanian	U. Jur.	Choffat	1885	Lusitania, Portugal	Faun. Jur. Portugal	Corallian
Lutetian	M. Eoc.	de Lapparent	1883	Lutetia, Paris	Géol., p.989	
Maastrichtian	U. Cret.	Dumont	1849	Maastricht, Holland	Bull. Acad. sc. Belg.	Pt. Danian
Magdalenian	Pleisto.	Mortillet	1878	Madelaine, France	Congr. géol., Paris, p.179	
Manresian	U. Eoc.	Vézian	1858	Manresa, Spain	Bull. Soc. géol. Fr. (2) xv, p.439	Bartonian
Mareniscan	Pampal.	Van Hise	1892	Marenisco, N. Amer.	Bull. U.S. Geol. Surv., No. 86	
Marquettian	Pampal.	Winchell	1888	Marquette, America	Congr. géol. inter. Rep. Am. Com., p.14	
Marylandian	M. Mio.	Heilprin	1882	Maryland, America	Proc. Acad. sc. Philad.	
Maudunian	L. Eoc.	de Lapparent	1883	Meudon, France	Géol. p.989	Montian
Mayencian	M. Mio.	Mayer-Eymar	1857	Mayence, Germany	Verh. Schweiz. Nat. Ges., Trogen, Table	Burdigalian
Mecklenburgian	M. Pleisto.	J. Geikie	1895	Mecklenburg, Ger.	Journ. Geol., p.250	4th Glaciation
Melbournian	Up. Sil.	Gregory	1902	Melbourne, Victoria	Proc. R. Soc. Vict., n.s., xv, p.171	
Menevian	M. Camb.	Salter and Hicks	1865	Menevia, St. David's Wales	Q. J. G. S., xxiv	
Messinian	L. Plio.	Mayer-Eymar	1867	Messina, Sicily	Cat. Foss. Mus. Zurich, p.13	
Mississippian	L. Carb.	H. S. Williams	1891	Mississippi, America	Bull. U.S. Geol. Surv., No. 80	
Modenian	L. Olig.	Pareto	1865	Modena, Italy	Bull. Soc. géol. Fr. (2), xxii, p.216	Tongrian
Mohawkian	M. Ord.	Hall	1842	R. Mohawk, N. Y.	Am. Jour. Sci., xliii. p.52	= Llandeilian
Monian	L. Camb.	Blake	1888	Mona, Anglesey	Congr. géol. inter. London, p.36	
Monongahelan	U. Carb.	Prosser	1901	R. Monongahela, U.S.A.	Am. Jour. Sci. (4), xi, p.199	Pt. Uralian
Montian	L. Eoc.	Dewalque	1868	Mons, Belgium	Prodr. Géol. Belgique, p.185	
Morfontian	U. Eoc.	Dollfus.	1880	Mortefontaine, France	Expos. géol. Havre, p.592	Bartonian
Mornasian	U. Cret.	Coquand	1862	Mornas, France	Bull. Soc. géol. Fr. (2), xx, p.50	Pt. Senonian
Moscovian	M. Carb.	Nikitin	1890	Moscow, Russia	Mem. Com. géol. Russ., v, p.147	
Mousterian	Pleisto.	Mortillet	1878	Moustier, France	Congr. géol. inter. Paris. p.179	
Naivashan	Plio.	Gregory	1896	Naivasha, Kenya Col.	Great Rift Valley, p.235	
Nemausian	L. Cret.	Saruan	1875	Nemausum, Nimes, France	Bull. sc. nat. Nimes	Valanginian
Neocomian	L. Cret.	Thurmann	1835	Neocomium, Neuchatel	Bull. Soc. géol. Fr., vii, p.209	
Neptodunian	M. Eoc.	Dollfus	1850	Neptodunum, Paris	Expos. géol. Havre, p.592	Lutetian
Nervian	U. Cret.	Dumont	1849	Nervians, people of Belgium	Bull. Acad. Belg., xvi, p.360	Pt. Senonian
Neudeckian	Pleisto.	J. Geikie	1895		Jour. Geol., p.249	3rd Interglacial
Newark System	U. Tr. and L. Jur.	Redfield	1856	Newark, N. Y.	*Cf.* J. E. Russell, 1892, Bull. U. S. G. S., 85	
Newbournian	L. Plio.	Harmer	1900	Newbourne, Suffolk	Q. J. G. S., lvi, p.270	= M. Astian
Niagaran	Silur.	Sterry-Hunt		Niagara R., N. Amer.		M. Silur.
Norfolkian	Pleisto.	J. Geikie	1895	Norfolk, England	Jour. Geol. p.247	1st Interglacial
Norian	Pampal.	Sterry-Hunt	1870		Congr. géol. inter. London, p.73	Lower Trias
Norian	U. Trias.	Mojsisovics	1869	Alps	Verh. geol. Reichs., p.65	
Nyasan	Olig.	Gregory	1896	L. Nyasa	Great Rift Valley, p.235	
Oranian	U. Mio.	Welsch	1895	Oran, Algeria	Bull. Soc. géol. Fr. (3), xxiii, P. v, p.60	Pontian
Oriskanian	L. Dev.	Clark and Schuchert	1898	Oriskany	Science, n.s., x, pp.874–78	
Oswegan	L. Sil.	Clark and Schuchert	1898	Oswego, N. Y.	Science, x, p.876	
Oxfordian	M. Jur.	Brogniart	1829	Oxford, England	Tabl. Terr.	
Paleocene	L. Eoc.	Schimper	1874	*Palaois* and Eocene	Pal. veget., iii, p.680	
Paleolithic	Pleisto.	Lubbock	1865	Ancient stone	Prehistoric Times	
Paniselian	M. Eoc.	Dumont	1851	Mt. Panisel, Belgium	Bull. Acad. sc. Belg.	Lutetian

Name	System	Author	Year	Locality or derivation	Reference	Use
Parisian	Eoc. and L. Olig.	Brogniart	1820	Paris		Eocene and Tongrian
Parnian	M. Eoc.	Dollfus	1880	Parnes, near Paris	Expos. géol. Havre, p.591	Lutetian
Patagonian	Mio.	d'Orbigny	1842	Patagonia, S. Amer.	Voy. Amér. Mérid., iii	
Pauletian	U. Cret	Dumas	1852	St. Paulet, France	Carte géol. d'Uzès 1874	Pt. Cenomanian
Pebidian	Pampal.	Hicks	1878	Pebidia, Wales	Rep. Brit. Assoc., 1878	
Pennsylvanian	Mid. Carb.	H. S. Williams	1898	Pennsylvania, U.S.A.	Bull. U. S. G. S., No. 80	
Peorian	Pleisto.	Chamberlin and Salisbury	1906		Geol. iii, p.383	4th Amer. Interglacial
Petchorian	L. Cret.	Nikitin	18??	Petchora, Russia		Berriasian
Pilatan	M. Eoc.	Kaufmann	1872	M. Pilatus, Switz.	Mat. Carte Suisse 11th Livre, p.158	Lutetian
Plaisancian	L. Plio.	Mayer-Eymar	1857	Plaisance, Italy	Verh. Nat. Ges. Trogen, Table	
Platian	Pleisto.	Ameghino	1889	La Plata, S. Amer.	Mam. Foss. Arg., p.106	
Pliensbachian	L. Jur.	Oppel	1858	Pliensbach, Germany	Juraformation, p.815	Charmouthian
Poecilitic	Permian and Trias	Conybeare	1832		Rep. Brit. Ass., 1832, p.379	
Poederlian	U. Plio.	Vincent	1889	Poederlé, Belgium		
Polandian	Pleisto.	J. Geikie	1895	Poland	Jour. Geol. p.249	3rd Glaciation
Pontian	U. Mio.	Marny	1869	Pontus Euxinus	Géol. de Cherson	
Portlandian	U. Jur.	Brongniart	1829	Portland, England	Tabl. Terr.	
Priabonian	U. Eoc.	de Lapparent	1893	Priabona, Italy	Géol. 3rd ed., p.149	Bartonian
Proterozoic	L. Paleoz.	Lapworth	1888	*Proteros* = first	Intro.Textbk.Geol.,p.152	
Punjabian	M. Perm.	de Lapparent	1893	Punjab, India	Géol., 3rd ed., p.886	
Purbeckian	U. Jur.	Brongniart	1829	Purbeck, England	Tabl. Terr.	
Radstockian	M. Carb.	Kidston	1905	Radstock, England	Q. J. G. S., lxi, pp.308–21	Pt. Moscovian
Raiblian	U. Trias.	Stoppani	1860	Raibl, Austria	Pal. Lomb., pp.226, 229	
Rauracian	M. Jur.	Gressly	1867	Rauracia, Jura	Essai sur Jura, p.72	Corallian
Revinian	M. Carb.	Dumont	1847	Revin, France	Bull. Acad. Belg.	Acadian
Rhenan	L. Dev.	Dumont	1848	R. Rhine, Germany	*Ibid.*	Siegenian
Rhaetian	L. Jur.	Guembel	1861	Rhaetian Alps	Bay. Alp., p.122	L. Jurassic or U. Trias
Rhodanian	U. Cret.	Renevier	1854	Perte-du-Rhône, France	Mem. sur Perte-du-Rhône, p.68	
Rognacian	U. Cret.	Caziot	1890	Rognac, France	Bull. Soc. géol. Fr. (3), xviii, p.227	Pt. Danian
Rotomagian	U. Cret.	Coquand	1857	Rothomagus, Rouen	*Ibid.* (2), xiv, p.882	Pt. Cenomanian
Roubian	L. Olig.	Vézian	1858	Rubio, Spain	*Ibid.* (2), xv, p.440	Tongrian
Rupelian	M. Olig.	Dumont	1849	Rupel, Belgium	Bull. Acad. Belg., xvi, p.367	Regarded by Continental authors as Upper Olig.
Sahelian	U. Mio.	Pomel	1858	Sahel, Algeria	C. R. Acad. Sci., xlvii, p.479	Pontian
Saliferian	U. Trias	d'Orbigny	1852	Salt bearing	Cours élém., p.404	Keuper
Sallomacian	M. Mio.	Fallot	1893	Salles, France	Bull. Soc. géol. Fr. (3), xxi, Pr. v, p.77	Helvetian
Salmian	U. Camb.	Dumont	1847	Salm, France	Bull. Acad. Belg.	Potsdamian
Salopian	M. Silur.	Lapworth	1879	Salop, England	Ann. Mag. Nat. Hist. (5), iii, opp. p.455	Wenlock
Sangamon	Pleisto.	Leverett	1899	Sangamon Co., Ill.	Mon. U.S. Geol. S., xxxviii	3rd Amer. Interglac.
Sannoisian	L. Olig.	de Lapparent	1893	Sannoise, France	Géol., 3rd ed., p.1263	Tongrian
Santonian	U. Cret.	Coquand	1857	Santes, France	Bull. Soc. géol. Fr. (2), xiv, p.882	Pt. Senonian
Sarmatian	U. Mio.	Barbot de Marny	1869	Sarmatia	Esq. géol. de Cherson	Phase of Pontian
Saxonian	Pleisto.	J. Geikie	1895	Saxony, Germany	Journ. Géol., p.247	2nd Glaciation
Saxonian	M. Perm.	de Lapparent	1893	Saxony, Germany	Géol., 3rd ed., p.886	
Scaldisian	U. Plio.	Dumont	1849		Bull Acad. Belg.	Astian
Scandinavian	U. Camb.	de Lapparent	1883	Scandinavia	Géol., p.732	
Scanian	Pleisto.	J. Geikie	1895	Scania, Sweden	Journ. Géol., p.246	1st Glaciation
Senecan	U. Dev.	Clarke and Schuchert	1898	Seneca, U.S.A.	Science, n.s., x, pp.874–8	
Senonian	U. Cret	d'Orbigny	1843	Senones, a tribe in France	Pal. fr., Crét., ii, Table, pl. 236 bis.	
Sequanian	M. Jur.	Thurmann and Marcou	1848	Sequania, Jura	Mem. Soc. géol. Fr., iii, p.96	Corallian
Sicilian	L. Pleisto	Döderlein	1872	Sicily	Nat. sur. Carte géol. de Moden, p.14	
Siderolitique	L. Olig.	Gressly	1841	Ferruginous	Jur. sol., p.251	Tongrian
Sinemurian	L. Jur.	d'Orbigny	1849	Sémur, France	Pal. fr., Jur., i, p.604	
Sinian	Camb.	Richthofen	1882	China	Nord. Chin	U. Pampalozoic
Skiddavian	L. Ord.	Marr	1905	Skiddaw	Q. J. G. S., lxi, Proc. p.lxxxvi	
Skytian	L. Trias	Waagen and Diener	1895		Ak. Wis. Wien, civ	
Snowdonian	U. Ord.	S. P. Woodward	1856	Snowdon, Wales	Mon. Moll., p.409	= Bala
Soissonian	L. Eoc.	Mayer-Eymar	1857	Soissons, France	Verh. Schweiz. Nat. Ges., Trogen, Table	
Solutrian	Pleisto.	Mortillet	1878	Solutré, France	Congr. géol. inter. Paris, p.179	

Name	System	Author	Year	Locality or derivation	Reference	Use
Sparnacian	L. Eoc.	Dollfus	1880	Epernay, France	Bull. Soc. géol. Norm., vi, p.588	
Staffordian	M. Carb.	Kidston	1905	Stafford, England	Q. J. G. S., lxi, pp.308–21	Pt. Moscovian
Stampian	L. Olig.	Rouville	1853	Etampes, France	Géol. de Montpel., p.180	Tongrian
Stephanian	U. Carb.	Mayer-Eymar	1878	St. Etienne, France	Class. inter., 1881	Uralian
Suessonian	L. Eoc.	d'Orbigny	1852	Soissons, France	Cours, élém. pal., p.712	
Taconian	L. Camb.	Emmons	1842	Taconic Mts., Amer.	Geol. N. Y., pp.135–64	Lower Ordovician
Taunusian	L. Dev.	Dumont	1848	Taunus, Belgium	Mem. Ter. Arden., p.183	Siegenian
Tennessean	L. Carb.	Ulrich	1911	Tennessee	Geol. Soc. Am. Bull., vol. xxii, pp.581–2	Visean
Thanetian	L. Eoc.	Renevier	1873	Thanet, England	Tabl. Terr.	
Thurgovian	U. Mio.	Rollier	1892	Thurgovia, Switz.	Eclog. geol. Helv., iii, p.83	Tortonian
Thuringian	U. Perm.	Renevier	1874	Thuringia, Germany	Tabl. Terr.	
Toarcian	L. Jur.	d'Orbigny	1849	Thouars, France	Pal. fr., Jur., i, p.106	
Tongrian	L. Olig.	Dumont	1839	Tongres, Belgium	Bull. Acad. sc. Belg., vi, p.773	
Tortonian	U. Mio.	Mayer-Eymar	1857	Tortona, Italy	Verh. Schweiz. Nat. Ges., Trogen, Table	
Tournaisian	L. Carb.	Koninck	1872	Tournai, Belgium	Mem. Ac. R. Sc. L.B.A. Belge., xxxix, pp.1–178	
Tremadocian	U. Camb.	Renevier	1874	Tremadoc, Wales	Tabl. Terr.	Potsdamian
Turbasian	Pleisto.	J. Geikie	1895		Jour. Geol.	Geikie's 5th and 6th Glaciations
Turonian	U. Cret.	d'Orbigny	1843	Touraine, France	Pal. fr., Crét, ii, Table, pl. 236 bis.	Pt. Senonian
Tyrolian	U. Trias.	de Lapparent	1885	Tyrol, Austria	Geol., 2nd ed., p.905	Keuper
Ulsterian	M. Dev.	Clarke and Schuchert	1898		Science, n.s., x, pp.874–8	
Uralian	U. Carb.	de Lapparent	1893	Ural Mountains	Géol., 3rd ed., p.819	
Urgonian	L. Cret.	d'Orbigny	1850	Orgon, France	Cours. élém., ii, p.606	Barremian and Aptian
Valanginian	L. Cret.	Desor	1854	Valangin, Switzerland	Bull. Sc. nat. Neuchat., iii, p.177	Pt. Neocomian
Valdonnian	U. Cret.	Matheson	1878	Valdonne, France	Rech. pal. Midi	Pt. Senonian
Valentian	L. Silur.	Lapworth	1879	Valentia, S. Scotland	Ann. Mag. Nat. Hist. (5), iii, opp. p.455	
Vasconian	L. Mio.	Fallot	1893	Vasconia, France	Bull. Soc. géol. Fr. (3), xxi, Pr. v, p.79	Burdigalian
Vectian	L. Cret.	Jukes-Brown	1885	Vectium, I.O.W.	Geol. Mag., iii, p.298	Urgonian
Vesulian	M. Jur.	Marcou	1848	Vesoul, France	Mem. Soc. géol. Fr., iii, p.73	L. Bathonian
Villafranchian	L. Pleisto.	Pareto	1865	Villafranca, Italy	Bull. Soc. géol. Fr. (2), xxii, p.262	Sicilian
Vindobonian	U. Mio.	Depéret	1895	Vindobona, Vienna	*Ibid.* (3), xxiii, Pr. v, p.34	Helvetian and Tortonian
Virginian	M. Mio.	Heilprin	1882	Virginia, America	Acad. Nat. Sc. Philad.	
Virglorian	L. Trias	Renevier	1874	Virgloria, Switzerland	Tabl. Terr.	Pt. Bunter
Virgulian	U. Jur.	Thurmann	1852	*Exogyra virgula*	Mitth. Bern. naturf. Ges., p.217	Kimeridgian
Virtonian	L. Jur.	Mourlon	1880	Virton, Belgium	Geol. Belg., i, p.143	Charmouthian
Visean	M. Carb.	Dupont	1883	Visé, Belgium	Bull. Acad. Belg., xv, p.212	Moscovian
Volgian	U. Jur.	Nikitin	1881	Volga, Russia	Mem. Acad. Imp. Sci. St. Petersb., 7th Ser., xxviii, p.98	Portlandian
Vosgian	L. Trias	de Lapparent (after de Beaumont)	1885	Vosges, Germany	Géol., 2nd ed., p.905	Bunter
Vraconnian	U. Cret.	Renevier	1867	Vraconne, Jura	Faun. Chevill, p.201	Pt. Cenomanian
Waltonian	U. Plioc.	Harmer	1900	Walton, Essex	Q. J. G. S., lvi, p.709	=
Waucoban	L. Camb.	Walcott	1912	Waucoba Springs, Cal.	Smithson. Misc. Coll., vol. 57, No. 10, pp.305–6	
Waverlian	L. Carb.	C. Briggs, jr. (Mather)	1838		Ohio Geol. Surv., 1st Ann. Rep., pp.74, 79–80	Tournaisian
Wealden	L. Cret.	Mantell		Weald, England	Foss. S. Downs	Pt. Neocomian
Wenlock	M. Silur.	Murchison	1839	Wenlock, England	Silurian System	
Werfenian	L. Trias	Renevier	1874	Werfen, Austria	Tabl. Terr.	Bunter
Westphalian	M. Carb.	de Lapparent	1893	Westphalia, Germany	Géol., 3rd ed. p.819	= Moscovian
Weybournian	U. Plio.	Harmer	1900	Weybourne, Suffolk	Q. J. G. S., lvi, p.724	
Wisconsin	Pleisto.	Chamberlin and Salisbury	1906	Wisconsin, U.S.A.	Geol., iii, p.383	= Wurmian
Yarmouthian	Pleisto.	Chamberlin and Salisbury	1906	Yarmouth, Iowa	Geol., iii, p.383	2nd Amer. Interglac.
Yeovilian	L. Ool.	Buckman		Yeovil		Midford Sands
Yeringian	U. Sil.	Gregory	1902	Yering, Victoria	Proc. R. Soc. Victoria, n.s., xv, p.172	
Yorkian	M. Carb.	Watts		Ypres, Belgium		Pt. Moscovian
Ypresian	L. Eoc.	Dumont	1849	Ypres, Belgium	Bull. Acad. sc. Belg., xvi, p.369	= London Clay
Zechstein	U. Perm.	German miners' term		? from Zähe, tough		Thuringian

APPENDIX 3

List of preferred abbreviations of chronostratic names

Aalenian		Aal
Actonian		Act
Albertian		Abt
Albian		Alb
Alexandrian		Alx
Alportian		Alp
Anisian		Ans
Aphebian		Aph
Aptian		Apt
Aquitanian		Aqt
Archean	Ar	
Arenig		Arg
Arnsbergian		Arn
Artinskian		Art
Arundian		Aru
Asbian		Asb
Ashgill		Ash
Asselian		Ass
Atdabanian		Atb
Bajocian		Baj
Bala		Bal
Barremian		Brm
Bartonian		Brt
Bashkirian		Bsh
Bathonian		Bth
Berriasian		Ber
Brigantian		Bri
Burdigalian		Bur
Burzyan	B	
Caerfai		Crf
Calabrian		Clb
Callovian		Clv
Cambrian	€	
Campanian		Cmp
Canadian		Cnd
Cantabrian		Ctb
Caradoc		Crd
Carboniferous	C	
Carnian		Crn
Cautleyan		Cau
Cayugan		Cay
Cenomanian		Cen
Cenozoic	Cz	
Chadian		Chd
Chamovnicheskian		Chv
Champalian		Chp
Chattian		Cht
Cheremshanskian		Che
Chokierian		Cho
Cincinnatian		Cin
Comley		Com
Coniacian		Con
Costonian		Cos
Courceyan		Cor
Couvinian		Cov
Cretaceous	K	
Croixian		Crx
Danian		Dan
Devonian	D	
Dienerian		Die
Dinantian		Din
Dogger		Dog
Dolgellian		Dol
Dorogomilovskian		Dor
Downtonian		Dow
Dzhulfian		Dzh
Early	E	
Ediacaran		Edi
Eifelian		Eif
Emsian		Ems
Eocene		Eoc
Famennian		Fam
Frasnian		Frs
Fronian		Fro
Gedinnian		Ged
Givetian		Giv
Gleedon		Gle
Gorstian		Gor
Griesbachian		Gri
Guadelupian		Gua
Gzelian		Gze
Hadean		Hde
Hadrynian		Hdy
Harnagian		Har
Hastarian		Has
Hauterivian		Hau
Helikian		Hel
Hettangian		Het
Hirnantian		Hir
Holkerian		Hlk
Holocene		Hol
Homerian		Hom
Huronian	H	
Idwian		Idw
Induan		Ind
Isuan	I	
Ivorian		Ivo
Jurassic	J	
Kashirskian		Ksk
Kasimovian		Kas
Kazanian		Kaz
Kimmeridgian		Kim
Kinderscoutian		Kin
Klasminskian		Kla
Krevyakinskian		Kre
Kungurian		Kun
Ladinian		Lad
Langhian		Lan
Late	L	
Latdorfian		Lat
Lenian		Len
Leonardian		Leo
Lias		Lia
Llandeilo		Llo
Llandovery		Lly
Llanvirn		Lln
Lochkovian		Lok
Longvillian		Lon
Ludfordian		Ldf
Ludlow		Lud
Lutetian		Lut
Maastrichtian		Maa
Maentwrogian		Mnt
Malm		Mlm
Marsdenian		Mrd
Marshbrookian		Mrb
Melekesskian		Mel
Menevian		Men
Merioneth		Mer
Mesozoic	Mz	
Messinian		Mes
Middle	M	

Miocene			Mio
Mississippian			Mis
Mortensnes			Mor
Moscovian			Mos
Myachkovskian			Mya
Namurian			Nam
Neocomian			Neo
Neogene	Ng		
Niagaran			Nia
Noginskian			Nog
Norian			Nor
Olenekian			Olk
Oligocene			Oli
Onnian			Onn
Ordovician	O		
Orenburgian			Orn
Oxfordian			Oxf
Paleocene			Pal
Paleogene	Pg		
Paleozoic	Pz		
Pendleian			Pnd
Pennsylvanian			Pen
Permian	P		
Phanerozoic	Ph		
Piacenzian			Pia
Pleistocene			Ple
Pleistogene	Q	or	Ptg
Pliensbachian			Plb
Pliocene			Pli
Podolskian			Pod
Portlandian			Por
Poundian			Pou
Precambrian	P-Є		
Priabonian			Prb
Pridoli			Prd
Priscoan	Pr		
Proterozoic	Pt		
Purbeckian			Pur
Pusgillian			Pus
Quaternary	Q		
Randian			Ran
Rawtheyan			Raw
Rhaetian			Rht
Rhuddanian			Rhu
Riphean	R		
Rupelian			Rup
Ryazanian			Ryz
Sakmarian			Sak
Santonian			San
Scythian			Scy
Senonian			Sen
Serpukhovian			Spk
Serravallian			Srv
Sheinwoodian			She
Siegenian			Sig
Silesian			Sls
Silurian	S		
Sinemurian			Sin
Smalfjord			Sma
Smithian			Smi
Solva			Sol
Soudleyan			Sou
Spathian			Spa
Stampian			Sta
St David's			StD
Stephanian			Ste
Strunian			Str
Sturtian	U	or	Stu
Swazian	Sw		
Tatarian			Tat
Telychian			Tel
Tertiary	TT		
Thanetian			Tha
Tithonian			Tth
Toarcian			Toa
Tommotian			Tom
Tortonian			Tor
Tournaisian			Tou
Tremadoc			Tre
Triassic	Tr	or	TR
Turonian			Tur
Ufimian			Ufi
Valanginian			Vlg
Valdaian			Vld
Varangian			Var
Vendian	V		
Vereiskian			Vrk
Vetternian			Vet
Visean			Vis
Volgian			Vol
Waucoban			Wau
Wenlock			Wen
Westphalian			Wes
Whitwell			Whi
Witwatersrand			Wit
Wonokan			Won
Woolstonian			Woo
Yeadonian			Yea
Ypresian			Ypr
Yurmatin	Y		
Zanclian			Zan

APPENDIX 4

The unit of time

Introduction

The statement of age of an event on a chronometric scale is given as the number of units of time that lapsed between the instant of that event and a defined datum 'Present'. It is thus a periodic scale compounded by repetition of identical units and so is not a calendar date as might be derived from so many circuits of the Earth round the Sun in relation to the stars because each circuit varies in duration. Thus the standardisation of the chronometric scale requires only that the unit of time be defined. However two competing standards for this unit have been defined and the IUGS has not yet decided which to follow (George *et al.* 1969). That the difference in numerical terms is insignificant for our purposes does not eliminate the need to state the observational principle by which geologic time is calibrated.

Mean solar second

Formerly the unit of time was the mean solar second, defined as 1/86 400 of the mean solar day.

Ephemeris second

The International Astronomical Union (IAU) recommended in 1957 that in astronomy and related sciences the ephemeris second be adopted as the fundamental invariable unit of time. It was defined as 1/31 556 925.9747 of the tropical year at 1900 January 0 days 12 hours ephemeris time. The IAU meeting in Prague in 1967 recommended its continuation in face of the competing standards for a second (references in George *et al.* 1969, especially Sadler 1968).

Atomic second

The following definition was adopted on 13 October 1967 at the meeting of the Thirteenth General Conference on Weights and Measures: 'The second is the duration of 9 192 631 770 periods of the radiation of the atom of Cesium 133'. The frequency (9 192 631 770 Hz) which the definition assigns to the cesium radiation was carefully chosen to make it impossible, by any existing experimental evidence, to distinguish the atomic second from the ephemeris second based on the Earth's motion. No changes were anticipated in data stated in terms of the old standard to convert them to the new one (Weast 1969 *et seq.*).

Discussion

The advantages claimed were that it is theoretically possible for anyone anywhere to build an atomic clock with a precision of ±1 part per 10^{11} (or better), controlled by the cesium radiation, and calibration can be achieved in the laboratory in a few hours without the long time necessary to make astronomical observations. In practice clocks are calibrated against broadcast time signals.

Chronometric calibration of the age of geologic events is based fundamentally on these two kinds of observation. (1) The Earth's motion is reflected in sedimentary phenomena such as varves, or biologic phenomena such as variations in growth, (2) isotopic decay rates are determined in the first place in real time in the laboratory using a clock.

Conventions

Although the time standard was initially based on the year the fundamental unit of time is the second (s). The Système International d'Unités (SI) of the General Conference on Weights and Measures also allows the use of the year (a). Geoscientists could have used seconds rather than years.

The conversion of years to seconds (useful for some physical calculations) is based on the above values. Thus 1 year (a) = 31.56 teraseconds (Ts) and a million years (Ma) = 3.156 x 10^4 petaseconds (Ps). Were seconds to be used extensively there would thus be a need for a name and symbol for 10^{21}. The presently available SI conventions at 10^3 intervals are:

10^{18}	exa	E	10^{-3}	milli	m
10^{15}	peta	P	10^{-6}	micro	μ
10^{12}	tera	T	10^{-9}	nano	n
10^{9}	giga	G	10^{-12}	pico	p
10^{6}	mega	M	10^{-15}	femto	f
10^{3}	kilo	k	10^{-18}	atto	a
10^{0}	unity	1			

Ages are given in years before present (BP) rather than BC. To avoid a constantly changing datum (as in ^{14}C determinations) present = AD 1950.

References

To keep the work within bounds references have been severely restricted; however, a number of references used in our internal BP and CASP reports have been repeated here even if they are not cited in the text because we found them useful. They are distinguished by asterisks.

* Afanas'yev, G.D. & Rubinstein, M.M. (1964). Explanatory note to the geochronological scale in the absolute chronological system. In *Absolute age of geologic formations – papers presented by Soviet geologists at 22nd International Geological Congress (Delhi, India)*, pp. 287–324. Dokl. Sov. Geol., Problem 3, Nauka.

Alvarez, W. & Lowrie, W. (1978). Upper Cretaceous paleomagnetic stratigraphy at Moria (Umbrian Apennines, Italy): verification of the Gubbio section. *Geophys. J. R. Astr. Soc.*, **55**, 1–17.

Alvinerie, J., *et al.* (1973). A propos de la limite oligo-miocène; résultats préliminaires d'une recherche collective sur les gisements d'Escornebéou (Saint Geours de Maremme, Landes, Aquitaine méridionale). Présence de *Globigerinoides* dans les faunes de l'Oligocène supérieur. *C.R. Soc. Geol. France*, **15**, (3–4), 75–6.

* Anderton, R., Bridges, P.H., Leeder, M.R. & Sellwood, B.W. (1979). *A dynamic stratigraphy of the British Isles, a study in crustal evolution.* London: George Allen & Unwin, 301 pp.

Anglada, R. (1971a). Sur la position du datum à *Globigerinoides* (Foraminiferida) la zone N4 (Blow 1967) et la limite oligo-miocène en Méditerranée. *Acad. Sci. Comptes Rendues*, **272**, 1067–70.

Anglada, R. (1971b). Sur la limite Aquitanien–Burdigalien, sa place dans l'échelle des Foraminifères planctoniques et sa signification dans le Sud-Est de la France. *Acad. Sci. Comptes Rendues*, **272**, 1948–50.

Anhaeusser, C.R. & Wilson, J.F. (1981). The granitic–gneiss greenstone shield. In Hunter (1981), pp. 423–99.

Anonymous (1880). *Congrès International de Géologie – Comptes Rendues de la 2me Session, Bologne, 1881.* Bologna: Fava & pp.

Anonymous (1882). *Congrès Géologique International – Comptes Rendues de la 2me Session, Bologne, 1881.* Bologne: Fava & Garagni, 661 pp.

Anonymous (1979). Magnetostratigraphic polarity units, a supplementary chapter of the International Subcommission on Stratigraphic Classification International Stratigraphic Guide. *Geology*, **7**, 578–83.

Arkell, W.J. (1933). *The Jurassic System in Great Britain.* Oxford: Clarendon Press, 681 pp.

Arkell, W.J. (1956). *Jurassic geology of the world.* Edinburgh: Oliver & Boyd , 806 pp.

Armstrong, R.L. (1978). Pre-Cenozoic Phanerozoic time scale – computer file of critical dates and consequences of new and in-progress decay-constant revisions. In Cohee *et al.* (1978), pp. 73–91.

Ascoli, P. (1976). Foraminiferal and ostracod biostratigraphy of the Mesozoic–Cenozoic, Scotian Shelf, Atlantic Canada, 1st International Symposium on Benthonic Foraminifera of Continental Margins. Part B: Paleoecology and biostratigraphy. *Maritime Sediments Spec. Pub.* **1**, 653–771.

* Baadsgaard, H., Lerbekmo, J.F. & Evans, M.E. (1978). Geochronology and magnetostratigraphy of fluvial–deltaic sediments embracing the Cretaceous–Tertiary boundary, Red Deer Valley, Alberta, Canada. In Zartman (1978), pp. 17–18.

* Bamber, E.W. *et al.* (1970). Biochronology: standard of Phanerozoic time. In *Geology and economic minerals of Canada*, 5th edn, ed. R.J.W. Douglas, pp. 591–674. Ottawa: Geological Survey of Canada, Economic Geology Report No. 1.

Banerjee, S.K., Lund, S.P. & Levi, S. (1979). Geomagnetic record in Minnesota lake sediments – absence of the Gothenburg and Erieau excursions, *Geology*, **7**, 588–91.

* Banks, P.O. (1973). Permian–Triassic radiometric time scale. In *The Permian and Triassic Systems and their mutual boundary*, ed. A. Logan & L.V. Hills, pp. 669–77. Calgary: Canadian Society Petroleum Geologists Memoir 2.

Banner, F.T. & Blow, W.H. (1965). Progress in the planktonic foraminiferal biostratigraphy of the Neogene. *Nature, Lond.*, **208** (5016), 1164–6.

Barbier, R. & Thieuloy, J. P. (1965). Étage Valanginien, *Mémoires du Bureau de Recherches Géologiques et Minières*, No. 34, 79–84.

* Bassett, M.G. (ed.) (1976). *The Ordovician System. Proceedings of a Palaeontological Association Symposium, Birmingham, September 1974.* Cardiff: University of Wales Press/National Museum of Wales (for Palaeontological Association), 696 pp.

Bassett, M.G., Cocks, L.R.M., Holland, C.H., Rickards, R.B. & Warren, P.T. (1975). *The type Wenlock Series.* Institute of Geological Sciences Report No. 75/13. London: Her Majesty's Stationery Office, 19 pp.

Bates, R.L. & Jackson, J.A. (eds.) (1980). *Glossary of geology*, 2nd edn. Falls Church, Virginia: American Geological Institute, 749 pp.

Beard, J.H., Sangree, J.B. & Smith, L.A. (1982). Quaternary chronology, palaeoclimate, depositional sequences and eustatic cycles. *Bull. Am. Assoc. Petrol. Geol.*, **66**, 158–69.

Benedek, P.N. von & Müller, C. (1974). Nannoplankton–Phytoplankton Korrelation im Mittel und Ober-Oligozän von NW-Deutschland. *Neues Jb. Geol. Paläont. Mh.*, **7**, 385–97.

Berggren, W.A. (1971). Tertiary boundaries and correlations. In *The micropalaeontology of oceans*, ed. B.M. Funnell & W.R. Riedel, pp. 693–809. Cambridge University Press.

Berggren, W.A. (1972). A Cenozoic time-scale – some implications for regional geology and paleobiogeography. *Lethaia*, **5**, 195–215.

Berggren, W.A. (1973). The Pliocene time-scale: calibration of planktonic foraminiferal and calcareous nannoplankton zones. *Nature, Lond.*, **243** (5407), 391–7.

* Berggren, W.A. (1978). Recent advances in Cenozoic planktonic foraminiferal biostratigraphy, biochronology, and biogeography: Atlantic Ocean. *Micropaleontology*, **24** (4), 337–70.

Berggren, W.A. *et al.* (1980). Towards a Quaternary time scale. *Quaternary Research*, **13**, 277–302.

Berggren, W.A. & Van Couvering, J.A. (1974). The Late Neogene. *Palaeogeogr., Palaeoclimatol., Palaeoecol.*, **16** (1/2), 1–260.

Berggren, W.A. & Van Couvering, J.A. (1978). Biochronology. In Cohee *et al.* (1978), pp. 39–55.

Blackwelder, E. (1912). United States of America. In *Handbuch der Regionalen Geologie*, ed. G. Steinmann & O. Wilckens, vol. 8, part 2, pp. 1–258.

Blow, W.H. (1969). Late Middle Eocene to Recent planktonic foraminiferal biostratigraphy. In *Proceedings of the First International Conference on Planktonic Microfossils, Geneva, 1967.* Leiden: Brill.

Blow, W.H. (1979). *The Cainozoic Globigerinida.* Leiden: Brill.

Bond, G. (1978). Speculations on real sea-level changes and vertical motions of continents at selected times in the Cretaceous. *Geology*, **6**, 247–50.

* Bonhommet, N. & Zahringer, J. (1969). Paleomagnetism and potassium–argon age determinations of the Laschamp geomagnetic polarity event. *Earth Planet. Sci. Lett.*, **6**, 43–6.

Bouché, P.M. (1962). Nannofossiles calcaires du Lutétien du Bassin de Paris. *Revue Micropaléont.*, **5**(2), 75–103.

Boucot, A.J. (1975). *Evolution and extinction rate controls.* Amsterdam: Elsevier, 427 pp.

Bouroz, A. (1978). Report on isotopic dating of rocks in the Carboniferous System. In Cohee *et al.* (1978), pp. 323–6.

Bowen, D.Q. (1978). *Quaternary Geology*, Oxford: Pergamon Press, 221 pp.

Bramlette, M.N. & Sullivan, F.R. (1961). Coccolithophorids and related nannoplankton of the early Tertiary in California. *Micropaleontology*, **7**, 129–88.

Breistroffer, M. (1947). Sur les zones d'Ammonites de l'Albien de France et d'Angleterre *Trav. Lab. Géol. Grenoble Mém.*, **26**. 1–88.

Brown, L.E., Jr & Fisher, W.L. (1979). *Principles of seismic stratigraphic interpretation.* In AAPG–SEG Stratigraphic interpretation of seismic data school notes. Austin, Texas: American Association Petroleum Geologists Education Department.

Bukry, D. & Kennedy, M.P. (1969). *Cretaceous and Eocene coccoliths at San Diego, California.* San Francisco: California Division Mines and Geology Special Report 100.

Busnardo, R. (1965). Le stratotype du Barrémien: 1 – Lithologie et macrofaune. *Mémoires du Bureau de Recherches Géologiques et Minières*, No.34, 99–116.

* Butler, R.F. & Coney, P.J. (1981). A revised magnetic polarity time scale for the Paleocene and Early Eocene and implications for Pacific plate motion. *Geophys. Res. Lett.*, **8**, 301–4.

Butler, R.F., Lindsay, E.H., Jacobs, L.L. & Johnson, N.M. (1977). Magnetostratigraphy of the Cretaceous–Tertiary boundary in the San Juan basin, New Mexico. *Nature, Lond.*, **267**, 318–23.

Button, A. *et al.* (1981). The cratonic environment. In Hunter (1981), 501–639.

Cande, S.C., Larson, R.L. & LaBrecque, J.L. (1978). Magnetic lineations in the Pacific Jurassic quiet zone. *Earth Planet. Sci. Lett.*, **41**, 434–40.

* Carloni, G.C., Marks, P., Rutsch, R.F. & Selli, R. (1971). Stratotypes of Mediterranean Neogene stages, *G. Geol.*, ser.2, **37**, fasc. 2.

Casey, R. (1963). The dawn of the Cretaceous Period in Britain. *Bull. S.-East. Un. Sci. Socs.*, No. 117, 1–15.

Casey, R. (1967). The position of the Middle Volgian in the English Jurassic. *Proc. Geol. Soc. Lond.*, No. 1640, 246–7.

* Challinor, J. (1978). *A dictionary of geology*, 5th edn. Cardiff: University of Wales Press, 365 pp.

Champion, D.E., Dalrymple, G.B. & Kuntz, M.A. (1981). Radiometric and paleomagnetic evidence for the Emperor Reversed Polarity Event at 0.46 ± 0.05 m.y. in basalt lava flows from the eastern Snake River Plain, Idaho. *Geophys. Res. Lett.*, **8**, 1055–8.

Chen, C.-c. (1974). The Triassic System. In *Handbook of the stratigraphy and palaeontology in southwest China.* pp. 58–65. Beijing: Nanjing Institute of Geology and Palaeontology Academia Sinica, (in Chinese).

Chlupáč, I., Jaeger, H. & Zikmundova, J. (1972). The Silurian–Devonian boundary in the Barrandian. *Bull. Can. Petrol. Geol.*, **20** (1), 104–74.

Choubert, G. & Faure-Muret, A. (General co-ordinators), Chanteux, P. (Cartographic art) (Commission for the Geological Map of the World) (1976). *Geological world atlas 1/10 000 000*, Paris: Unesco.

* Churkin, M., Jr, Carter, C. & Johnson, B.R. (1977). Subdivision of Ordovician and Silurian time scale using accumulation rates of graptolitic shale. *Geology*, **5** (8), 452–6.

Cita, M.B. & Elter, G. (1960). La posizione stratigrafica delle marne a Pteropodi della Langhe della Collina di Torino ed il significato cronologico del Langhiano. *Accad. Naz. Lincei*, ser.8, **29** (5), 360–9.

Cita, M.B. & Premoli Silva, I. (1960). Pelagic foraminifera from the type Langhian. International Geological Reports, XXI Sess. Part XXII, *Proc. Internat. Paleont. Union*, 39–50, Copenhagen.

Cloud, P. (1972). A working model of the primitive earth. *Am. J. Sci.*, **272**, 537–48.

Cloud, P. (1976). *Major features of crustal evolution, 13th A.L. du Toit Memorial Lecture.* Geol. Soc. South Africa.

Coats, R.P. (1981). Late Proterozoic (Adelaidean) tillites of the Adelaide Geosyncline. In Hambrey & Harland (1981), pp. 537–48 (D21).

Cobban, W.A. & Reedside, J.B. Jr (1952). Correlation of the Cretaceous formations of the western interior of the United States. *Bull. Geol. Soc. Am.*, **63**, 1011–44.

Cocks, L.R.M., Holland, C.H., Rickards, R.B. & Strachan, I. (1971). *A correlation of Silurian rocks in the British Isles.* Geol. Soc. Lond., Special Paper No. 1., 136 pp.

Cocks, L.R.M., Toghill, P. & Ziegler, A.M. (1970). Stage names within the Llandovery Series. *Geol. Mag.*, **107** (1), 79–87.

Cogley, N.G. (1981). Late Phanerozoic extent of dry land. *Nature, Lond.*, **291**, 56–8.

Cohee, G.V. (1970). *Generally recognised European stages.* Paper issued April 15, 1970 by George V. Cohee, Chairman AAPG Advisory Committee on Stratigraphic Coding.

Cohee, G.V. *et al.* (1967). Standard stratigraphic code adopted by AAPG. *Bull. Am. Assoc. Petrol. Geol.*, **51** (10), 2146–50.

Cohee, G.V., Glaessner, M.F. & Hedberg, H.D. (eds.) (1978). *Contributions to the geologic time scale*, papers given at the Geological Time Scale Symposium 106.6, 25th IGC Sydney, Australia, August 1976. Tulsa: American Association of Petroleum Geologists, Studies in Geology No. 6, 388 pp.

Colalongo, M.L. (1970). Appunti biostratigrafical sul Messiniano. *Gionale di Geologia ser.2*, **36**, 515–42.

* Conil, R., Groessens, E. & Pirlet, H. (1977). Nouvelle charte stratigraphique du Dinantian type de la Belgique. *Annls. Soc. Géol. N.*, **96**, 363–71.

Conybeare, W.D. & Phillips, W. (1822). *Outlines of the geology of England and Wales, with an introductory compendium of the general principles of that science, and comparative view of the structure of foreign countries, Part I*, London: Phillips, 470 pp.

* Cooke, C.W., Gardner, J. & Woodring, W.P. (1943). Correlation of the Cenozoic formations of the Atlantic Coastal Plain and Caribbean region, *Bull. Geol. Soc. Am.*, **53**, 569–98.

* Cooper, G.A. *et al.* (1942). Correlation of the Devonian sedimentary formations of North America. *Bull. Geol. Soc. Am.*, **53**, 1729–94.

Cooper, M.R. (1977). Eustacy during the Cretaceous: its implication and importance, *Palaeogeogr., Palaeoclimatol., Palaeoecol.*, **22**, 1–60.

Cope, J.C.W., Duff, K.L., Parsons, C.F., Torrens, H.S., Wimbledon, W.A. & Wright, J.K. (1980a). *A correlation of Jurassic rocks in the British Isles. Part two: Middle and Upper Jurassic.* Geol. Soc. Lond., Special Report No. 15, 109 pp.

Cope, J.C.W., Getty, T.A., Howarth, M.K., Morton, N. & Torrens, H.S. (1980b). *A correlation of Jurassic rocks in the British Isles. Part one: Introduction and Lower Jurassic*. Geol. Soc. London., Special Report No. 14, 73 pp.

Cowie, J.W. (in press). In *Geology of England and Wales*, ed. A.J. Smith & McDuff. London: Academic Press.

Cowie, J.W. & Cribb, S.J. (1978). The Cambrian System. In Cohee *et al.* (1978), pp. 355–62.

Cowie, J.W., Rushton, A.W.A. & Stubblefield, C.J. (1972). *A correlation of Cambrian rocks in the British Isles.* Geol. Soc. Lond., Special Report No. 2, 42 pp.

Cox, A.V. (1968). Lengths of geomagnetic polarity intervals. *J. Geophys. Res.*, 73 (10), 3247–60.

Cox, A.V. (1981). A stochastic approach towards understanding the frequency and polarity bias of geomagnetic reversals. *Phys. Earth Planet. Interiors*, **24**, 178–90.

Cox, A.V. & Dalrymple, G.B. (1967). Statistical analysis of geomagnetic reversal data and the precision of potassium–argon dating. *J. Geophys. Res.*, 72 (10), 2603–14.

Creer, K.M., Readman, P.W. & Jacobs, A.M. (1980). Paleomagnetic and paleontological dating of a section of Gioia Tauro, Italy: identification of the Blake Event. *Earth Planet. Sci. Lett.*, **50**, 289–300.

Curry, D., Adams, C.G., Boulter, M.C., Dilley, F.C., Eames, F.E., Funnell, B.M. & Wells, M.K. (1978). *A correlation of Tertiary rocks in the British Isles.* Geol. Soc. Lond., Special Report No. 12, 72 pp.

* Dalrymple, G.B. (1979). Critical tables for conversion of K–Ar ages from old to new constants. *Geology*, 7 (11), 558–60.

Debelmas, J. & Thieuloy, J.P. (1965). Étage Hauterivien. *Mémoires du Bureau de Recherches Géologiques et Minières*, No. 34, 85–96.

* Denham, C.R., Anderson, R.F. & Bacon, M.P. (1977). Paleomagnetism and radiochemical age estimates for Late Bruhnes polarity episodes. *Earth Planet. Sci. Lett.*, **35**, 384–97.

* de Rouville, P.G. (1853). *Description géologique des environs de Montpellier.* Montpellier: Boehm, 185 pp.

Dietl, G. & Etzold, A. (1977). The Aalenian at the type locality. *Beitr. Naturk. Stuttgart*, ser. B, No. 30, 1–13.

d'Onofrio, S. (1964). IForaminiferi del neostratotipo del Messiniano. *G. Geol.*, ser.2, **32** (2), 409–61.

Donovan, D.T. & Jones. E.J.W. (1979). Causes of world-wide changes in sea level. *J. Geol. Soc. Lond.*, **136** (2), 187–92.

Douglas, R.J.W. (1980). *Proposals for time classification and correlation of Precambrian rocks and events in Canada and adjacent areas of the Canadian Shield.* Geol. Surv. Can. Paper, 80–24, 19 pp.

Drooger, C.W. (1964). Problems of mid-Tertiary stratigraphic interpretation. *Micropaleontology*, **10**(3), 369–74.

* Drury, S.A. *et al.* (1976). *Lunar geology case study.* Earth Science Topics and Methods. Milton Keynes: Open University Press, 116 pp.

Dunbar, C.O. (1960). Correlation of the Permian formations of North America. *Bull. Geol. Soc. Am.*, **71**, 1763–1806.

* Dunbar, C.O. *et al.* (1942). Correlation charts prepared by the Committee on Stratigraphy of the National Research Council. *Bull. Geol. Soc. Am.*, **53**, 429–34.

Edwards, M.B. & Føyn, S. (1981). Late Proterozoic tillites in Finnmark, North Norway. In Hambrey & Harland (1981), pp. 606–10 (E12).

Egyed, L. (1956). Change of earth dimensions as determined from palaeogeographical data. *Geofisica Pura e Applicata*, **33**, 42–8.

El-Naggar, Z.R. (1966a). Stratigraphy and planktonic foraminifera of the Upper Cretaceous–Lower Tertiary succession in the Esna–Idfu region, Nile Valley, Egypt, UAR. *Bull. Br. Mus. Nat. Hist. (Geology, Supplement)*, **2**, 1–279.

El-Naggar, Z.R. (1966b). Stratigraphy and classification of type Esna Group of Egypt. *Bull. Am. Ass. Petrol. Geol.*, **50** (7), 455–77.

Emiliani, C. (1965). Pleistocene temperatures. *J. Geol.*, **63**, 538–78.

Emiliani, C. (1966). Paleotemperature analysis of Caribbean cores P6304–8 and P6304–9 and a generalized temperature curve for the past 425 000 years. *J. Geol.*, **74**, 109–26.

Evans, P. (1971). Towards a Pleistocene time-scale. In Harland *et al.* (1971), pp. 123–356.

Fairbridge, R.W. (1961). Eustatic changes in sea-level. In *Physics and chemistry of the earth*, vol. 4, ed. L.H. Ahrens *et al.*, pp. 99–185. London: Pergamon Press.

Fewtrell Smith, M. (compiler) (1981). Open University Handbook and Wall Chart, s364, *Evolution.* Milton Keynes: Open University.

Flemming, N.C. & Roberts, D.G. (1973). Tectono-eustatic changes in sea-level and sea-floor spreading. *Nature, Lond.*, **243**, 19–22.

Flint, R.F. (1971). *Glacial and Quaternary geology.* New York: Wiley, 892 pp.

* Frebold, H. (1953). Correlation of the Jurassic formations of Canada. *Bull. Geol. Soc. Am.*, **64**, 1229–46.

* Frith, R.A. (1979). Precambrian division. *Geology*, 8 (3), 19. (Also summarised in *Open Earth*, **5**, 13.)

* Gale, N.H., Beckinsale, R.D. & Wadge, A.J. (1979). A Rb–Sr whole-rock isochron for the Stockdale Rhyolite of the English Lake District and a revised mid-Palaeozoic time-scale. *J. Geol. Soc. Lond.*, **136** (2), 235–42.

Gartner, S. (1977). Calcareous nannofossil biostratigraphy and revised zonation of the Pleistocene. *Marine Micropaleontology*, **2**, 1–25.

* Geological Society of London (1968). International Geological Correlation Programme – United Kingdom Contribution – *Recommendations on stratigraphical classification.* London: The Royal Society, 43 pp. (Typewritten report.)

George, T.N. *et al.* (1967). The stratigraphical code – report of the stratigraphical code sub-committee. *Proc. Geol. Soc. Lond.*, No. 1638, 75–87.

George, T.N. *et al.* (1969). Recommendations on stratigraphical usage. *Proc. Geol. Soc. Lond.*, No. 1656, 139–66.

George, T.N., Johnson, G.A.L., Mitchell, M., Prentice, J.E., Ramsbottom, W.H.C., Sevastopulo, G.D. & Wilson, R.B. (1976). *A correlation of Dinantian rocks in the British Isles.* Geol. Soc. Lond., Special Report No. 7., 87 pp.

George, T.N. & Wagner, R.H. (1972). IUGS Subcommission on Carboniferous Stratigraphy, *C.R. 7me Cong. int. Strat. Géol. Carb. Krefeld 1971*, 139–47.

Gerasimov, P., Kuznetsova, K., Mikchailov, N.P. & Uspenskaya, E.A. (1975). Correlation of the Volgian, Portlandian and Tithonian stages. *Mémoires du Bureau de Recherches Géologiques et Minières*, (Colloque sur La Limite Jurassique–Crétacé, 1973, Lyon, Neuchâtel).

Geyer, O.F. (1964). Die Typuslokalität des Pliensbachium in Württemberg Südwestdeutschland. *Colloque du Jurassique, Luxembourg, Vol. des C.R. et Mém.*, 161–7.

* Gignoux, M. (1955). *Stratigraphic geology.* London: Freeman, 682 pp.

Gino, G.F. *et al.* (1953). Studi stratigrafice e micropaleontologiche sull'Apennino Tortonese. In *Observazione geologiche sui Dintorne di Sant'Agata Fossili (Tortona Alessandria)*, pp. 7–24, Milan: Memoria Rivista Italiana di Paleontologia e Stratigrafia VI.

Glaessner, M. (1977). The Ediacara fauna and its place in the evolution of the Metazoa. In *Correlation of the Precambrian*, vol. 1, ed. A.V. Sidorenko, pp. 257–68.

* Gordon, M. Jr & Mamet, B.L. (1978). Moscow: Academy of Sciences USSR Committee for IGCP. The Mississippian–Pennsylvanian boundary. In Cohee *et al.* (1978), pp. 327–35.

* Gradstein, F.M. (1978). A revision of the Mesozoic–Cenozoic time-scale. *Geology*, 7 (3), 34. (Also summarised in *Open Earth*, 3, 13.)

Gregory, J.W. & Barrett, B.H. (1931). *General stratigraphy*. London: Methuen, 285 pp.

*Hall, C.M. & York, D. (1978). K–Ar and ^{40}Ar/^{39}Ar age of the Laschamp geomagnetic polarity reversal. *Nature, Lond.*, **274**, 462–4.

*Hallam, A. (1975). *Jurassic environments*. Cambridge University Press, 269 pp.

Hallam, A. (1977). Secular changes in marine inundation of USSR and North America through the Phanerozoic, *Nature, Lond.*, **269**, 769–72.

Hallam, A. (1978). Eustatic cycles in the Jurassic, *Palaeogeogr., Palaeoclimatol., Palaeoecol.*, **23**, 1–32.

Hallam, A. (1981). A revised sea-level curve for the early Jurassic. *J. Geol. Soc. Lond.*, **138**, 735–43.

Hallberg, J.A. & Glikson, A.Y. (1981). Archaean granite–greenstone terranes of western Australia. In Hunter (1981), pp. 33–103.

Hambrey, M.J. & Harland, W.B. (eds.) (1981). *Earth's pre-Pleistocene glacial record.* Cambridge University Press, 1004 pp.

Hancock, J.M. & Kauffman, E.G. (1979). The great transgression of the Late Cretaceous. *J. Geol. Soc. Lond.*, **136**, 175–86.

Hansen, H.J. (1970). Danian foraminifera from Nugssuaq, West Greenland, *Bull. Grønlands Geol. Undersøgelse*, No. 93, 1–132.

Haq. B., Berggren, W.A. & Van Couvering, J.A. (1977). Corrected age of the Pliocene/Pleistocene boundary. *Nature, Lond.*, **269**, 483–8.

Hardenbol, J. (1968). The Priabon type section (France). *Mémoires du Bureau du Recherches Géologiques et Minières*, No. 58, 629–35.

Hardenbol, J. & Berggren, W.A. (1978). A new Paleogene numerical time scale. In Cohee *et al.* (1978), pp. 213–34.

Harland, W.B. (1974). The Pre-Cambrian – Cambrian boundary. In *Cambrian of the British Isles, Norden and Spitsbergen, vol. 2. Lower Palaeozoic rocks of the world*, ed. C.H. Holland, pp. 15–42. London: Wiley.

Harland, W.B. (1975). The two geological time scales, *Nature, Lond.*, **253**, 505–7.

Harland, W.B. (1978). Geochronologic scales. In Cohee *et al.* (1978), pp. 9–32.

Harland, W.B. (in press). The Proterozoic glacial record. *Mem. Geol. Soc. Am.*

*Harland, W.B. *et al.* (eds.) (1967). *The fossil record*. London: Geological Society of London, 827 pp.

Harland, W.B. *et al.* (eds.) (1971). *The Phanerozoic time-scale – a supplement.* London: Geological Society of London Special Publication No. 5, 356 pp.

Harland, W.B. & Herod, K.M. (1975). Glaciations through time. In *Ice ages: ancient and modern*, ed. A.E. Wright & F. Moseley, pp. 189–216, Geological Journal Special Issue 6. Liverpool: Steel Horse Press.

Harland, W.B., Smith, A.G. & Wilcock, B. (eds.) (1964). *The Phanerozoic time-scale.* (A symposium dedicated to Professor Arthur Holmes). Quarterly Journal Geological Society of London 120s, 458 pp.

* Harris, A.L., Shackleton, R.M., Watson, J., Downie, C., Harland, W.B. & Moorbath, S. (1975). *A correlation of Precambrian rocks in the British Isles.* Geol. Soc. Lond., Special Report No. 6., 136 pp.

Harrison, C.G.A., Brass, G.W., Saltzman, E., Sloan, J., Southam, J. & Whitman, J.M. (1981). Sea level variations, global sedimentation rates and the hypsographic curve. *Earth Planet. Sci. Lett.*, **54**, 1–16.

Hay, W.W. (1967). Calcareous nannoplankton zonation of the Cenozoic of the Gulf Coast and Caribbean–Antillean area and transoceanic correlation. *Trans. Gulf Cst. Ass. Geol. Socs.*, 17, 428–80.

Hay, W.W. & Mohler, H.P. (1967). Calcareous nannoplankton from early Tertiary rocks at Pont Labau, France, and Paleocene – early Eocene correlations, *J. Paleont.*, **41**, 1505–41.

Hays, J.D., Imbrie, J. & Shackleton, N.J. (1976). Variations in the Earth's orbit: pacemaker of the ages. *Science*, **194**, 1121–32.

Hays, J.D. & Pitman, W.C. (1973). Lithospheric plate motion, sea-level changes, and climatic and ecological consequences. *Nature, Lond.*, **246**, 18–22.

Hedberg, H.D. (1976). *International stratigraphic guide*, New York: Wiley, 200 pp.

Heirtzler, J.R., Dickson, G.O., Herron, E.M., Pitman, W.C.III & Le Pichon, X. (1968). Marine magnetic anomalies, geomagnetic field reversals, and motions of the ocean floor and continents. *J. Geophys. Res.*, **73**, 2119–36.

Helsley, C.E. & Steiner, M.B. (1969). Evidence for long intervals of normal polarity during the Cretaceous period, *Earth Planet. Sci. Lett.*, **5**, 325–32.

Hill, D. (1967). Devonian of Eastern Australia. In *International Symposium on the Devonian System*, ed. D.H. Oswald, vol. 1, pp. 613–30. Calgary: Alberta Society of Petroleum Geologists.

Holland, C.H. (1980). Silurian series and stages: decisions concerning chronostratigraphy. *Lethaia*, **13** (3), 238.

Holland, C.H. *et al.* (1978). *A guide to stratigraphical procedure.* Geol. Soc. Lond., Special Report No. 11., 18 pp.

Holland, C.H., Lawson, J.D. & Walmsley, V.G. (1963). The Silurian rocks of the Ludlow district, Shropshire. *Bull. Br. Mus. Nat. Hist. (Geol.)*, 8 (3), 93–171.

*Holland, C.H., Lawson, J.D., Walmsley, V.G. & White, D.E. (1980). Ludlow stages. *Lethaia*, **13** (3), 268.

Holmes, A. (1937). *The age of the earth.* London: Nelson, 263 pp.

Holmes, A. (1947). The construction of a geological time-scale. *Trans. Geol. Soc. Glasg.*, **21**, 117–52.

Holmes, A. (1959). A revised geological time-scale. *Trans. Edinb. Geol. Soc.*, **17** (3), 183–216.

House, M.R., Richardson, J.B., Chaloner, W.G., Allen, J.R.L., Holland, C.H., & Westoll, T.S. (1977). *A correlation of Devonian rocks in the British Isles.* Geol. Soc. Lond., Special Report No. 8, 110 pp.

Howarth, M.K. (1955). Domerian of the Yorkshire coast. *Proc. Yorks. Geol. Soc.*, **30**, 147–75.

Hughes, N.F., Williams, D.B., Cutbill, J.L. & Harland, W.B. (1967). A use of reference points in stratigraphy. *Geol. Mag.*, **104**, 634–5.

Hunter, D.R. (ed.) (1981). *Precambrian of the southern hemisphere. Developments in Precambrian geology 2.* Amsterdam: Elsevier, 882 pp.

Hunter, D.R. & Pretorius, D.A. (1981). Structural framework. In Hunter (1981), pp 397–422.

* Imlay, R.W. (1944). Correlation of the Cretaceous formations of the Greater Antilles, Central America and Mexico. *Bull. Geol. Soc. Am.*, **55**, 1005–45.

* Imlay, R.W. & Reedside, J.B., Jr (1954). Correlation of the Cretaceous formations of Greenland and Alaska. *Bull. Geol. Soc. Am.*, **65**, 223–46.

Irving, E. & Couillard, R.W. (1973). Cretaceous normal polarity interval. *Nature Physical Science*, **244** (131), 10–11.

Irving, E. & McGlynn, J.C. (1976). Proterozoic magnetostratigraphy and the tectonic evolution of Laurentia, *Phil. Trans. R. Soc. A*, **280**, 433–68.

Irving, E. & Parry, L.G. (1963). The magnetism of some Permian rocks from New South Wales. *Geophys. J. R. Astr. Soc.*, 7, 395–411.

Irving, E. & Pullaiah, G. (1976). Reversals of the geomagnetic field, magnetostratigraphy, and relative magnitude of paleosecular variation in the Phanerozoic, *Earth Sci. Rev.*, 12, 35–64.

James, H.L. (1972). Subdivision of Precambrian: an interior scheme to be used by US Geological Survey. *Bull. Am. Assoc. Petrol. Geol.*, 56, 1026–30.

* James, H.L. (1979). Precambrian subdivided. *Episodes*, 1979, No. 4, 34.

Jenkins, R.J.F. (in press). The concept of an 'Ediacaran Period' and its stratigraphic significance in Australia. *Trans. R. Soc. S. Aust.*, 105.

* Johnsen, S.J., Dansgaard, W., Clausen, H.B. & Langway, C.C. (1972). Oxygen isotope profile through the Antarctic and Greenland ice sheets. *Nature, Lond.*, 235, 429–34.

Johnson, C.D. & Hills, L.V. (1973). Microplankton zones of the Savik Formation (Jurassic), Axel Heiberg and Ellesmere Islands, District of Franklin. *Bull. Can. Petrol. Geol.*, 21, 178–218.

Keller, B.M. (1979). Precambrian stratigraphic scale of the USSR. *Geol. Mag.*, 116 (6), 419–29.

Kent, L.E. & Hugo, P. J. (1978). Aspects of the revised South African stratigraphic classification and a proposal for the chronostratigraphic subdivision of the Precambrian. In Cohee *et. al.* (1978), pp. 367–79.

* Khramov, A.N. (1963). Paleomagnitnoye izucheniye rasrezov verkhney permi i nizhnego triasa severa i vostoke Russkoy platformy (Paleomagnetic study of sections of Upper Permian and Lower Triassic of the northern and eastern Russian platform). In *Paleomagnitnyye Stratigrafichecheskiye Issledovaniya, Vses. Neft. Nauchno-Issled. Geol. Razved. Inst. Trudy*, no. 204, pp. 145–74.

Khramov, A.N. (1967). The earth's magnetic field in the late Paleozoic, (Akad. Nauk SSSR Izvestiya). *Phys. Solid Earth*, 1, 50–63.

Kilian, W. (1887). Note géologique sur la chaîne de Lure (Basses-Alpes). *Feuille Jeun. Nat.*, 17, 53.

Kiparisova, L.D. & Popov, Yu.N. (1956). Subdivision of the lower series of the Triassic system into stages. *Dokl. Akad. Sci. USSR*, 109 (4), 842–5. (in Russian).

Kiparisova, L.D. & Popov, Yu.N. (1964). The project of subdivision of the Lower Triassic into stages. *Int. Geol. Congress, XXII Session, Reports Soviet Geologists, Problem 16A*, 91–9 (in Russian).

Kiparisova, L.D., Radchenko, G.P. & Gorskiy, V.P. (eds.) (1973). *Triasovaya sistema (Triassic System)*, 537 pp. In *Stratigrafiya SSSR (Stratigraphy of the USSR)*, 14 vols, ed. D.V. Nalivkin. Moscow: Nedra Press.

Kotlyar, G.V. (ed.) (1977). Karbon. Perm. (The Carboniferous and Permian), 535 pp. In series *'Stratigraficheskiy slovar' SSSR. (A stratigraphical dictionary of the USSR)*, ed V.N. Vereschagin. Leningrad: Nedra Press.

* Kröner, A. (ed.) (1981). *Precambrian plate tectonics, Developments in Precambrian Geology 4.* Amsterdam: Elsevier, 781 pp.

Krymgol'ts, G. Ya (ed.) (1972). *Yurskaya sistema. (Jurassic System)*, 524 pp. In *Stratigrafiya SSSR (Stratigraphy of the USSR)*, 14 vols, ed. D.V. Nalivkin. Moscow: Nedra Press.

* Kukla, G.J. (1977). Pleistocene land-sea correlations. I. Europe. *Earth Sci. Rev.*, 13, 307–74.

Kulling, O. (1951). Spar av Varangeristiden i Norbotten. Eocambriska Varvskiffrar i Nordbottens fjällens ostra rand i nordligaste Sverige. *Sver. Geol. Unders. Afh.*, ser. C, 43 1–44.

Kulp, J.L. (1961). Geologic time-scale. *Science*, 133, 1105–14.

Kummel, B. (1961). *History of the earth.* San Francisco: Freeman, 610 pp.

Kummel, B. & Teichert, C. (1966). Relations between the Permian and Triassic formations in the Salt Range and Trans-Indus ranges, West Pakistan. *Neues Jb. Geol. Paläont. Abh.*, 125, 297–333.

LaBrecque, J.L., Kent, D.V. & Cande, S.C. (1977). Revised magnetic polarity time scale for Late Cretaceous and Cenozoic time. *Geology*, 5 (6), 330–5.

Lamb, J.L. & Stainforth, R.M. (1976). Unreliability of *Globigerinoides* datum, *Bull. Am. Assoc. Petrol Geol.*, 60 (9), 1564–9.

Lambert, R.St. J. (1971). The pre-Pleistocene Phanerozoic time-scale – a review. In Harland *et al.* (1971), pp. 9–34.

Lang, W.D. (1928). The Belemnite Marls of Charmouth, a series in the Lias of the Dorset Coast. *Quart. J. Geol. Soc. Lond.*, 74, 179–257.

* Lanphere, M.A., Churkin, M. Jr & Eberlein, G.D. (1977). Radiometric age of the *Monograptus cyphus* graptolite zone in southeastern Alaska – an estimate of the age of the Ordovician–Silurian boundary. *Geol. Mag.*, 114 (1), 15–24.

Lanphere, M.A. & Jones, D.L. (1978). Cretaceous time scale from North America. In Cohee *et al.* (1978), pp. 259–68.

Larcher, C., Rat, P. & Malapris, M. (1965). Documents paléontologiques et stratigraphiques sur l'Albien de l'Aube. *Mémoires du Bureau de Recherches Géologiques et Minières*, No. 34, 237–53.

Larson, R.L., Golovchenko, X. & Pitman, W.C. III (1981). Geomagnetic polarity time scale. In *Plate tectonic map of the circum-Pacific region, northeast quadrant*, ed. Ch.K.J. Drummond. Tulsa: The American Association of Petroleum Geologists.

Larson, R.L. & Hilde, T.W.C. (1975). A revised time scale of magnetic reversals for the Early Cretaceous and Late Jurassic. *J. Geophys. Res.*, 80, 2586–94.

Larson, R.L. & Pitman, W.C. III (1972). World-wide correlation of Mesozoic magnetic anomalies and its implications. *Bull. Geol. Soc. Am.*, 83 (12), 3645–62.

Le Pichon, X. & Heirtzler, J.R. (1968). Magnetic anomalies in the Indian Ocean and sea-floor spreading. *J. Geophys. Res.*, 73, 2101–17.

Likharev, B.K. (ed.) (1966). *The Permian System*, 536 pp. In *Stratigrafiya SSSR (Stratigraphy of the USSR)*, 14 vols, ed. D.V. Nalivkin. Moscow: Nedra Press. (in Russian)

* Lowman, P.D., Jr (1972). The geologic evolution of the moon. *J. Geol.* 80 (2), 125–66.

Lowrie, W. & Alvarez, W. (1981). One hundred million years of geomagnetic polarity history. *Geology*, 9 (9), 392–7.

Lowrie, W., Channell, J.E.T. & Alvarez, W. (1980). A review of magnetic stratigraphy investigations in Cretaceous pelagic carbonate rocks. *J. Geophys. Res.*, 85, 3597–605.

* Mankinen, E.A. & Dalrymple, G.B. (1978). Revised Late Cenozoic geomagnetic polarity time-scale. *USGS Prof. Paper* 1100 (Research reports), 167.

Mankinen, E.A. & Dalrymple, G.B. (1979). Revised geomagnetic polarity time scale for the interval 0–5 m.y. BP *J. Geophys. Res.*, 84 (B2), 615–26. (Also summarised in *Open Earth*, 4, 23–4.)

Marks, P. (1967). *Rotalipora* et *Globotruncana* dans la Craie de Théligny (Cénomanien, Dépt. de la Sarthe). *Proc. K. Ned. Akad. Wet.*, ser. B, 70, 264–75.

* Marks, P. (1977). Micropaleontology and the Cenomanian-Turonian boundary problem. *Proc. K. Ned. Akad. Wet.*, ser. B. 80, 1–6.

Martini, E. (1971). Standard Tertiary and Quaternary calcareous nannoplankton zonation. In *Proceedings of the II Planktonic Conference Roma, 1969*, pp. 739–85. Rome: Edizioni Tecnoscienza.

Martini, E. & Müller, C. (1975). Calcareous nannoplankton from the type Chattian (upper Oligocene). *6th Congr. Reg. Comm. Mediterranean Neogene Stratigraphy Proc.*, 37–41.

Martinsson, A. (1974). The Cambrian of Norden. In *Cambrian of the British Isles, Norden and Spitsbergen, vol. 2 of Lower Palaeozoic rocks of the world*, ed C.H. Holland, pp. 185–283. London: Wiley.

* Martinsson, A., Bassett, M.G. & Holland, C.H. (1981). Ratification of standard chronostratigraphical divisions and stratotypes for the Silurian System. *Episodes*, 1981 (2), 36.

McDougall, I. (1978). Revision of the geomagnetic polarity time scale for the last 5 m.y. In Zartman (1978), pp. 287–9.

McDougall, I., Watkins, N.D., Walker, G.P.L. & Kristjansson, L.

(1976). Potassium-argon and paleomagnetic analysis of Icelandic lava flows: limits on the age of anomaly 5. *J. Geophys. Res.*, 81, 1505–12.

McElhinny, M.W. (1971). Geomagnetic reversals during the Phanerozoic. *Science*, 172, 157–9.

McElhinny, M.W. (1978). The magnetic polarity time scale: prospects and possibilities in magnetostratigraphy. In Cohee *et al.* (1978), pp. 57–65.

McElhinny, M.W. & Burek, P.J. (1971). Mesozoic palaeomagnetic stratigraphy. *Nature, Lond.*, **232**, 98–102.

McKerrow, W.S., Lambert, R.St J. & Chamberlain, V.E. (1980). The Ordovician, Silurian and Devonian time scales. *Earth Planet. Sci. Lett.*, **51** (1), 1–8.

McLaren, D.J. (1977). The Silurian–Devonian Boundary Committee. In *The Silurian–Devonian boundary*. IUGS Series A, No. 5, ed. A. Martinsson, pp. 1–34. Stuttgart: Schweizerbart'sche Verlagsbuchhandlung.

* McLean, F.H. (1953). Correlation of the Triassic formations of Canada. *Bull. Geol. Soc. Am.*, **64**, 1206–28.

Mitchell, G.F., Penny, L.F., Shotton, F.W. & West, R.G. (1973). *A correlation of Quaternary deposits in the British Isles.* Geol. Soc. Lond., Special Report No. 4, 99 pp.

* Mitchum, R.M., Jr, Vail, P.R. & Thompson, S., III. (1977). Seismic stratigraphy and global changes of sea level, Part 2: The depositional sequence as a basic unit for stratigraphic analysis. In Payton (1977), pp. 53–97.

* Moore, R.C. *et al.* (1944). Correlation of Pennsylvanian formations of North America. *Bull. Geol. Soc. Am.*, **55**, 657–706.

Mörner, N.-A. (1976). The Pleistocene/Holocene boundary: proposed boundary stratotype in Gothenburg, Sweden. *Boreas*, 5, 193–275.

Mörner, N.-A. (1981). Revolution in Cretaceous sea-level analysis. *Geology*, 9, 344–6.

Morton, N. (1971). The definition of standard Jurassic stages. *Mémoires du Bureau de Recherches Géologiques et Minières*, No. 75, 83–93.

Murray, G.E. (1961). *Geology of the Atlantic and Gulf Coastal Province of North America.* New York: Harper.

* Naeser, C.W., Ross, R.J. & Izett, G.A. (1978). Fission-track dating of the type Ordovician and Silurian. *USGS Prof. Paper* 1100 (Research Reports), 191.

Nalivkin, D.V. (1973). *Geology of the USSR.* Edinburgh: Oliver & Boyd, 855 pp. (First published by The Academy of Sciences, USSR, Moscow–Leningrad.)

Ness, G., Levi, S. & Couch, R. (1980). Marine magnetic anomaly time-scales for the Cenozoic and Late Cretaceous: a precis, critique, and synthesis. *Rev. Geophys. Space Phys.*, No. 18, 753–70.

Nikiforova, K.V. (1978). Status of the boundary between Pliocene and Pleistocene. In Cohee *et al.* (1978), pp. 171–8.

Nisbet, E.G., Wilson, J.F. & Bickle, M.J. (1981). Evolution of the Rhodesian and adjacent Archaean terrain: tectonic models. In Kröner (1981), pp. 161–83.

Norford, B.S., Bolton, T.E., Copeland, L.M., Cumming, L.M. & Sinclair, G.W. (1970). Ordovician and Silurian faunas. In *Geology and economic minerals of Canada*, 5th edn, ed. R.J.W. Douglas, pp. 601–13 Ottawa: Geological Survey of Canada, Economic Geology No. 1.

* Obradovich, J.D. & Cobban, W.A. (1975). A time-scale for the Late Cretaceous of the Western Interior of North America. In *The Cretaceous System in the Western Interior of North America* (proceedings of symposium, Saskatchewan, May 1973), ed. W.G.E. Caldwell, pp. 31–54. Geological Association Canada Special Paper 13.

Obradovich, J.D. & Cobban, W.A. (1978). K–Ar dating of the Albian. *USGS Prof. Paper* 1100 (Research Reports), 191.

Odin, G.S. (1978a). Results of dating Cretaceous, Paleogene sediments, Europe. In Cohee *et al.* (1978), pp. 127–41.

Odin, G.S. (1978b). Isotopic dates for a Paleogene time scale. In Cohee *et al.* (1978), pp. 247–57.

Oliver, W.A.Jr, De Witt, W.Jr, Dennison, J.M., Hoskins, D.M. & Huddle, J.W. (1967). Devonian of the Appalachian Basin, United States. In *International Symposium on the Devonian System*, Vol. 1, ed. D.H. Oswald, pp. 1001–40. Calgary: Alberta Society of Petroleum Geologists.

Oradovskaya, M.M. & Sobolevskaya, R.F. (compilers) (1979). *Guidebook to field excursion to the Omulev Mountains, Tour VIII.* Problem: The Ordovician–Silurian boundary. Pacific Science Association XIV, Pacific Science Congress, Magadan, 50 pp.

Paproth, E. (1980). The Devonian–Carboniferous boundary. *Lethaia*, **13**, 287.

Pareto, M.F. (1865). Note sur la subdivision, que l'on pourrait établir dans les terrains tertiaires de l'Appenin septentrional. *Bull. Soc. Géol. Fr.*, sér.2, **22**, 210–77.

Payton, C.E. (ed.) (1977). *Seismic stratigraphy – applications to hydrocarbon exploration.* Tulsa: Memoir American Association Petroleum Geologists 26.

Pechersky, D.M. & Khramov, A.N. (1973). Mesozoic palaeomagnetic scale of the USSR. *Nature, Lond.*, **244**, 499–501.

Perch-Nielsen, K. (1972). Les nannofossiles calcaires de la limite Crétace-Tertiaire (France). *Mémoires du Bureau de Recherches Géologiques et Minières*, No. 77, 181–8.

Pitman, W.C. (1978). Relationship between eustacy and stratigraphic sequences of passive margins. *Bull. Geol. Soc. Am.*, 89, 1389–1403.

Pitman, W.C., III, Herron, E.M. & Heirtzler, J.R. (1968). Magnetic anomalies in the Pacific and sea floor spreading. *J. Geophys. Res.*, **73**, 2069–85.

* Pomerol, C. (1978). Critical review of isotopic dates in relation to Paleogene stratotypes. In Cohee *et al.* (1978), pp. 235–45.

* Pomerol, C. (1981). *The Cenozoic Era.* Chichester: Wiley.

* Popenoe, W.P., Imlay, R.W. & Murphy, M.A. (1960). Correlation of the Cretaceous formations of the Pacific coast (United States and northwestern Mexico). *Bull. Geol. Soc. Am.*, 71, 1491–1540.

Postuma, J.A. (1971). *Manual of planktonic foraminifera.* Amsterdam: Elsevier.

Rampino, M.R. (1981). Revised age estimates of Bruhnes palaeomagnetic events: support for a link between geomagnetism and eccentricity. *Geophys. Res. Lett.*, 8, 1047–50.

Ramsbottom, W.H.C. (ed.) (1981). *Field guide to the boundary stratotypes of the Carboniferous stages of Britain.* IUGS Subcommission on Carboniferous Stratigraphy.

Ramsbottom, W.H.C., Calver, M.A., Eagar, R.M.C., Hodson, F., Holliday, D.W., Stubblefield, C.J. & Wilson, R.B. (1978). *A correlation of Silesian rocks in the British Isles.* Geol. Soc. Lond., Special Report No. 10, 81 pp.

Rauser-Chernousova, D.M. & Shchegolev, A.K. (1979). The Carboniferous–Permian boundary in the USSR. In Wagner *et al.* (1979), pp. 175–95.

Rawson, P.F., Curry, D., Dilley, F.C., Hancock, J.M., Kennedy, W.J., Neale, J.W., Wood, C.J. & Worssam, B.C. (1978). *A correlation of Cretaceous rocks in the British Isles.* Geol. Soc. Lond., Special Report No. 9, 70 pp.

Rea, D.K. & Blakely, R.J. (1975). Short-wavelength magnetic anomalies in a region of rapid seafloor spreading, *Nature, Lond.*, **225**, 126–8.

* Reedside, J.B., Jr *et al.* (1957). Correlation of the Triassic formations of North America exclusive of Canada. *Bull. Geol. Soc. Am.*, 68, 1451–1514.

* Rénevier, E. (1867). Notices géologiques et paléontologiques sur les Alpes vandoises et les régions environnantes: V. Complément de la faune de Cheville. *Bull. Soc. Vaud. Sci. Nat.*, 9, 115–208.

Rénevier, E. (1897). Chronologie géologique. *Congrès géologique international, IV, Sess., Zürich, 1894*, 523–695.

* Reyment, R.A. & Mörner, N.A. (1977). Cretaceous transgressions and regressions exemplified by the South Atlantic. *Paleont. Soc. Japan, Spec. Papers*, No. 21, 217–61.

* Richards, J.R. (1978). The length of the Devonian Period. In Zartman (1978), p. 351.

Richter, R. (1942). Geschichte und Aufgabe des Wetteldorfer Richtschnittes. *Senckenbergiana,* **25**, 357–61.

Riedel, W.R. & Sanfilippo, A. (1971). Cenozoic radiolaria from the Western Tropical Pacific, Leg 7, *Initial Reports of the Deep-sea Drilling Project,* vol. VII. Washington D.C., 1529–1672.

* Ross, J.R.P. (1981). Biogeography of Carboniferous ectoproct bryozoa. *Palaeontology,* **24**, (2), 313–41.

* Ross, R.J., Jr, Naeser, C.W. & Lambert, R.S. (1978a). Ordovician geochronology. In Cohee *et al.* (1978), pp. 347–54.

* Ross, R.J., Jr *et al.* (1978b). Fission-track dating of Lower Paleozoic volcanic ashes in British stratotypes. In Zartman (1978), pp. 363–5.

Ross, R.J. *et al.* (1982). Fission-track dating of British Ordovician and Silurian stratotypes. *Geol. Mag.,* **119**, 135–53.

Rotai, A.P. (1979). Carboniferous stratigraphy of the USSR: proposal for an international classification. In: Wagner *et al.* (1979), pp 225–47.

Roth, H.P., Baumann, P. & Bertolino, V. (1971). Late Eocene-Oligocene calcareous nannoplankton from central and northern Italy. *2nd International Conference on Planktonic Microfossils Rome 1970, Proc.,* pp. 1069–97.

Rudwick, M.J.S. (1979). The Devonian: a system born from conflict. In: *The Devonian System (Special Papers in Palaeontology No. 23),* ed. M.R. House, C.T. Scrutton & M.G. Bassett, pp. 9–21. London: Palaeontological Association.

Rutland, R.W.R., Parker, A.J., Pitt, G.M., Preiss, W.V. & Murrell, B. (1981). *The Precambrian of South Australia.* In Hunter (1981), pp. 309–60.

* Ryan, W.B.F., Cita, M.B., Rawson, M.D., Burckle, L.H. & Saito, T. (1974). A paleomagnetic assignment of Neogene stage boundaries and the development of isochronous datum planes between the Mediterranean, the Pacific and Indian Oceans in order to investigate the response of the world ocean to the Mediterranean 'salinity crisis'. *Riv. Ital. Paleont. Stratigr.,* 80 (4), 631–87.

Sachs, V.N. & Strelkov, S.A. (1961). Mesozoic and Cenozoic of the Soviet Arctic. In *Geology of the Arctic. Proceedings of First International Symposium on Arctic Geology,* ed. G.O. Raasch, vol. 1, pp. 48–67. University of Toronto Press.

Sadler, D.H. (1968). Astronomical measures of time. *Quart. J. R. Astr. Soc.,* **9**, 281–93.

Sarjent, W.A.S. (1979). Middle and Upper Jurassic dinoflagellate cysts: the world excluding North America. In *Contributions of stratigraphic palynology (with emphasis on North America),* vol. 2. Mesozoic palynology, pp. 133–57. AASP Contribution Series, No. 5B.

Sasajima, S. & Shimada, M. (1966). Paleomagnetic studies of the Cretaceous volcanic rocks in southwest Japan – an assumed drift of the Honshu Island. *J. Geol. Soc. Japan* 72, 503–14 (in Japanese with English abstract).

* Schopf, J.W. (1970). Precambrian micro-organisms and evolutionary events prior to the origin of vascular plants. *Biol. Rev.,* **45**, 319–52.

Scott, G.H. (1972). *Globigerinoides* from Escornebéou (France) and the basal Miocene *Globigerinoides* datum. *N. Z. J. Geol. Geophys.,* **15** (2), 287–95.

Seguenza, G. (1868). La formation zancléenne, ou recherches sur une nouvelle formation tertiaire. *Bull. Soc. Géol. Fr., sér.* 2, **25**, 465–86.

* Seguenza, G. (1879). Le formazioni terziare nella provincia di Reggio Calabria. *Mem. Acad. R. Lince, Cl.Sci.Fis.Mat.Nat.,* 3(6), 1–446.

Selli, R. (1960). Il Messiniano Mayer-Eymar, 1867. Proposta di un neostratotipo. *G. Geol.,* ser. 2, 28, 1–33.

Selli, R. *et al.* (1977). The Vrica section (Calabria, Italy). A potential Neogene/Quaternary boundary stratotype. *G. Geol.,* ser. 2, **42**, 181–204.

Shackleton, N.J. & Opdyke, N.D. (1973). Oxygen isotope and palaeomagnetic stratigraphy of equatorial Pacific core V28–238: oxygen isotope temperatures and ice volumes on a 10^5 and 10^6 year scale. *Quaternary Research,* **3**, 39–55.

Shackleton, N.J. & Opdyke, N.D. (1976). Oxygen-isotope and paleomagnetic stratigraphy of Pacific core V28–239, Late Pliocene to Latest Pleistocene. *Mem. Geol. Soc. Am.,* **145**, 449–64.

Sheng Shen-Fu (1980). *The Ordovician System in China. Correlation chart and explanatory notes.* Ottawa: International Union of Geological Sciences, Publication No. 1., 7 pp + charts.

Sherlock, R.L. (1948). The Permo-Triassic formations: a world review. London: Hutchinson, 367 pp.

Shibata, K., Matsumota, T., Yanagi, T. & Hamamoto, R. (1978). Isotopic ages and stratigraphic control of Mesozoic igneous rocks in Japan. In Cohee *et al.* (1978), pp. 143–64.

Šibrava, V. (1978). Isotopic methods in Quaternary geology. In Cohee *et al.* (1978), pp. 165–9.

Sigal. J. (1977). Essai de zonation du Grétacé Méditerranéen à l'aide des foraminifères planctoniques. *Géologie Méditerranéenne,* 4(2), 99–108.

Silberling, N.J. & Tozer, E.T. (1968). *Biostratigraphic correlation of the marine Triassic in North America.* Boulder, Colorado: Special Papers Geological Society of America, 110, 63 pp.

Sims, P.K. (1980). Subdivision of the Proterozoic and Archean Eons: recommendations and suggestions by the International Subcommission on Precambrian Stratigraphy. *Precambrian Research,* **13**, 379–80.

Sissingh, W. (1977). Biostratigraphy of Cretaceous calcareous nannoplankton. *Geologie en Mijnbouw,* **56** (1), 37–56.

Smith, A.G., Hurley, A.M. & Briden, J.C. (1981). *Phanerozoic paleocontinental world maps.* Cambridge University Press, 102 pp.

Smith, D.B., Brunstrom, R.G.W., Manning, P.I., Simpson, S. & Shotton, F.W. (1974). *A correlation of Permian rocks in the British Isles.* Geol. Soc. Lond., Special Report No. 5, 45 pp.

Spath, L.F. (1923). On the ammonite horizons of the Gault and contiguous deposits. *Summ. Progr. Geol. Surv. Lond.* (for 1922), 139–49.

* Spath, L.F. (1932). The invertebrate faunas of the Bathonian-Callovian deposits of Jameson Land (East Greenland). *Meddr Grønland,* 87 (7), 1–158.

Spath, L.F. (1956). The Liassic ammonite faunas of the Stowell Park Borehole. *Bull. Geol. Surv. Gt. Br.,* **11**, 140–64.

* Spjeldnaes, N. (1978). The Silurian System. In Cohee *et al.* (1978), pp. 341–5.

Stainforth, R.M. *et al.* (1975). Cenozoic planktonic foraminiferal zonation and characteristics of index forms. *Kansas Univ. Paleont. Contr.,* article 62, 1–425.

Steiger, R.H. & Jäger, E. (1977). Subcommission on Geochronology: convention on the use of decay constants in geo- and cosmochronology. *Earth Planet. Sci. Lett.,* **36**, 359–62.

* Steiger, R.H. & Jäger, E. (1978). Subcommission on Geochronology: convention on the use of decay constants in geochronology and cosmochronology. In Cohee *et al.* (1978), pp. 67–71.

* Stephenson, L.W., King, P.B., Monroe, W.H. & Imlay, R.W. (1942). Correlation of the outcropping Cretaceous formations of the Atlantic and Gulf Coastal Plain and Trans-Pecos, Texas. *Bull. Geol. Soc. Am.,* **53**, 435–48.

Stevens, C.H., Wagner, D.B. & Sumsion, R.S. (1979). Permian fusulinid biostratigraphy, Central Cordilleran miogeosyncline. *J. Palaeont.,* **53**(1), 29–36.

Stevens, G.R. (compiler) (1980). *Geological time scale.* Geological Survey of New Zealand.

Stille, H. (1924). *Grundfragen der vergleichenden Tektonik.* Berlin: Borntraeger, 443 pp.

Stockwell, C.H. (1964). Fourth report on structural provinces, orogenies and time classification of rocks of the Canadian Precambrian shield. *Geol. Surv. Can. Paper,* pp. 64–17.

Stubblefield, C.J. (1956). Cambrian palaeogeography in Britain. In

El Sistema cámbrico, su paleogeografiá y el problema du su base I. XX International Geological Congress, Mexico, pp. 1–43.

Sturani, C. (1967). Ammonites and stratigraphy of the Bathonian in the Digne-Barrême area (SE France). *Boll. Soc. Paleont. Ital.,* 5, 1–55, pl. 1–24.

Suggate, R.P., Stevens, G.R. & Te Punga, M.T. (eds) (1978). *The geology of New Zealand,* 2 vols. Wellington: Government Printer, 820 pp.

Surlyk, F. (1977). Stratigraphy, tectonics and palaeogeography of the Jurassic sediments of the areas north of Kong Oscars Fjord, East Greenland. *Bull. Grønlands. Geol. Undersøgelse,* No. 123, 1–56.

* Swartz, C.K. *et al.* (1942). Correlation of the Silurian formations of North America. *Bull. Geol. Soc. Am.,* **53**, 533–8.

Sweet, W.C. & Bergström, S.M. (1976). Conodont biostratigraphy of the Middle and Upper Ordovician of the United States Midcontinent. In Bassett (1976), pp. 121–51.

Takai, F., Matsumoto, T. & Toriyama, R. (eds) (1963). *The geology of Japan.* University of Tokyo Press, 279 pp.

Tappan, H. (1980). *The paleobiology of plant protists.* Reading: Freeman, 1028 pp.

* Tappan, H. & Loeblich, A.R. (1971). Geobiologic implications of fossil phytoplankton evolution and timespace distribution. *Spec. Pap. Geol. Soc. Am.,* **127**, 247–340. (Reprinted from 'Symposium on palynology of the Late Cretaceous and Early Tertiary' (1971) ed. Kosanke & Cross.)

Theyer, F. & Hammond, S.R. (1974). Cenozoic magnetic time scale in deep sea cores: completion of the Neogene, *Geology,* **2**(10), 487–92.

* Toriyama, R. (1963). The Permian. In Takai *et al.* (1963), pp. 43–58.

Toucas, A. (1888). Note sur le Jurassique supérieur et le Crétacé inférieur de la vallée du Rhône. *Bull. Soc. Géol. Fr.,* **16**(3), 903.

Tozer, E.T. (1967). A standard for Triassic time. *Bull. Geol. surv. Can.,* **156**, 1–103.

Tozer, E.T. (1979). Latest Triassic ammonoid faunas and biochronology, Western Canada. *Geol. Surv. Can. Paper,* 79-1B, 127–35.

* Twenhofel, W.H. *et al.* (1954). Correlation of the Ordovician formations of North America. *Bull. Geol. Soc. Am.,* **65**, 247–98.

Vail, P.R., Mitchum, R.M., Jr & Thompson, S. III. (1977). Seismic stratigraphy and global changes of sea level, Part 4: Global cycles of relative changes of sea level. In Payton (1977), pp. 83–97.

Vail, P.R. & Todd, R.G. (1981). Northern North Sea Jurassic unconformities, chronostratigraphy and sea-level changes from seismic stratigraphy. In *Petroleum geology of continental shelf of north-west Europe,* ed. J.V. Cling & C.D. Hobson, pp. 216–35. London: Institute of Petroleum.

Van Couvering, J.A. & Berggren, W.A. (1977). Biostratigraphical basis of the Neogene time-scale. In *Concepts and methods of biostratigraphy.* J. Hasel & E. Kaufmann, pp. 283–305. Stroudsberg, Pa: Dowden, Hutchinson & Ross.

Van Donk, J. (1976). O record of the Atlantic Ocean for the entire Pleistocene Epoch. *Mem. Geol. Soc. Am.,* **145**, 147–63.

Van Eysinga, F.W.B. (compiler) (1975). *Geological timetable* (3rd edn). Amsterdam: Elsevier.

Van Hinte, J.E. (1978a). A Cretaceous time scale. In Cohee *et al.* (1978), pp. 269–87.

Van Hinte, J.E. (1978b). A Jurassic time scale. In Cohee *et al.* (1978), pp. 289–97.

* Vass, D. (1975). Report of the working group on radiometric age and palaeomagnetism. *6th Congr. Reg. Comm. Mediterranean Neogene Stratigraphy Proc.,* 103–17.

* Verbeek, J.W. (1977). Calcareous nannoplankton biostratigraphy of Middle and Upper Cretaceous deposits in Tunisia, Southern Spain and France. *Utrecht Micropaleontological Bulletins,* No. 16, 1–157.

Verosub, K.L. & Banerjee, S.K. (1977). Geomagnetic excursions and their palaeomagnetic record. *Rev. Geophys. Space Phys.,* **15**, 145–55.

Vervloet, C.C. (1966). Stratigraphical and micropalaeontological data on the Tertiary of southern Piemont (northern Italy). Utrecht: Schotanus & Jens.

Vidal, G. (1979). *Acritarchs and the correlation of the Upper Proterozoic.* Publications from the Institutes of Mineralogy, Paleontology, and Quaternary Geology, University of Lund, Sweden, No. 219, 21 pp.

Vidal, G. (1981). Micropalaeontology and biostratigraphy of the Upper Proterozoic and Lower Cambrian sequence in East Finnmark, northern Norway. *Norg. Geog. Unders.,* **362**, 1–53.

Vine, F.J. (1966). Spreading of the ocean floor: new evidence. *Science,* **154**, 1405–15.

Vojacek, H.J. (1979). UNESCO geological world atlas. *Cartography,* **11**(1), 32–9.

Wagner, R.H., Higgins, A.C. & Meyen, S.V. (eds.) (1979). *The Carboniferous of the USSR (Reports to IUGS Subcommission on Carboniferous stratigraphy and geology, 1975).* Yorkshire Geol. Soc. Occasional Publication, 4, 22 pp.

Wang Hong-zhen & Liu Ben-pei (1980). *Text book of historical geology.* Beijing, 352 pp. (in Chinese).

Wang Yuelun, Lu Songnian, Gao Zhenjia, Lin Weixing & Ma Guogan (1981). Sinian tillites of China. In Hambrey & Harland (1981), pp. 386–401 (C33).

Warrington, G., Audley-Charles, M.G., Elliott, R.E., Evans, W.B., Ivimey-Cook, H.C., Kent, P.E., Robinson, P.L., Shotton, F.W. & Taylor, F.M. (1980). *A correlation of Triassic rocks in the British Isles.* Geol. Soc. Lond., Special Report No. 13., 78 pp.

Waterhouse, J.B. (1978). Chronostratigraphy for the world Permian. In Cohee *et al.* (1978), pp. 299–322.

Watts, A.B. (1982). Tectonic subsidence, flexure and global changes of sea level. *Nature, Lond.,* **297**, 469–74.

Weast, R.C. (ed.) (1969). Definitions and formulas. In *Handbook of Chemistry and Physics,* 49th edn, F65–103. Cleveland, Ohio: Chemical Rubber Co. Press.

* Weaver, C.E. *et al.* (1944). Correlation of the marine Cenozoic formations of western North America. *Bull. Geol. Soc. Am.,* **55**, 569–98.

Webby, B.D. *et al.* (1981). *The Ordovician System in Australia, New Zealand and Antarctica.* Ottawa: International Union of Geological Sciences Publication No. 3, 64 pp. + figures and charts.

* Weller, J.M. *et al.* (1948). Correlation of the Mississippian formations of North America. *Bull. Geol. Soc. Am.,* **59**, 91–196.

Whittard, W.F. (1961). *Lexique stratigraphique international. Vol. I. Europe fasc. 3a, Angleterre, Pays de Galles, Écosse: V. Silurien.* Paris: Centre National de la Recherche Scientifique, 273 pp.

Whittington, H.B. & Williams, A. (1964). The Ordovician period. In Harland, Smith & Wilcock (1964), pp. 241–54.

Williams, A., Strachan, I., Bassett, D.A., Dean, W.T., Ingham, J.K., Wright, A.D. & Whittington, H.B. (1972). *A correlation of Ordovician rocks in the British Isles.* Geol. Soc. Lond., Special Report No. 3, 74 pp.

Williams, H.S. (1893). Elements of the geological time scale, *J. Geol.* **1**, 283–95.

Wilmarth, M.G. (1925). The geologic time classification of the United States Geological Survey compared with other classifications accompanied by the original definitions of era, period and epoch terms – a compilation. *Bull. U.S. Geol. Surv.,* **769**, 1–138.

Wilson, D.S. & Hey, R.N. (1981). The Galapagos axial magnetic anomaly: evidence for the Emperor event within the Bruhnes and for a two-layer magnetic source. *Geophys. Res. Lett.,* 8, 1051–4.

Wise, D.U. (1974). Continental margins, freeboard and the volumes of continents and oceans through time. In *The geology of continental margins,* ed. C.A. Burk & C.L. Drake, pp. 45–58. New York: Springer-Verlag.

Zartman, R.E. (ed.) (1978). *Short papers of the 4th International Conference, Geochronology, Cosmochronology, Isotope Geology, 1978.* USGS Open-File Report, 78–701, 476 pp.

Zeuner, F.E. (1945). *The Pleistocene Period, its climate, chronology and faunal successions.* London: Royal Society, 322 pp.

Ziegler, W. (1978). Devonian. In Cohee *et al.* (1978), pp. 337–9.

Ziegler, W. (1979). Historical subdivisions of the Devonian. In *The Devonian System,* (Special Papers in Palaeontology No. 23), ed. M.R. House, C.T. Scrutton & M.G. Bassett, pp. 23–47. London: Palaeontological Association.

Zittel, K.A. von (1901). *History of geology and palaeontology to the end of the nineteenth century.* London: Walter Scott, 562 pp. (translated from 1899 German edition by M.M. Ogilvie-Gordon).

Index

This index provides full reference to both the divisions that appear in the summary time scale and particular topics discussed in the text. Index fossils are not catalogued and local stratigraphic divisions (as correlated in Charts 2.1 to 2.17) are included only in selected cases, mostly for widely used names or those particularly referred to in the text. Separate reference is not made to the alphabetical 'List of formations' which appears on pages 110–15 or to the 'List of preferred abbreviations' on pages 116–17.

Italics denote entry on a chart, figure or table.